Ma
2002

To Alec.

In memory of the
visit to Albany

March 8, 04   Boris K.

Graduate Texts in Mathematics 199

Springer
New York
Berlin
Heidelberg
Barcelona
Hong Kong
London
Milan
Paris
Singapore
Tokyo

Graduate Texts in Mathematics

(continued after index)

Haakan Hedenmalm
Boris Korenblum
Kehe Zhu

Theory of Bergman Spaces

With 4 Illustrations

 Springer

Haakan Hedenmalm
Department of Mathematics
Lund University
Lund, S-22100
Sweden

Boris Korenblum
Kehe Zhu
Department of Mathematics
State University of New York at Albany
Albany, NY 12222-0001
USA

Mathematics Subject Classification (2000): 47-01, 47A15, 32A30

Library of Congress Cataloging-in-Publication Data
Hedenmalm, Haakan.
  Theory of Bergman spaces / Haakan Hedenmalm, Boris Korenblum, Kehe Zhu.
     p. cm. — (Graduate texts in mathematics ; 199)
  Includes bibliographical references and index.
  ISBN 0-387-98791-6 (alk. paper)
     1. Bergman kernel functions. I. Korenblum, Boris. II. Zhu, Kehe, 1961– III. Title.
  IV. Series.
  QA331 .H36 2000
  515—dc21   99-053568

Printed on acid-free paper.

Production managed by Jenny Wolkowicki; manufacturing supervised by Jeffrey Taub.
Photocomposed copy prepared from the authors' LaTeX files.
Printed and bound by R.R. Donnelley and Sons, Harrisonburg, VA.
Printed in the United States of America.

9 8 7 6 5 4 3 2 1

ISBN 0-387-98791-6 Springer-Verlag New York Berlin Heidelberg   SPIN 10715348

Preface

Their memorials are covered by sand,
their rooms are forgotten.
But their names live on by the books they wrote,
for they are beautiful.

(Egyptian poem, 1500–1000 BC)

The theory of Bergman spaces experienced three main phases of development during the last three decades.

The early 1970's marked the beginning of function theoretic studies in these spaces. Substantial progress was made by Horowitz and Korenblum, among others, in the areas of zero sets, cyclic vectors, and invariant subspaces. An influential presentation of the situation up to the mid 1970's was Shields' survey paper "*Weighted shift operators and analytic function theory*".

The 1980's saw the thriving of operator theoretic studies related to Bergman spaces. The contributors in this period are numerous; their achievements were presented in Zhu's 1990 book "*Operator Theory in Function Spaces*".

The research on Bergman spaces in the 1990's resulted in several breakthroughs, both function theoretic and operator theoretic. The most notable results in this period include Seip's geometric characterization of sequences of interpolation and sampling, Hedenmalm's discovery of the contractive zero divisors, the relationship between Bergman-inner functions and the biharmonic Green function found by

Duren, Khavinson, Shapiro, and Sundberg, and deep results concerning invariant subspaces by Aleman, Borichev, Hedenmalm, Richter, Shimorin, and Sundberg.

Our purpose is to present the latest developments, mostly achieved in the 1990's, in book form. In particular, graduate students and new researchers in the field will have access to the theory from an almost self-contained and readable source.

Given that much of the theory developed in the book is fresh, the reader is advised that some of the material covered by the book has not yet assumed a final form.

The prerequisites for the book are elementary real, complex, and functional analysis. We also assume the reader is somewhat familiar with the theory of Hardy spaces, as can be found in Duren's book *"Theory of H^p Spaces"*, Garnett's book *"Bounded Analytic Functions"*, or Koosis' book *"Introduction to H^p Spaces"*.

Exercises are provided at the end of each chapter. Some of these problems are elementary and can be used as homework assignments for graduate students. But many of them are nontrivial and should be considered supplemental to the main text; in this case, we have tried to locate a reference for the reader.

We thank Alexandru Aleman, Alexander Borichev, Bernard Pinchuk, Kristian Seip, and Sergei Shimorin for their help during the preparation of the book. We also thank Anders Dahlner for assistance with the computer generation of three pictures, and Sergei Treil for assistance with one.

January 2000

Haakan Hedenmalm
Boris Korenblum
Kehe Zhu

Contents

1

The Bergman Spaces

In this chapter we introduce the Bergman spaces and concentrate on the general aspects of these spaces. Most results are concerned with the Banach (or metric) space structure of Bergman spaces. Almost all results are related to the Bergman kernel. The Bloch space appears as the image of the bounded functions under the Bergman projection, but it also plays the role of the dual space of the Bergman spaces for small exponents ($0 < p \le 1$).

1.1 Bergman Spaces

Throughout the book we let $\mathbb{C}$ be the complex plane, let

$$\mathbb{D} = \{z \in \mathbb{C} : |z| < 1\}$$

be the open unit disk in $\mathbb{C}$, and let

$$\mathbb{T} = \{z \in \mathbb{C} : |z| = 1\}$$

be the unit circle in $\mathbb{C}$. Likewise, we write $\mathbb{R}$ for the real line. The normalized area measure on $\mathbb{D}$ will be denoted by dA. In terms of real (rectangular and polar) coordinates, we have

$$dA(z) = \frac{1}{\pi}\,dx\,dy = \frac{1}{\pi}\,r\,dr\,d\theta, \qquad z = x + iy = re^{i\theta}.$$

We shall freely use the Wirtinger differential operators

$$\frac{\partial}{\partial z} = \frac{1}{2}\left(\frac{\partial}{\partial x} - i\,\frac{\partial}{\partial y}\right), \qquad \frac{\partial}{\partial \overline{z}} = \frac{1}{2}\left(\frac{\partial}{\partial x} + i\,\frac{\partial}{\partial y}\right),$$

where again $z = x + iy$. The first acts as differentiation on analytic functions, and the second has a similar action on antianalytic functions.

The word *positive* will appear frequently throughout the book. That a function f is positive means that $f(x) \geq 0$ for all values of x, and that a measure μ is positive means that $\mu(E) \geq 0$ for all measurable sets E. When we need to express the property that $f(x) > 0$ for all x, we say that f is *strictly positive*. These conventions apply – *mutatis mutandis* – to the word *negative* as well. Analogously, we prefer to speak of increasing and decreasing functions in the less strict sense, so that constant functions are both increasing and decreasing.

We use the symbol $\sim$ to indicate that two quantities have the same behavior asymptotically. Thus, $A \sim B$ means that A/B is bounded from above and below by two positive constants in the limit process in question.

For $0 < p < +\infty$ and $-1 < \alpha < +\infty$, the *(weighted) Bergman space* $A_\alpha^p = A_\alpha^p(\mathbb{D})$ of the disk is the space of analytic functions in $L^p(\mathbb{D}, dA_\alpha)$, where

$$dA_\alpha(z) = (\alpha + 1)(1 - |z|^2)^\alpha \, dA(z).$$

If f is in $L^p(\mathbb{D}, dA_\alpha)$, we write

$$\|f\|_{p,\alpha} = \left[\int_{\mathbb{D}} |f(z)|^p \, dA_\alpha(z) \right]^{1/p}.$$

When $1 \leq p < +\infty$, the space $L^p(\mathbb{D}, dA_\alpha)$ is a Banach space with the above norm; when $0 < p < 1$, the space $L^p(\mathbb{D}, dA_\alpha)$ is a complete metric space with the metric defined by

$$d(f, g) = \|f - g\|_{p,\alpha}^p.$$

Since $d(f, g) = d(f - g, 0)$, the metric is invariant. The metric is also p-homogeneous, that is, $d(\lambda f, 0) = |\lambda|^p d(f, 0)$ for scalars $\lambda \in \mathbb{C}$. Spaces of this type are called *quasi-Banach spaces*, because they share many properties of the Banach spaces.

We let $L^\infty(\mathbb{D})$ denote the space of (essentially) bounded functions on $\mathbb{D}$. For $f \in L^\infty(\mathbb{D})$ we define

$$\|f\|_\infty = \text{ess sup} \left\{ |f(z)| : z \in \mathbb{D} \right\}.$$

The space $L^\infty(\mathbb{D})$ is a Banach space with the above norm. As usual, we let H^∞ denote the space of bounded analytic functions in $\mathbb{D}$. It is clear that H^∞ is closed in $L^\infty(\mathbb{D})$ and hence is a Banach space itself.

PROPOSITION 1.1 *Suppose $0 < p < +\infty$, $-1 < \alpha < +\infty$, and that K is a compact subset of $\mathbb{D}$. Then there exists a positive constant $C = C(n, K, p, \alpha)$ such that*

$$\sup \left\{ |f^{(n)}(z)| : z \in K \right\} \leq C \|f\|_{p,\alpha}$$

for all $f \in A_\alpha^p$ and all $n = 0, 1, 2, \ldots$. In particular, every point-evaluation in $\mathbb{D}$ is a bounded linear functional on A_α^p.

Proof. Without loss of generality we may assume that

$$K = \{z \in \mathbb{C} : |z| \le r\}$$

for some $r \in (0, 1)$. We first prove the result for $n = 0$.

Let $\sigma = (1 - r)/2$ and let $B(z, \sigma)$ denote the Euclidean disk at z with radius σ. Then by the subharmonicity of $|f|^p$,

$$|f(z)|^p \le \frac{1}{\sigma^2} \int_{B(z,\sigma)} |f(w)|^p \, dA(w)$$

for all $z \in K$. It is easy to see that for all $z \in K$ we have

$$1 - |z|^2 \ge 1 - |z| \ge (1 - r)/2.$$

Thus, we can find a positive constant C (depending only on r) such that

$$|f(z)|^p \le C \int_{B(z,\sigma)} |f(w)|^p \, dA_\alpha(w) \le C \int_{\mathbb{D}} |f(w)|^p \, dA_\alpha(w)$$

for all $z \in K$. This proves the result for $n = 0$.

By the special case we just proved, there exists a constant $M > 0$ such that $|f(\zeta)| \le M \|f\|_{p,\alpha}$ for all $|\zeta| = R$, where $R = (1 + r)/2$. Now if $z \in K$, then by Cauchy's integral formula,

$$f^{(n)}(z) = \frac{n!}{2\pi i} \int_{|\zeta|=R} \frac{f(\zeta) \, d\zeta}{(\zeta - z)^{n+1}}.$$

It follows that

$$|f^{(n)}(z)| \le \frac{n! M R}{\sigma^{n+1}} \|f\|_{p,\alpha}$$

for all $z \in K$ and $f \in A_\alpha^p$.   ∎

As a consequence of the above proposition, we show that the Bergman space A_α^p is a Banach space when $1 \le p < +\infty$ and a complete metric space when $0 < p < 1$.

PROPOSITION 1.2 *For every $0 < p < +\infty$ and $-1 < \alpha < +\infty$, the weighted Bergman space A_α^p is closed in $L^p(\mathbb{D}, dA_\alpha)$.*

Proof. Let $\{f_n\}_n$ be a sequence in A_α^p and assume $f_n \to f$ in $L^p(\mathbb{D}, dA_\alpha)$. In particular, $\{f_n\}_n$ is a Cauchy sequence in $L^p(\mathbb{D}, dA_\alpha)$. Applying the previous proposition, we see that $\{f_n\}_n$ converges uniformly on every compact subset of $\mathbb{D}$. Combining this with the assumption that $f_n \to f$ in $L^p(\mathbb{D}, dA_\alpha)$, we conclude that $f_n(z) \to f(z)$ uniformly on every compact subset of $\mathbb{D}$. Therefore, f is analytic in $\mathbb{D}$ and belongs to A_α^p.   ∎

In many applications, we need to approximate a general function in the Bergman space A_α^p by a sequence of "nice" functions. The following result gives two commonly used ways of doing this.

PROPOSITION 1.3 *For an analytic function f in $\mathbb{D}$ and $0 < r < 1$, let f_r be the dilated function defined by $f_r(z) = f(rz)$, $z \in \mathbb{D}$. Then*

(1) For every $f \in A_\alpha^p$, we have $\|f_r - f\|_{p.\alpha} \to 0$ as $r \to 1^-$.

(2) For every $f \in A_\alpha^p$, there exists a sequence $\{p_n\}_n$ of polynomials such that $\|p_n - f\|_{p.\alpha} \to 0$ as $n \to +\infty$.

Proof. Let f be a function in A_α^p. To prove the first assertion, let δ be a number in the interval $(0, 1)$ and note that

$$
\int_{\mathbb{D}} |f_r(z) - f(z)|^p \, dA_\alpha(z) \leq \int_{|z| \leq \delta} |f_r(z) - f(z)|^p \, dA_\alpha(z)
$$
$$
+ \int_{\delta < |z| < 1} \left(|f_r(z)| + |f(z)| \right)^p \, dA_\alpha(z).
$$

Since f is in $L^p(\mathbb{D}, dA_\alpha)$, we can make the second integral above arbitrarily small by choosing δ close enough to 1. Once δ is fixed, the first integral above clearly approaches 0 as $r \to 1^-$.

To prove the second assertion, we first approximate f by f_r and then approximate f_r by its Taylor polynomials. ∎

Although any function in A_α^p can be approximated (in norm) by a sequence of polynomials, it is not always true that a function in A_α^p can be approximated (in norm) by its Taylor polynomials. Actually, such approximation is possible if and only if $1 < p < +\infty$; see Exercise 4.

We now turn our attention to the special case $p = 2$. By Proposition 1.2 the Bergman space A_α^2 is a Hilbert space. For any nonnegative integer n, let

$$
e_n(z) = \sqrt{\frac{\Gamma(n + 2 + \alpha)}{n! \, \Gamma(2 + \alpha)}} \, z^n, \qquad z \in \mathbb{D}.
$$

Here, $\Gamma(s)$ stands for the usual Gamma function, which is an analytic function of s in the whole complex plane, except for simple poles at the points $\{0, -1, -2, \ldots\}$. It is easy to check that $\{e_n\}_n$ is an orthonormal set in A_α^2. Since the set of polynomials is dense in A_α^2, we conclude that $\{e_n\}_n$ defined above is an orthonormal basis for A_α^2. It follows that if

$$
f(z) = \sum_{n=0}^{+\infty} a_n z^n \qquad \text{and} \qquad g(z) = \sum_{n=0}^{+\infty} b_n z^n
$$

are two functions in A_α^2, then

$$
\|f\|_2^2 = \sum_{n=0}^{+\infty} \frac{n! \, \Gamma(2 + \alpha)}{\Gamma(n + 2 + \alpha)} |a_n|^2
$$

and

$$\langle f, g \rangle_\alpha = \sum_{n=0}^{+\infty} \frac{n!\,\Gamma(2+\alpha)}{\Gamma(n+2+\alpha)}\, a_n \overline{b}_n,$$

where $\langle \cdot, \cdot \rangle_\alpha$ is the inner product in A_α^2 inherited from $L^2(\mathbb{D}, dA_\alpha)$.

PROPOSITION 1.4 *For $-1 < \alpha < +\infty$, let $\mathbf{P}_\alpha$ be the orthogonal projection from $L^2(\mathbb{D}, dA_\alpha)$ onto A_α^2. Then*

$$\mathbf{P}_\alpha f(z) = \int_{\mathbb{D}} \frac{f(w)\, dA_\alpha(w)}{(1 - z\overline{w})^{2+\alpha}}, \qquad z \in \mathbb{D},$$

for all $f \in L^2(\mathbb{D}, dA_\alpha)$.

Proof. Let $\{e_n\}_n$ be the orthonormal basis of A_α^2 defined a little earlier. Then for every $f \in L^2(\mathbb{D}, dA_\alpha)$ we have

$$\mathbf{P}_\alpha f = \sum_{n=0}^{+\infty} \langle \mathbf{P}_\alpha f, e_n \rangle_\alpha\, e_n.$$

In particular,

$$\mathbf{P}_\alpha f(z) = \sum_{n=0}^{+\infty} \langle \mathbf{P}_\alpha f, e_n \rangle_\alpha\, e_n(z)$$

for every $z \in \mathbb{D}$ and the series converges uniformly on every compact subset of $\mathbb{D}$. Since

$$\langle \mathbf{P}_\alpha f, e_n \rangle_\alpha = \langle f, \mathbf{P}_\alpha e_n \rangle_\alpha = \langle f, e_n \rangle_\alpha,$$

we have

$$\begin{aligned}
\mathbf{P}_\alpha f(z) &= \sum_{n=0}^{+\infty} \frac{\Gamma(n+2+\alpha)}{n!\,\Gamma(2+\alpha)} \int_{\mathbb{D}} f(w)(z\overline{w})^n\, dA_\alpha(w) \\
&= \int_{\mathbb{D}} f(w) \left[\sum_{n=0}^{+\infty} \frac{\Gamma(n+2+\alpha)}{n!\,\Gamma(2+\alpha)}(z\overline{w})^n \right] dA_\alpha(w) \\
&= \int_{\mathbb{D}} \frac{f(w)\, dA_\alpha(w)}{(1 - z\overline{w})^{2+\alpha}}.
\end{aligned}$$

The interchange of integration and summation is justified, because for each fixed $z \in \mathbb{D}$, the series

$$\sum_{n=0}^{+\infty} \frac{\Gamma(n+2+\alpha)}{n!\,\Gamma(2+\alpha)}(z\overline{w})^n$$

converges uniformly in $w \in \mathbb{D}$.   ∎

The operators $\mathbf{P}_\alpha$ above are called the *(weighted) Bergman projections* on $\mathbb{D}$. The functions

$$K_\alpha(z, w) = \frac{1}{(1 - z\overline{w})^{2+\alpha}}, \qquad z, w \in \mathbb{D},$$

are called the (weighted) Bergman kernels of $\mathbb{D}$. These kernel functions play an essential role in the theory of Bergman spaces.

Although the Bergman projection $\mathbf{P}_\alpha$ is originally defined on $L^2(\mathbb{D}, dA_\alpha)$, the integral formula

$$\mathbf{P}_\alpha f(z) = \int_\mathbb{D} \frac{f(w) \, dA_\alpha(w)}{(1 - z\overline{w})^{2+\alpha}}$$

clearly extends the domain of $\mathbf{P}_\alpha$ to $L^1(\mathbb{D}, dA_\alpha)$. In particular, we can apply $\mathbf{P}_\alpha$ to a function in $L^p(\mathbb{D}, dA_\alpha)$ whenever $1 \le p < +\infty$.

If f is a function in A_α^2, then $\mathbf{P}_\alpha f = f$, so that

$$f(z) = \int_\mathbb{D} \frac{f(w) \, dA_\alpha(w)}{(1 - z\overline{w})^{2+\alpha}}, \qquad z \in \mathbb{D}.$$

Since this is a pointwise formula and A_α^2 is dense in A_α^1, we obtain the following.

COROLLARY 1.5 *If f is a function in A_α^1, then*

$$f(z) = \int_\mathbb{D} \frac{f(w) \, dA_\alpha(w)}{(1 - z\overline{w})^{2+\alpha}}, \qquad z \in \mathbb{D},$$

and the integral converges uniformly for z in every compact subset of $\mathbb{D}$.

This corollary will be referred to as the reproducing formula. The Bergman kernels are special types of reproducing kernels.

On several occasions later on theorems will hold only for the unweighted Bergman spaces. Thus, we set $A^p = A_0^p$ and call them the ordinary Bergman spaces. The corresponding Bergman projection will be denoted by $\mathbf{P}$, and the Bergman kernel in this case will be written as

$$K(z, w) = \frac{1}{(1 - z\overline{w})^2}.$$

The Bergman kernel functions are intimately related to the Möbius group Aut $(\mathbb{D})$ of the disk. To see this, let $z \in \mathbb{D}$ and consider the Möbius map φ_z of the disk that interchanges z and 0,

$$\varphi_z(w) = \frac{z - w}{1 - \overline{z}w}, \qquad w \in \mathbb{D}.$$

We list below some basic properties of φ_z, which can all be checked easily.

PROPOSITION 1.6 *The Möbius map φ_z has the following properties:*

(1) $\varphi_z^{-1} = \varphi_z$.

(2) The real Jacobian determinant of φ_z at w is $|\varphi_z'(w)|^2 = \dfrac{(1-|z|^2)^2}{|1-z\overline{w}|^4}.$

(3) $1 - |\varphi_z(w)|^2 = \dfrac{(1-|z|^2)(1-|w|^2)}{|1-z\overline{w}|^2}.$

As a simple application of the properties above, we mention that the formula for the Bergman kernel function $K_\alpha(z, w)$ can be derived from a simple change of variables, instead of using an infinite series involving the Gamma function. More specifically, if $f \in A_\alpha^1$, then the rotation invariance of dA_α gives

$$f(0) = \int_{\mathbb{D}} f(w)\, dA_\alpha(w).$$

Replacing f by $f \circ \varphi_z$, making an obvious change of variables, and applying properties (2) and (3) above, we obtain

$$f(z) = (1-|z|^2)^{2+\alpha} \int_{\mathbb{D}} \frac{f(w)\, dA_\alpha(w)}{(1-w\overline{z})^{2+\alpha}(1-z\overline{w})^{2+\alpha}}.$$

Fix $z \in \mathbb{D}$, and replace f by the function $w \mapsto (1-w\overline{z})^{2+\alpha} f(w)$. We then arrive at the reproducing formula

$$f(z) = \int_{\mathbb{D}} \frac{f(w)}{(1-z\overline{w})^{2+\alpha}}\, dA_\alpha(w), \qquad z \in \mathbb{D},$$

for $f \in A_\alpha^1$. From this we easily deduce the integral formula for the Bergman projection $\mathbf{P}_\alpha$.

1.2 Some L^p Estimates

Many operator-theoretic problems in the analysis of Bergman spaces involve estimating integral operators whose kernel is a power of the Bergman kernel. In this section, we present several estimates for integral operators that have proved very useful in the past. In particular, we will establish the boundedness of the Bergman projection $\mathbf{P}_\alpha$ on certain L^p spaces.

THEOREM 1.7 *For any $-1 < \alpha < +\infty$ and any real β, let*

$$I_{\alpha,\beta}(z) = \int_{\mathbb{D}} \frac{(1-|w|^2)^\alpha}{|1-z\overline{w}|^{2+\alpha+\beta}}\, dA(w), \qquad z \in \mathbb{D},$$

and

$$J_\beta(z) = \int_0^{2\pi} \frac{d\theta}{|1-ze^{-i\theta}|^{1+\beta}}, \qquad z \in \mathbb{D}.$$

Then we have

$$I_{\alpha,\beta}(z) \sim J_{\beta}(z) \sim \begin{cases} 1 & \text{if } \beta < 0, \\ \log \dfrac{1}{1 - |z|^2} & \text{if } \beta = 0, \\ \dfrac{1}{(1 - |z|^2)^{\beta}} & \text{if } \beta > 0, \end{cases}$$

as $|z| \to 1^-$.

Proof. The condition $-1 < \alpha < +\infty$ ensures that the integral $I_{\alpha,\beta}(z)$ is convergent for every $z \in \mathbb{D}$. The integral $J_{\beta}(z)$ clearly converges for all $z \in \mathbb{D}$.

Let $\lambda = (2 + \alpha + \beta)/2$. If λ is a nonpositive integer, then clearly $\beta < 0$ and $J_{\alpha,\beta}(z)$ is bounded. In what follows, we assume that λ is not a nonpositive integer. In this case, we make use of the following power series:

$$\frac{1}{(1 - z\overline{w})^{\lambda}} = \sum_{n=0}^{+\infty} \frac{\Gamma(n + \lambda)}{n! \, \Gamma(\lambda)} (z\overline{w})^n.$$

Since the measure $(1 - |w|^2)^{\alpha} \, dA(w)$ is rotation invariant, we have

$$\begin{aligned} I_{\alpha,\beta}(z) &= \int_{\mathbb{D}} \frac{(1 - |w|^2)^{\alpha} \, dA(w)}{|1 - z\overline{w}|^{2\lambda}} \\ &= \sum_{n=0}^{+\infty} \frac{\Gamma(n + \lambda)^2}{(n!)^2 \Gamma(\lambda)^2} |z|^{2n} \int_{\mathbb{D}} (1 - |w|^2)^{\alpha} |w|^{2n} \, dA(w) \\ &= \frac{\Gamma(\alpha + 1)}{\Gamma(\lambda)^2} \sum_{n=0}^{+\infty} \frac{\Gamma(n + \lambda)^2}{n! \, \Gamma(n + \alpha + 2)} |z|^{2n}. \end{aligned}$$

By Stirling's formula,

$$\frac{\Gamma(n + \lambda)^2}{n! \, \Gamma(n + \alpha + 2)} \sim (n + 1)^{\beta - 1}, \qquad n \to +\infty.$$

If $\beta < 0$, then the series

$$\sum_{n=0}^{+\infty} \frac{|z|^{2n}}{(n + 1)^{1 - \beta}}$$

clearly defines a bounded function on $\mathbb{D}$, and so $I_{\alpha,\beta}(z)$ is bounded on $\mathbb{D}$.

If $\beta = 0$, then we have

$$I_{\alpha,0}(z) \sim \sum_{n=0}^{+\infty} \frac{|z|^{2n}}{n + 1} \sim \log \frac{1}{1 - |z|^2}$$

as $|z| \to 1^-$.

If $\beta > 0$, then we have

$$I_{\alpha,\beta}(z) \sim \sum_{n=0}^{+\infty} (n + 1)^{\beta - 1} |z|^{2n} \sim \frac{1}{(1 - |z|^2)^{\beta}}$$

as $|z| \to 1^-$, because

$$\frac{1}{(1 - |z|^2)^\beta} = \sum_{n=0}^{+\infty} \frac{\Gamma(n + \beta)}{n! \, \Gamma(\beta)} \, |z|^{2n}$$

and

$$\frac{\Gamma(n + \beta)}{n! \, \Gamma(\beta)} \sim (n + 1)^{\beta - 1}$$

by Stirling's formula again.

The estimate for $J_\beta(z)$ is similar; we omit the details. ∎

The following result, usually called Schur's test, is a very effective tool in proving the L^p-boundedness of integral operators.

THEOREM 1.8 *Suppose X is a measure space and μ a positive measure on X. Let $T(x, y)$ be a positive measurable function on $X \times X$, and $\mathbf{T}$ the associated integral operator*

$$\mathbf{T}f(z) = \int_X T(x, y) \, f(y) \, d\mu(y), \qquad x \in X,$$

defined wherever the integral converges. If, for some $1 < p < +\infty$, there exists a strictly positive measurable function h on X and a positive constant M such that

$$\int_X T(x, y) \, h(y)^q \, d\mu(y) \leq M \, h(x)^q, \qquad x \in X,$$

and

$$\int_X T(x, y) \, h(x)^p \, d\mu(x) \leq M \, h(y)^p, \qquad y \in X,$$

where $p^{-1} + q^{-1} = 1$, then $\mathbf{T}$ is bounded on $L^p(X, d\mu)$ with $\|\mathbf{T}\| \leq M$.

Proof. Fix a function f in $L^p(X, d\mu)$. Applying Hölder's inequality to the integral below,

$$|\mathbf{T}f(x)| \leq \int_X h(y) \, h(y)^{-1} \, |f(y)| \, T(x, y) \, d\mu(y),$$

we obtain

$$|\mathbf{T}f(x)| \leq \left[\int_X T(x, y) \, h(y)^q \, d\mu(y) \right]^{\frac{1}{q}} \left[\int_X T(x, y) \, h(y)^{-p} |f(y)|^p \, d\mu(y) \right]^{\frac{1}{p}}.$$

Using the first inequality in the assumption, we have

$$|\mathbf{T}f(x)| \leq M^{1/q} \, h(x) \left[\int_X T(x, y) h(y)^{-p} |f(y)|^p \, d\mu(y) \right]^{\frac{1}{p}}.$$

Using Fubini's theorem and the second inequality in the assumption, we easily arrive at the following:

$$\int_X |\mathbf{T}f(x)|^p \, d\mu(x) \le M^p \int_X |f(y)|^p \, d\mu(y).$$

Thus, $\mathbf{T}$ is a bounded operator on $L^p(X, d\mu)$ of norm less than or equal to M. ∎

We now prove the main result of this section.

THEOREM 1.9 *Suppose a, b, and c are real numbers and*

$$d\mu(z) = (1 - |z|^2)^c \, dA(z).$$

Let T and S be the integral operators defined by

$$Tf(z) = (1 - |z|^2)^a \int_{\mathbb{D}} \frac{(1 - |w|^2)^b}{(1 - z\overline{w})^{2+a+b}} f(w) \, dA(w)$$

and

$$Sf(z) = (1 - |z|^2)^a \int_{\mathbb{D}} \frac{(1 - |w|^2)^b}{|1 - z\overline{w}|^{2+a+b}} f(w) \, dA(w).$$

Then for $1 \le p < +\infty$ the following conditions are equivalent:

(1) T is bounded on $L^p(\mathbb{D}, d\mu)$.

(2) S is bounded on $L^p(\mathbb{D}, d\mu)$.

(3) $-pa < c + 1 < p(b + 1)$.

Proof. It is obvious that the boundedness of S on $L^p(\mathbb{D}, d\mu)$ implies that of T. Now, assume that T is bounded on $L^p(\mathbb{D}, d\mu)$. Apply T to a function of the form $f(z) = (1 - |z|^2)^N$, where N is sufficiently large. An application of Theorem 1.7 then yields the inequality $c + 1 > -pa$. To prove the inequality $c + 1 < p(b + 1)$, we first assume $p > 1$ and let q be the conjugate exponent. Let T^* be the adjoint operator of T with respect to the dual action induced by the inner product of $L^2(\mathbb{D}, d\mu)$. It is given explicitly by

$$T^*f(z) = (1 - |z|^2)^{b-c} \int_{\mathbb{D}} \frac{(1 - |w|^2)^{a+c} f(w)}{(1 - z\overline{w})^{2+a+b}} \, dA(w),$$

must be bounded on $L^q(\mathbb{D}, d\mu)$. Again, by looking at the action of T^* on a function of the form $f(z) = (1 - |z|^2)^N$, where N is sufficiently large, and applying Theorem 1.7, we obtain the inequality $c + 1 < p(b + 1)$. If $p = 1$, then T^* is bounded on $L^\infty(\mathbb{D})$, and the desired inequality becomes $c < b$. Let T^* act on the constant function 1. We see that $c \le b$. To see that strict inequality must occur, we consider functions of the form

$$f_z(w) = \frac{(1 - z\overline{w})^{2+a+b}}{|1 - z\overline{w}|^{2+a+b}}, \qquad z, w \in \mathbb{D}.$$

Clearly, $\|f_z\|_\infty = 1$ for every $z \in \mathbb{D}$. If $b = c$, then

$$T^* f_z(z) = \int_{\mathbb{D}} \frac{(1 - |w|^2)^{a+c} \, dA(w)}{|1 - z\overline{w}|^{2+a+c}} \sim \log \frac{1}{1 - |z|^2}, \qquad |z| \to 1^-,$$

by Theorem 1.7. This implies $\|T^* f_z\|_\infty \to +\infty$ as $|z| \to 1^-$, a contradiction to the boundedness of T^* on $L^\infty(\mathbb{D})$. Thus, the boundedness of T on $L^p(\mathbb{D}, d\mu)$ implies the inequalities $-pa < c + 1 < p(b + 1)$.

Next, assume $-pa < c + 1 < p(b + 1)$. We want to prove that the operator S is bounded on $L^p(\mathbb{D}, d\mu)$. The case $p = 1$ is a direct consequence of Theorem 1.7 and Fubini's theorem. When $p > 1$, we appeal to Schur's test. Thus, we assume $1 < p < +\infty$ and seek a positive function $h(z)$ on $\mathbb{D}$ that will satisfy the assumptions in Schur's test. It turns out that such a function exists in the form $h(z) = (1 - |z|^2)^s$, where s is some real number. In fact, if we rewrite

$$Sf(z) = \int_{\mathbb{D}} \frac{(1 - |z|^2)^a (1 - |w|^2)^{b-c}}{|1 - z\overline{w}|^{2+a+b}} f(w) \, d\mu(w),$$

then the conditions that the number s has to satisfy become

$$\int_{\mathbb{D}} \frac{(1 - |w|^2)^{b+qs} \, dA(w)}{|1 - z\overline{w}|^{2+a+b}} \le \frac{C}{(1 - |z|^2)^{a-qs}}, \qquad z \in \mathbb{D},$$

and

$$\int_{\mathbb{D}} \frac{(1 - |z|^2)^{a+ps+c} \, dA(z)}{|1 - z\overline{w}|^{2+a+b}} \le \frac{C}{(1 - |w|^2)^{b-ps-c}}, \qquad w \in \mathbb{D},$$

where q is the conjugate exponent of p and C is some positive constant. According to Theorem 1.7, these estimates are correct if

$$b + qs > -1, \qquad a - qs > 0,$$

and

$$a + ps + c > -1, \qquad b - ps - c > 0.$$

We rewrite these inequalities as

$$-\frac{b+1}{q} < s < \frac{a}{q}, \qquad -\frac{a+c+1}{p} < s < \frac{b-c}{p}.$$

It is easy to check that the inequalities $-pa < c + 1 < p(b + 1)$ are equivalent to

$$-\frac{b+1}{q} < \frac{b-c}{p}, \qquad -\frac{a+c+1}{p} < \frac{a}{q},$$

which clearly imply that the intersection of intervals

$$\left(-\frac{b+1}{q}, \frac{q}{q} \right) \cap \left(-\frac{a+c+1}{p}, \frac{b-c}{p} \right)$$

is nonempty. This shows that the desired s exists, and so the operator S is bounded on $L^p(\mathbb{D}, d\mu)$.   ∎

One of the advantages of the theory of Bergman spaces over that of Hardy spaces is the abundance of analytic projections. For example, it is well known that there is no bounded projection from L^1 of the circle onto the Hardy space H^1, while there exist a lot of bounded projections from $L^1(\mathbb{D}, dA)$ onto the Bergman space A^1, as the following result demonstrates.

THEOREM 1.10 *Suppose* $-1 < \alpha, \beta < +\infty$ *and* $1 \le p < +\infty$. *Then* $\mathbf{P}_\beta$ *is a bounded projection from* $L^p(\mathbb{D}, dA_\alpha)$ *onto* A_α^p *if and only if* $\alpha + 1 < (\beta + 1)p$.

Proof. This is a simple consequence of Theorem 1.9. ∎

Two special cases are worth mentioning. First, if $\alpha = \beta$, then $\mathbf{P}_\alpha$ is a bounded projection from $L^p(\mathbb{D}, dA_\alpha)$ onto A_α^p if and only if $1 < p < +\infty$. In particular, the (unweighted) Bergman projection $\mathbf{P}$ maps $L^p(\mathbb{D}, dA)$ onto A^p if and only if $1 < p < +\infty$. Second, if $p = 1$, then $\mathbf{P}_\beta$ is a bounded projection from $L^1(\mathbb{D}, dA_\alpha)$ onto A_α^1 if and only if $\beta > \alpha$. In particular, $\mathbf{P}_\beta$ is a bounded projection from $L^1(\mathbb{D}, dA)$ onto A^1 when $\beta > 0$.

PROPOSITION 1.11 *Suppose* $1 \le p < +\infty$, $-1 < \alpha < +\infty$, *and that* n *is a positive integer. Then an analytic function* f *in* $\mathbb{D}$ *belongs to* A_α^p *if and only if the function* $(1 - |z|^2)^n f^{(n)}(z)$ *is in* $L^p(\mathbb{D}, dA_\alpha)$.

Proof. First assume $f \in A_\alpha^p$. Fix any $\beta > \alpha$. Then, by Corollary 1.5,

$$f(z) = (\beta + 1) \int_\mathbb{D} \frac{(1 - |w|^2)^\beta}{(1 - z\overline{w})^{2+\beta}} f(w) \, dA(w), \qquad z \in \mathbb{D}.$$

Differentiating under the integral sign n times, we obtain

$$(1 - |z|^2)^n f^{(n)}(z) = C (1 - |z|^2)^n \int_\mathbb{D} \frac{(1 - |w|^2)^\beta}{(1 - z\overline{w})^{2+n+\beta}} \overline{w}^n f(w) \, dA(w),$$

where C is the constant

$$C = (\beta + 1)(\beta + 2) \cdots (\beta + n + 1).$$

By Theorem 1.9, the function $(1 - |z|^2)^n f^{(n)}(z)$ is in $L^p(\mathbb{D}, dA_\alpha)$.

Next, assume that f is analytic in $\mathbb{D}$ and the function $(1 - |z|^2)^n f^{(n)}(z)$ is in $L^p(\mathbb{D}, dA_\alpha)$. We show that f belongs to the weighted Bergman space A_α^p. Without loss of generality, we may assume that the first $2n + 1$ Taylor coefficients of f are all zero. In this case, the function φ defined by

$$\varphi(z) = C \frac{(1 - |z|^2)^n f^{(n)}(z)}{\overline{z}^n}, \qquad z \in \mathbb{D},$$

is in $L^p(\mathbb{D}, dA_\alpha)$, for any constant C. Fix β, $\alpha < \beta < +\infty$, and let $g = \mathbf{P}_\beta \varphi$. By Theorem 1.10, the function g belongs to A_α^p. The explicit formula for g is

$$g(z) = (\beta + 1) \int_\mathbb{D} \frac{(1 - |w|^2)^\beta}{(1 - z\overline{w})^{2+\beta}} \varphi(w) \, dA(w), \qquad z \in \mathbb{D}.$$

If we set the constant C to be

$$C = \frac{1}{(\beta + 1)(\beta + 2) \cdots (\beta + n + 1)},$$

then differentiating n times in the formula for g yields

$$g^{(n)}(z) = (n + \beta + 1) \int_{\mathbb{D}} \frac{(1 - |w|^2)^{n+\beta}}{(1 - z\overline{w})^{2+n+\beta}} f^{(n)}(w) \, dA(w), \qquad z \in \mathbb{D}.$$

Applying Corollary 1.5 again, we find that $g^{(n)} = f^{(n)}$, so that f and g differ only by a polynomial. Since g is in A_α^p, we have $f \in A_\alpha^p$. ∎

1.3   The Bloch Space

An analytic function f in $\mathbb{D}$ is said to be in the Bloch space $\mathcal{B}$ if

$$\|f\|_{\mathcal{B}} = \sup \left\{ (1 - |z|^2)|f'(z)| : z \in \mathbb{D} \right\} < +\infty.$$

It is easy to check that the seminorm $\| \cdot \|_{\mathcal{B}}$ is Möbius invariant. The little Bloch space $\mathcal{B}_0$ is the subspace of $\mathcal{B}$ consisting of functions f with

$$\lim_{|z| \to 1^-} (1 - |z|^2)|f'(z)| = 0.$$

The Bloch space plays the same role in the theory of Bergman space as the space BMOA does in the theory of Hardy spaces. When normed with

$$\|f\| = |f(0)| + \|f\|_{\mathcal{B}},$$

the Bloch space $\mathcal{B}$ is a Banach space, and the little Bloch space $\mathcal{B}_0$ is the the closure of the set of polynomials in $\mathcal{B}$.

If f is an analytic function in $\mathbb{D}$ with $\|f\|_\infty \leq 1$, then by Schwarz's lemma,

$$(1 - |z|^2)|f'(z)| \leq 1 - |f(z)|^2, \qquad z \in \mathbb{D}.$$

It follows that $H^\infty \subset \mathcal{B}$ with $\|f\|_{\mathcal{B}} \leq \|f\|_\infty$.

Let $C(\overline{\mathbb{D}})$ be the space of continuous functions on the closed unit disk $\overline{\mathbb{D}}$. Denote by $C_0(\mathbb{D})$ the subspace of $C(\overline{\mathbb{D}})$ consisting of functions vanishing on the unit circle $\mathbb{T}$. It is clear that both $C(\overline{\mathbb{D}})$ and $C_0(\mathbb{D})$ are closed subspaces of $L^\infty(\mathbb{D})$.

THEOREM 1.12 Suppose $-1 < \alpha < +\infty$ and that $\mathbf{P}_\alpha$ is the corresponding weighted Bergman projection. Then

(1) $\mathbf{P}_\alpha$ maps $L^\infty(\mathbb{D})$ boundedly onto $\mathcal{B}$.

(2) $\mathbf{P}_\alpha$ maps $C(\overline{\mathbb{D}})$ boundedly onto $\mathcal{B}_0$.

(3) $\mathbf{P}_\alpha$ maps $C_0(\mathbb{D})$ boundedly onto $\mathcal{B}_0$.

Proof. First assume $g \in L^\infty(\mathbb{D})$ and $f = \mathbf{P}_\alpha g$, so that

$$f(z) = (\alpha + 1) \int_{\mathbb{D}} \frac{(1 - |w|^2)^\alpha g(w)}{(1 - z\overline{w})^{2+\alpha}} \, dA(w), \qquad z \in \mathbb{D}.$$

Differentiating under the integral sign and applying Theorem 1.7, we see that f belongs to $\mathcal{B}$ with

$$|f(0)| + \|f\|_\mathcal{B} \leq C \|g\|_\infty$$

for some positive constant C (independent of g). Thus, $\mathbf{P}_\alpha$ maps $L^\infty(\mathbb{D})$ boundedly into $\mathcal{B}$.

Next, assume $g \in C(\overline{\mathbb{D}})$. We wish to show that $f = \mathbf{P}_\alpha g$ is in the little Bloch space. By the Stone-Weierstrass approximation theorem, the function g can be uniformly approximated on $\mathbb{D}$ by finite linear combinations of functions of the form

$$g_{n,m}(z) = z^n \overline{z}^m, \qquad z \in \mathbb{D},$$

where n and m are nonnegative integers. Using the symmetry of $\mathbb{D}$, we easily check that each $\mathbf{P}_\alpha g_{n,m}$ belongs to the little Bloch space. Since $\mathbf{P}_\alpha$ maps $L^\infty(\mathbb{D})$ boundedly into $\mathcal{B}$, and $\mathcal{B}_0$ is closed in $\mathcal{B}$, we conclude that $\mathbf{P}_\alpha$ maps $C(\overline{\mathbb{D}})$ boundedly into $\mathcal{B}_0$.

Finally, for $f \in \mathcal{B}$ we write the Taylor expansion of f as

$$f(z) = a + bz + cz^2 + f_1(z), \qquad z \in \mathbb{D},$$

where $f_1(0) = f_1'(0) = 0$, and define a function g in $L^\infty(\mathbb{D})$ by

$$g(z) = (1 - |z|^2) \left[a + \frac{\alpha^2 + 5\alpha + 6}{(\alpha + 1)^2} bz + \frac{\alpha^2 + 7\alpha + 12}{2(\alpha + 1)^2} cz^2 + \frac{f_1'(z)}{\overline{z}(\alpha + 1)} \right].$$

It is clear that g is in $C_0(\mathbb{D})$ if f is in the little Bloch space. A direct calculation shows that $f = \mathbf{P}_\alpha g$. Thus, $\mathbf{P}_\alpha$ maps $L^\infty(\mathbb{D})$ onto $\mathcal{B}$; and it maps $C_0(\mathbb{D})$ (and hence $C(\overline{\mathbb{D}})$) onto $\mathcal{B}_0$. ∎

PROPOSITION 1.13 *Suppose n is a positive integer and f is analytic in $\mathbb{D}$. Then $f \in \mathcal{B}$ if and only if the function $(1 - |z|^2)^n f^{(n)}(z)$ is in $L^\infty(\mathbb{D})$, and $f \in \mathcal{B}_0$ if and only if the function $(1 - |z|^2)^n f^{(n)}(z)$ is in $C(\overline{\mathbb{D}})$ (or $C_0(\mathbb{D})$).*

Proof. If f is in the Bloch space, then by Theorem 1.12 there exists a bounded function g such that

$$f(z) = \int_{\mathbb{D}} \frac{g(w) \, dA(w)}{(1 - z\overline{w})^2}, \qquad z \in \mathbb{D}.$$

Differentiating under the integral sign and applying Theorem 1.7, we see that the function $(1 - |z|^2)^n f^{(n)}(z)$ is bounded.

If the function g above has compact support in $\mathbb{D}$, then clearly the function $(1 - |z|^2)^n f^{(n)}(z)$ is in $C_0(\mathbb{D})$ (and hence in $C(\overline{\mathbb{D}})$). If f is in the little Bloch space, then by Theorem 1.12 we can choose the function g in the previous paragraph to be in $C_0(\mathbb{D})$. Such a function g can then be uniformly approximated by continuous functions with compact support in $\mathbb{D}$. This shows that the function $(1-|z|^2)^n f^{(n)}(z)$ is in $C_0(\mathbb{D})$ (and hence in $C(\overline{\mathbb{D}})$) whenever f is in the little Bloch space.

To prove the "if" parts of the theorem, we may assume the first $2n+1$ Taylor coefficients of f are all zero. In this case, we can consider the function

$$g(z) = C \frac{(1-|z|^2)^n f^{(n)}(z)}{\overline{z}^n}, \qquad z \in \mathbb{D}.$$

By the proof of Proposition 1.11, the functions f and $\mathbf{P}g$ differ by a polynomial. The desired result then follows from Theorem 1.12. ∎

As a consequence of this result and Proposition 1.11, we see that $\mathcal{B}$ is contained in every weighted Bergman space A_α^p. We can then use this observation and the following result to construct nontrivial functions in weighted Bergman spaces. In particular, we see that every weighted Bergman space contains functions that do not have any boundary values.

Recall that a sequence $\{\lambda_n\}_n$ of positive integers is called a *gap sequence* if there exists a constant $\lambda > 1$ such that $\lambda_{n+1}/\lambda_n \geq \lambda$ for all $n = 1, 2, 3, \ldots$. In this case, we call a power series of the form $\sum_{n=0}^{+\infty} a_n z^{\lambda_n}$ a *lacunary series*.

THEOREM 1.14 *A lacunary series defines a function in $\mathcal{B}$ if and only if the coefficients are bounded. Similarly, a lacunary series defines a function in $\mathcal{B}_0$ if and only if the coefficients tend to 0.*

Proof. Suppose $\{a_n\}_n$ is a sequence of complex numbers with $|a_n| \leq M$ for all $n = 1, 2, 3, \ldots$, and suppose $\{\lambda_n\}_n$ is sequence of positive integers with $\lambda_{n+1}/\lambda_n \geq \lambda$ for all $n = 1, 2, 3, \ldots$, where $1 < \lambda < +\infty$ is a constant. Let

$$f(z) = \sum_{n=0}^{+\infty} a_n z^{\lambda_n}, \qquad z \in \mathbb{D}.$$

Clearly, f is analytic in $\mathbb{D}$ and

$$f'(z) = \sum_{n=0}^{+\infty} a_n \lambda_n z^{\lambda_n - 1}, \qquad z \in \mathbb{D}.$$

Let $C = \lambda/(\lambda - 1)$; then $1 < C < +\infty$. It is easy to check that

$$\lambda_{n+1} \leq C(\lambda_{n+1} - \lambda_n), \qquad n = 1, 2, 3, \ldots.$$

This implies that

$$\lambda_{n+1}|z|^{\lambda_{n+1}-1} \leq C(\lambda_{n+1} - \lambda_n)|z|^{\lambda_{n+1}-1}$$
$$\leq C\left(|z|^{\lambda_n} + \cdots + |z|^{\lambda_{n+1}-1}\right), \qquad n = 1, 2, 3, \ldots.$$

We also have, rather trivially,

$$\lambda_1 |z|^{\lambda_1 - 1} \leq 1 + |z| + \cdots + |z|^{\lambda_1 - 1} \leq C(1 + |z| + \cdots + |z|^{\lambda_1 - 1}).$$

It follows that

$$|f'(z)| \le MC \sum_{n=0}^{+\infty} |z|^n = \frac{MC}{1 - |z|}, \qquad z \in \mathbb{D},$$

and hence f is in the Bloch space.

A similar argument shows that if f is defined by a lacunary series whose coefficients tend to 0, then f must be in the little Bloch space.

Conversely, if

$$f(z) = \sum_{n=0}^{+\infty} a_n z^n, \qquad z \in \mathbb{D},$$

is any function in the Bloch space, we show that its Taylor coefficients must be bounded. By Corollary 1.5, we have

$$f'(z) = 2 \int_{\mathbb{D}} \frac{1 - |w|^2}{(1 - z\overline{w})^3} f'(w) \, dA(w), \qquad z \in \mathbb{D},$$

whence it follows that

$$a_n = \frac{f^{(n)}(0)}{n!} = (n + 1) \int_{\mathbb{D}} \overline{w}^n (1 - |w|^2) f'(w) \, dA(w), \qquad n = 1, 2, 3 \ldots .$$

This clearly implies that $\{a_n\}_n$ is bounded. Similarly, the formula above together with an obvious partition of the disk implies that $\{a_n\}_n$ converges to 0 if f is in the little Bloch space. ∎

Finally in this section we present a characterization of the Bloch space in terms of the Bergman metric. Recall that for every $z \in \mathbb{D}$, the function φ_z is the Möbius transformation that interchanges z and the origin. The *pseudohyperbolic metric* ρ on $\mathbb{D}$ is defined by

$$\rho(z, w) = |\varphi_z(w)| = \left| \frac{z - w}{1 - z\overline{w}} \right|, \qquad z, w \in \mathbb{D},$$

and the *hyperbolic metric* β, also called the *Bergman metric* or the *Poincaré metric*, is defined by

$$\beta(z, w) = \frac{1}{2} \log \frac{1 + \rho(z, w)}{1 - \rho(z, w)}, \qquad z, w \in \mathbb{D}.$$

It is easy to check that the pseudohyperbolic metric (and hence the hyperbolic metric) is Möbius invariant. The infinitesimal distance element for the Bergman metric on $\mathbb{D}$ is given by

$$\frac{|dz|}{1 - |z|^2}.$$

THEOREM 1.15 *An analytic function f in $\mathbb{D}$ belongs to the Bloch space if and only if there exists a positive constant C such that*

$$|f(z) - f(w)| \le C \beta(z, w)$$

holds for all z and w in $\mathbb{D}$.

Proof. If f is analytic in $\mathbb{D}$, then

$$f(z) - f(0) = z \int_0^1 f'(tz)\,dt$$

for all $z \in \mathbb{D}$. If f is in the Bloch space, then it follows that

$$\left| \frac{f(z) - f(0)}{z} \right| \leq \|f\|_{\mathcal{B}} \int_0^1 \frac{dt}{1 - |z|^2 t^2} = \|f\|_{\mathcal{B}}\, \beta(z, 0)$$

for all $z \in \mathbb{D}$. Replacing f by $f \circ \varphi_z$, replacing z by $\varphi_z(w)$, and applying the Möbius invariance of both $\| \cdot \|_{\mathcal{B}}$ and β, we arrive at

$$|f(z) - f(w)| \leq \|f\|_{\mathcal{B}}\, \beta(z, w)$$

for all $f \in \mathcal{B}$ and $z, w \in \mathbb{D}$.

The other direction follows from the identity

$$\lim_{w \to z} \frac{|f(w) - f(z)|}{\beta(w, z)} = (1 - |z|^2)|f'(z)|, \qquad z \in \mathbb{D},$$

which can easily be checked. ∎

Carefully examining the above proof, we find that

$$\|f\|_{\mathcal{B}} = \sup \left\{ \frac{|f(z) - f(w)|}{\beta(z, w)} : z, w \in \mathbb{D}, z \neq w \right\}.$$

With the help of functions of the type

$$f(z) = \frac{1}{2} \log \frac{1 + ze^{i\theta}}{1 - ze^{i\theta}}, \qquad z \in \mathbb{D},$$

we can also prove that

$$\beta(z, w) = \sup \left\{ |f(z) - f(w)| : \|f\|_{\mathcal{B}} \leq 1 \right\}.$$

These formulas exhibit the precise relationship between the Bloch space and the Bergman metric.

1.4 Duality of Bergman Spaces

Suppose $0 < p < +\infty$ and $-1 < \alpha < +\infty$. A linear functional F on A_α^p is called bounded if there exists a positive constant C such that $|F(f)| \leq C\|f\|_{\alpha,p}$ for all $f \in A_\alpha^p$, where

$$\|f\|_{\alpha,p} = \left[\int_{\mathbb{D}} |f(z)|^p \, dA_\alpha(z) \right]^{1/p}.$$

Recall that point evaluation at every $z \in \mathbb{D}$ is a bounded linear functional on every A_α^p. In particular, every weighted Bergman space A_α^p has nontrivial bounded linear functionals. We let A_α^{p*} denote the space of all bounded linear functionals. Then A_α^{p*} is a Banach space with the norm

$$\|F\| = \sup\left\{|F(f)| : \|f\|_{\alpha,p} \le 1\right\},$$

even though A_α^p is only a metric space when $0 < p < 1$.

THEOREM 1.16 *For* $1 < p < +\infty$ *and* $-1 < \alpha < +\infty$, *we have* $A_\alpha^{p*} = A_\alpha^q$ *under the integral pairing*

$$\langle f, g \rangle = \int_{\mathbb{D}} f(z)\overline{g(z)}\, dA_\alpha(z), \qquad f \in A_\alpha^p, \ g \in A_\alpha^q,$$

where q *is the conjugate exponent of* p: $p^{-1} + q^{-1} = 1$.

Note that the identification isomorphism $A_\alpha^{p*} = A_\alpha^q$ need not be isometric for $p \ne 2$.

Proof. By Hölder's inequality, every function g in A_α^q defines a bounded linear functional on A_α^p via the above integral pairing. Conversely, if F is a bounded linear functional on A_α^p, then by the Hahn-Banach extension theorem, F can be extended to a bounded linear functional (still denoted by F) on $L^p(\mathbb{D}, dA_\alpha)$ without increasing its norm. By the duality theory of L^p spaces, there exists a function φ in $L^q(\mathbb{D}, dA_\alpha)$ such that

$$F(f) = \int_{\mathbb{D}} f(z)\overline{\varphi(z)}\, dA_\alpha(z), \qquad f \in A_\alpha^p.$$

Writing $f = \mathbf{P}_\alpha f$ and using the fact that the operator $\mathbf{P}_\alpha$ is self-adjoint with respect to the inner product associated with dA_α, we obtain

$$F(f) = \int_{\mathbb{D}} f(z)\overline{\mathbf{P}_\alpha\varphi(z)}\, dA_\alpha(z), \qquad f \in A_\alpha^p.$$

Letting $g = \mathbf{P}_\alpha\varphi$ and using Theorem 1.10, we conclude that g is in A_α^q and that

$$F(f) = \int_{\mathbb{D}} f(z)\overline{g(z)}\, dA_\alpha(z)$$

for all $f \in A_\alpha^p$. ∎

In order to identify the dual space of A_α^p when $0 < p \le 1$, we first introduce a certain type of fractional differentiation and integration.

Let $H(\mathbb{D})$ denote the space of all analytic functions in $\mathbb{D}$ and equip $H(\mathbb{D})$ with the topology of "uniform convergence on compact subsets". Thus, a linear operator T on $H(\mathbb{D})$ is continuous if and only if $Tf_n \to Tf$ uniformly on compact subsets whenever $f_n \to f$ uniformly on compact subsets.

LEMMA 1.17 *For every* α, $-1 < \alpha < +\infty$, *there exists a unique linear operator* D^α *on* $H(\mathbb{D})$ *with the following properties:*

(1) D^α *is continuous on* $H(\mathbb{D})$.

(2) $D_z^\alpha\left[(1 - z\overline{w})^{-2}\right] = (1 - z\overline{w})^{-(2+\alpha)}$ *for every* $w \in \mathbb{D}$.

Proof. Recall that

$$\frac{1}{(1 - z\overline{w})^2} = \sum_{n=0}^{\infty} (n+1) z^n \overline{w}^n$$

and

$$\frac{1}{(1 - z\overline{w})^{2+\alpha}} = \sum_{n=0}^{\infty} \frac{\Gamma(n+2+\alpha)}{n!\,\Gamma(2+\alpha)} z^n \overline{w}^n.$$

If we define

$$D^\alpha(z^n) = \frac{\Gamma(n+2+\alpha)}{(n+1)!\,\Gamma(2+\alpha)} z^n$$

for all $n = 0, 1, 2, 3, \ldots$ and extend D^α linearly to the whole space $H(\mathbb{D})$, then the resulting operator D^α has the desired properties. The uniqueness also follows from the earlier series expansions. ∎

By Stirling's formula,

$$\frac{\Gamma(n+2+\alpha)}{(n+1)!\,\Gamma(2+\alpha)} \sim n^\alpha$$

as $n \to \infty$. Thus, the operator D^α can be considered a fractional differential operator of order α in the case $\alpha > 0$.

It is easy to see that for each $-1 < \alpha < +\infty$, the operator D^α can also be represented by

$$D^\alpha f(z) = \lim_{r \to 1^-} \int_{\mathbb{D}} \frac{f(rw)\,dA(w)}{(1 - z\overline{w})^{2+\alpha}}, \qquad z \in \mathbb{D},$$

for $f \in H(\mathbb{D})$. In particular, the limit above always exists. If f is in A^1, then

$$D^\alpha f(z) = \int_{\mathbb{D}} \frac{f(w)\,dA(w)}{(1 - z\overline{w})^{2+\alpha}}, \qquad z \in \mathbb{D}.$$

LEMMA 1.18 *For every* $-1 < \alpha < +\infty$, *the operator* D^α *is invertible on* $H(\mathbb{D})$.

Proof. Define an operator D_α on monomials by

$$D_\alpha(z^n) = \frac{(n+1)!\,\Gamma(2+\alpha)}{\Gamma(n+2+\alpha)} z^n$$

and extend D_α linearly to the whole space $H(\mathbb{D})$. Then D_α is a continuous linear operator on $H(\mathbb{D})$, and it is the inverse of D^α. ∎

It is easy to see that

$$D_\alpha f(z) = \lim_{r \to 1^-} (\alpha + 1) \int_{\mathbb{D}} \frac{(1 - |w|^2)^\alpha}{(1 - z\overline{w})^2} f(rw)\,dA(w), \qquad z \in \mathbb{D},$$

for every $f \in H(\mathbb{D})$. When $\alpha > 0$, the operator D_α is a fractional integral operator of order α.

We now proceed to identify the dual space of A_α^p when $0 < p \leq 1$. The following two lemmas will be needed for this purpose, but they are also of some independent interest.

LEMMA 1.19 *For every* $0 < p \leq 1$ *and* $-1 < \alpha < +\infty$, *there exists a constant* $C, 0 < C < +\infty$, *such that*

$$\int_{\mathbb{D}} |f(z)| (1 - |z|^2)^{-2+(2+\alpha)/p} \, dA(z) \leq C \, \|f\|_{\alpha,p}$$

for all $f \in A_\alpha^p$.

Proof. For $z \in \mathbb{D}$, we let $D(z)$ be the Euclidean disk centered at z with radius $(1 - |z|)/2$. By the subharmonicity of $|f|^p$, we have

$$|f(z)|^p \leq \frac{4}{(1 - |z|)^2} \int_{D(z)} |f(w)|^p \, dA(w).$$

Since $(1 - |w|) \sim (1 - |z|)$ for $w \in D(z)$, we can find a positive constant C such that

$$|f(z)| \leq C (1 - |z|^2)^{-(2+\alpha)/p} \|f\|_{\alpha,p}, \qquad z \in \mathbb{D},$$

for all $f \in A_\alpha^p$. For $0 < p \leq 1$, we can write

$$|f(z)| = |f(z)|^p \, |f(z)|^{1-p};$$

use the above inequality to estimate the second factor, and write out the remaining integral. What comes out is the desired result.   ∎

LEMMA 1.20 *Suppose* $-1 < \alpha < +\infty$ *and* f *is analytic in* $\mathbb{D}$. *If either* f *or the function* $(1 - |z|^2)^{-\alpha} f(z)$ *is bounded, then the function* $(1 - |z|^2)^\alpha D^\alpha f(z)$ *is area-integrable and*

$$\int_{\mathbb{D}} f(z) \overline{g(z)} \, dA(z) = (\alpha + 1) \int_{\mathbb{D}} D^\alpha f(z) \overline{g(z)} \, (1 - |z|^2)^\alpha \, dA(z),$$

for all $g \in H^\infty$.

Proof. The case $\alpha = 0$ is trivial. If $0 < \alpha < +\infty$, then by the integral representation of D^α and Theorem 1.7, the function $(1 - |z|^2)^\alpha D^\alpha f(z)$ is bounded.

If $-1 < \alpha < 0$ and f is bounded, then Theorem 1.7 and the integral representation of D^α imply that $D^\alpha f(z)$ is bounded, and hence the function $(1 - |z|^2)^\alpha D^\alpha f(z)$ is area-integrable.

If $-1 < \alpha < 0$ and $|f(z)| \leq C_1 (1 - |z|^2)^\alpha$, then by Theorem 1.7 and the integral representation of D^α, we have

$$(1 - |z|^2)^\alpha |D^\alpha f(z)| \leq C_2 (1 - |z|^2)^\alpha \log \frac{1}{1 - |z|^2}, \qquad \frac{1}{2} < |z| < 1,$$

and hence $(1 - |z|^2)^\alpha D^\alpha f(z)$ is area-integrable.

The desired identity now follows from the integral form of D^α, the reproducing property of $\mathbf{P}_\alpha$, and Fubini's theorem.   ∎

THEOREM 1.21 *Suppose $0 < p \leq 1$, $-1 < \alpha < +\infty$, and $\beta = (2+\alpha)/p - 2$. Then $A_\alpha^{p*} = \mathcal{B}$ under the integral pairing*

$$\langle f, g \rangle = \lim_{r \to 1^-} \int_{\mathbb{D}} f(rz)\overline{g(z)}(1 - |z|^2)^\beta \, dA(z),$$

where $f \in A_\alpha^p$ and $g \in \mathcal{B}$.

Proof. First assume $F \in A_\alpha^{p*}$ and $f \in A_\alpha^p$. Since $\|f - f_r\|_{\alpha, p} \to 0$ as $r \to 1^-$, we have

$$F(f) = \lim_{r \to 1^-} F(f_r), \qquad f \in A_\alpha^p.$$

Write

$$f_r(z) = \int_{\mathbb{D}} \frac{f_r(w) \, dA(w)}{(1 - z\overline{w})^2}, \qquad z \in \mathbb{D}.$$

Since the integral converges in A_α^p, the continuity of F implies that

$$F(f_r) = \int_{\mathbb{D}} f_r(w) \, F\left[\frac{1}{(1 - z\overline{w})^2}\right] dA(w).$$

where on the right hand side we think of F as acting with respect to the running variable z. Let

$$\overline{h(w)} = F\left[\frac{1}{(1 - z\overline{w})^2}\right], \qquad w \in \mathbb{D}.$$

Then h is analytic in $\mathbb{D}$ and

$$F(f_r) = \int_{\mathbb{D}} f_r(w) \, \overline{h(w)} \, dA(w).$$

Put

$$\beta = \frac{2+\alpha}{p} - 2$$

and apply Lemma 1.20, with the result

$$F(f_r) = (\beta + 1) \int_{\mathbb{D}} f_r(w) \, \overline{D^\beta h(w)} \, (1 - |w|^2)^\beta \, dA(w).$$

Let $g = (\beta + 1)D^\beta h$ and apply the second property of Lemma 1.17. Then

$$\overline{g(w)} = (\beta + 1) \, F\left[\frac{1}{(1 - z\overline{w})^{(2+\alpha)/p}}\right]$$

and

$$\overline{g'(w)} = \frac{(\beta + 1)(2 + \alpha)}{p} \, F\left[\frac{z}{(1 - z\overline{w})^{(2+\alpha)/p+1}}\right], \qquad w \in \mathbb{D}.$$

Using Theorem 1.7 and the boundedness of F, we easily check that g is in the Bloch space and that

$$F(f) = \lim_{r \to 1^-} \int_{\mathbb{D}} f(rw)\,\overline{g(w)}\,(1 - |z|^2)^\beta\,dA(w)$$

for every $f \in A_\alpha^p$.

Next, assume $g \in \mathcal{B}$. We show that the formula

$$F(f) = \lim_{r \to 1^-} \int_{\mathbb{D}} f_r(z)\,\overline{g(z)}\,(1 - |z|^2)^\beta\,dA(z), \quad f \in A_\alpha^p,$$

defines a bounded linear functional on A_α^p. By Theorem 1.12, there exists a function $\varphi \in L^\infty(\mathbb{D})$ such that

$$g(z) = \mathbf{P}_\beta\varphi(z) = (\beta + 1)\int_{\mathbb{D}} \frac{(1 - |w|^2)^\beta}{(1 - z\overline{w})^{2+\beta}}\,\varphi(w)\,dA(w), \quad z \in \mathbb{D}.$$

Using Fubini's theorem and the reproducing property of $\mathbf{P}_\beta$, we easily obtain

$$\int_{\mathbb{D}} f_r(z)\,\overline{g(z)}\,(1 - |z|^2)^\beta\,dA(z) = \int_{\mathbb{D}} f_r(w)\,\overline{\varphi(w)}\,(1 - |w|^2)^\beta\,dA(w).$$

By Lemma 1.19, we have

$$F(f) = \int_{\mathbb{D}} f(z)\,\overline{\varphi(z)}\,(1 - |z|^2)^\beta\,dA(z), \quad f \in A_\alpha^p,$$

and this defines a bounded linear functional on A_α^p. ∎

1.5 Notes

The notions of Bergman spaces, Bergman metric, and Bergman kernel are by now classical. General references include Bergman's book [19], Rudin's book [105], Dzhrbashian and Shamoyan's book [36], and Zhu's book [135]; see also Axler's treatise [14]. The classical reference for Bloch spaces is [9].

Theorems 1.7 and 1.10 were proved by Forelli and Rudin in [47] in the context of the open unit ball in $\mathbb{C}^n$. Proposition 1.11 should be attributed to Hardy and Littlewood [53]. That the Bergman projection maps $L^\infty(\mathbb{D})$ onto the Bloch space was first proved by Coifman, Rochberg, and Weiss [34]. The duality results in the case $1 \leq p < +\infty$ follow directly from the estimates of the Bergman kernel obtained by Forelli and Rudin [47]. The duality problem for $0 < p < 1$ has been studied by several authors, including [41] and [115]. Theorem 1.21 is from Zhu [136].

1.6   Exercises and Further Results

1. Suppose $1 < p < +\infty$. Show that $f_n \to 0$ weakly in A^p as $n \to +\infty$ if and only if $\{\|f_n\|_p\}_n$ is bounded and $f_n(z) \to 0$ uniformly on compact subsets of $\mathbb{D}$ as $n \to +\infty$.

2. For $-1 < \alpha < +\infty$, show that the dual space of the little Bloch space can be identified with A_α^1 under the integral pairing

$$\langle f, g \rangle = \lim_{r \to 1^-} \int_{\mathbb{D}} f(rz) \overline{g(z)} \, dA_\alpha(z), \quad f \in \mathcal{B}_0, \ g \in A_\alpha^1.$$

3. Show that $f_n \to 0$ in the weak-star topology of A_α^1 if and only if the sequence $\{f_n\}_n$ is bounded in norm and $f_n(z) \to 0$ uniformly on compact subsets of $\mathbb{D}$ as $n \to +\infty$.

4. For an analytic function f on $\mathbb{D}$, let f_n be the n-th Taylor polynomial of f. If $1 < p < +\infty$, $-1 < \alpha < +\infty$, and $f \in A_\alpha^p$, show that $f_n \to f$ in norm in A_α^p as $n \to +\infty$. Show that this is false if $0 < p \le 1$.

5. Prove Proposition 1.6.

6. If f is a function in the Bloch space, then there exists a positive constant C such that $|f(z)| \le C \log(1/(1 - |z|^2))$ for all z with $\frac{1}{2} \le |z| < 1$. Similarly, if f is in the little Bloch space, then for every $\varepsilon > 0$ there exists $\delta \in (0, 1)$ such that $|f(z)| < \varepsilon \log(1/(1 - |z|^2))$ for all z with $\delta < |z| < 1$.

7. For every $\delta \in (0, 1)$, there exists a positive constant $C = C(p, \delta)$ such that if f and g are analytic functions in $\mathbb{D}$ with $|f(z)| \le |g(z)|$ for $\delta < |z| < 1$, then

$$\int_{\mathbb{D}} |f(z)|^p \, dA(z) \le C \int_{\mathbb{D}} |g(z)|^p \, dA(z).$$

8. There exists an absolute constant σ, $0 < \sigma < 1$, such that

$$\int_{\mathbb{D}} |f(z)|^2 \, dA(z) \le \int_{\mathbb{D}} |g(z)|^2 \, dA(z)$$

whenever $|f(z)| \le |g(z)|$ on $\sigma < |z| < 1$, where f and g are analytic in $\mathbb{D}$. For details, see [87], [57], and [75].

9. For $1 < p < +\infty$, let B_p denote the space of analytic functions f in $\mathbb{D}$ such that

$$\int_{\mathbb{D}} (1 - |z|^2)^p |f'(z)|^p \, d\lambda(z) < +\infty,$$

where

$$d\lambda(z) = \frac{dA(z)}{(1 - |z|^2)^2}$$

is the Möbius-invariant measure on $\mathbb{D}$. These are called *analytic Besov spaces*. Show that the Bergman projection **P** maps $L^p(\mathbb{D}, d\lambda)$ onto B_p for all $1 < p < +\infty$. For details, see [135].

10. If $1 < p \leq 2$, $p^{-1} + q^{-1} = 1$, and

$$f(z) = \sum_{n=0}^{+\infty} a_n z^n$$

is in A^p, then

$$\sum_{n=0}^{+\infty} \frac{|a_n|^q}{(n+1)^{q-1}} < +\infty.$$

For problems 10–14, see [95].

11. Suppose $1 < p \leq 2$ and $p^{-1} + q^{-1} = 1$. If

$$\sum_{n=0}^{+\infty} \frac{|a_n|^p}{(n+1)^{p-1}} < +\infty,$$

then the function

$$f(z) = \sum_{n=0}^{+\infty} a_n z^n$$

belongs to A^q.

12. If $1 \leq p \leq 2$ and

$$f(z) = \sum_{n=0}^{+\infty} a_n z^n$$

belongs to A^p, then

$$\sum_{n=0}^{+\infty} \frac{|a_n|^p}{(n+1)^{3-p}} < +\infty.$$

13. If $1 \leq p \leq 2$ and the function

$$f(z) = \sum_{n=0}^{+\infty} a_n z^n$$

is in A^p, then the function

$$g(z) = \sum_{n=0}^{+\infty} \frac{a_n}{(n+1)^{1/p}} z^n$$

belongs to the Hardy space H^p.

14. If $2 \leq p < +\infty$ and the function

$$f(z) = \sum_{n=0}^{+\infty} a_n z^n$$

is in H^p, then the function

$$g(z) = \sum_{n=0}^{+\infty} (n+1)^{1/p} a_n z^n$$

belongs to A^p.

15. Suppose $0 < p < +\infty$ and f is analytic and bounded in $\mathbb{D}$. Then

$$\lim_{\alpha \to -1^+} \int_{\mathbb{D}} |f(z)|^p \, dA_\alpha(z) = \frac{1}{2\pi} \int_0^{2\pi} |f(e^{it})|^p \, dt.$$

16. Suppose φ is analytic in $\mathbb{D}$. Then $\varphi A_\alpha^p \subset A_\alpha^p$ if and only if $\varphi \in H^\infty$.

17. Suppose φ is analytic in $\mathbb{D}$. Show that $\varphi \mathcal{B} \subset \mathcal{B}$ if and only if $\varphi \in H^\infty$ and

$$\sup \left\{ (1 - |z|^2) |\varphi'(z)| \log[1/(1 - |z|^2)] : z \in \mathbb{D} \right\} < +\infty.$$

Formulate and prove a similar result for the little Bloch space. See [134].

18. Recall that $K_\alpha(z, w)$ is the reproducing kernel for the weighted Bergman space A_α^2. Show that

$$|K_\alpha(z, w)|^2 \leq K_\alpha(z, z) K_\alpha(w, w)$$

for all z and w in $\mathbb{D}$, and that

$$\sum_{j=1}^{N} \sum_{k=1}^{N} c_j \bar{c}_k K(z_j, z_k) \geq 0$$

for all $c_1, \ldots, c_N$ in $\mathbb{C}$ and all $z_1, \ldots, z_N$ in $\mathbb{D}$.

19. Let X be a linear space of analytic functions in $\mathbb{D}$. Suppose there exists a complete seminorm $\| \cdot \|$ on X such that:

   (1) $\|f \circ \varphi\| = \|f\|$ for any $f \in X$ and any Möbius map φ of the disk.
   (2) Point evaluations are bounded linear functionals on X.

   Then $X \subset \mathcal{B}$. See [104].

20. Let X be a linear space of analytic functions in $\mathbb{D}$. Suppose there exists a complete semi-inner product $\langle \cdot, \cdot \rangle$ on X such that:

   (1) $\langle f \circ \varphi, g \circ \varphi \rangle = \langle f, g \rangle$ for all f, g in X and any Möbius map φ of the disk.
   (2) Point evaluations are bounded linear functionals on X.

   Then $X = B_2$ (See Exercise 9). Note that B_2 is usually called the Dirichlet space and frequently denoted by $\mathcal{D}$. See [11].

21. Show that there exist infinite Blaschke products in the little Bloch space. See [23].

22. If $f \in A^p$ and $\varphi : \mathbb{D} \to \mathbb{D}$ is analytic, then $f \circ \varphi \in A^p$. See [135].

23. For $0 < p < +\infty$ and $-1 < \alpha < +\infty$, define

$$d_{p,\alpha}(z, w) = \sup \{|f(z) - f(w)| : \|f\|_{p,\alpha} \leq 1\}, \qquad z, w \in \mathbb{D}.$$

Show that

$$\lim_{w \to z} \frac{d_{p,\alpha}(w, z)}{|w - z|} = \sup \{|f'(z)| : \|f\|_{p,\alpha} \leq 1\},$$

for each $z \in \mathbb{D}$. See [137].

24. There exist functions in the little Bloch space whose Taylor series do not converge in norm.

25. Let B_1 consist of analytic functions f in $\mathbb{D}$ such that $f'' \in A^1$. Show that $f \in B_1$ if and only if there exists a sequence $\{c_n\}_n$ in l^1 and a sequence $\{a_n\}_n$ in $\mathbb{D}$ such that

$$f(z) = \sum_{n=0}^{+\infty} c_n \frac{a_n - z}{1 - \bar{a}_n z}, \qquad z \in \mathbb{D}.$$

26. Show that the Bergman projection $\mathbf{P}$ maps the space $L^1(\mathbb{D}, d\lambda)$ onto B_1, where $d\lambda$ is as in Exercise 9.

27. Show that for $f \in H(\mathbb{D})$ and $1 < p < +\infty$, we have $f \in B_p$ if and only if

$$\int_{\mathbb{D}} \int_{\mathbb{D}} \frac{|f(z) - f(w)|^p}{|1 - z\bar{w}|^4} \, dA(z) \, dA(w) < +\infty.$$

See [135].

28. For each $1 \leq p < +\infty$ and $-1 < \alpha < +\infty$, there exists a positive constant C such that

$$\|f\|_{p,\alpha} \leq C \|\operatorname{Re} f\|_{p,\alpha}$$

for all $f \in A_\alpha^p$ with $f(0) = 0$.

29. For each $1 \leq p < +\infty$ and $-1 < \alpha < +\infty$, there exists a positive constant C such that

$$\int_{\mathbb{D}} |\tilde{u}(z)|^p \, dA_\alpha(z) \leq C \int_{\mathbb{D}} |u(z)|^p \, dA_\alpha(z)$$

for all harmonic functions u in $\mathbb{D}$, where $\tilde{u}$ is the harmonic conjugate of u with $\tilde{u}(0) = 0$.

30. Solve the extremal problem

$$\inf \{\|f\|_{p,\alpha} : f \in A_\alpha^p, \; f(w) = 1\},$$

where w is any point in $\mathbb{D}$.

31. Try to extend Proposition 1.11 to the case $0 < p < 1$.

2

The Berezin Transform

In this chapter we consider an analogue of the Poisson transform in the context of Bergman spaces, called the Berezin transform. We show that its fixed points are precisely the harmonic functions. We introduce a space of BMO type on the disk, the analytic part of which is the Bloch space, and characterize this space in terms of the Berezin transform.

2.1 Algebraic Properties

Recall that one way to obtain the Poisson kernel is to start out with a harmonic function h in $\mathbb{D}$ that is continuous up to the boundary and apply the mean value property to get

$$h(0) = \frac{1}{2\pi} \int_0^{2\pi} h(e^{it}) \, dt.$$

Replace h by $h \circ \varphi_z$, where φ_z is the Möbius map interchanging 0 and z,

$$\varphi_z(w) = \frac{z - w}{1 - \bar{z}w}, \qquad w \in \mathbb{D},$$

and make a change of variables. Then

$$h(z) = \frac{1}{2\pi} \int_0^{2\pi} \frac{1 - |z|^2}{|1 - z \, e^{-it}|^2} h(e^{it}) \, dt.$$

This is the Poisson formula for harmonic functions. The integral kernel

$$P(e^{it}, z) = \frac{1 - |z|^2}{|1 - z\,e^{-it}|^2}$$

is the *Poisson kernel*, and the transform

$$f \in L^1(\mathbb{T}, dt) \mapsto \frac{1}{2\pi} \int_0^{2\pi} P(e^{it}, z)\, f(e^{it})\, dt$$

is the *Poisson transform*.

Now, let us start out with a bounded harmonic function h in $\mathbb{D}$ and apply the area version of the mean value property

$$h(0) = \int_{\mathbb{D}} h(w)\, dA(w).$$

Again replace h by $h \circ \varphi_z$ and make a change of variables. We get

$$h(z) = \int_{\mathbb{D}} \frac{(1 - |z|^2)^2}{|1 - z\overline{w}|^4} h(w)\, dA(w), \quad z \in \mathbb{D}.$$

By a simple limit argument, we see that the formula above also holds for every harmonic function h in $L^1(\mathbb{D}, dA)$.

For every function $f \in L^1(\mathbb{D}, dA)$, we define

$$\mathbf{B}f(z) = \int_{\mathbb{D}} \frac{(1 - |z|^2)^2}{|1 - z\overline{w}|^4} f(w)\, dA(w), \quad z \in \mathbb{D}.$$

The operator $\mathbf{B}$ will be called the *Berezin transform*.

Actually, we shall need to use a family of Berezin type operators. Recall that for $\alpha > -1$, we have

$$dA_\alpha(z) = (\alpha + 1)(1 - |z|^2)^\alpha\, dA(z).$$

Suppose h is a bounded harmonic function on $\mathbb{D}$. The mean value property together with the rotation invariance of dA_α implies that

$$h(0) = (\alpha + 1) \int_{\mathbb{D}} h(w)(1 - |w|^2)^\alpha\, dA(w).$$

Replacing h by $h \circ \varphi_z$ and making a change of variables, we get

$$h(z) = (\alpha + 1) \int_{\mathbb{D}} \frac{(1 - |z|^2)^{\alpha+2}(1 - |w|^2)^\alpha}{|1 - z\overline{w}|^{4+2\alpha}} h(w)\, dA(w), \quad z \in \mathbb{D}.$$

Thus, for $f \in L^1(\mathbb{D}, dA_\alpha)$ we write

$$\mathbf{B}_\alpha f(z) = (\alpha + 1) \int_{\mathbb{D}} \frac{(1 - |z|^2)^{\alpha+2}(1 - |w|^2)^\alpha}{|1 - z\overline{w}|^{4+2\alpha}} f(w)\, dA(w), \quad z \in \mathbb{D}.$$

A change of variables shows that we also have

$$\mathbf{B}_\alpha f(z) = \int_{\mathbb{D}} f \circ \varphi_z(w)\, dA_\alpha(w), \quad z \in \mathbb{D},$$

for every $f \in L^1(\mathbb{D}, dA_\alpha)$. Note that $\mathbf{B}_0 = \mathbf{B}$.

PROPOSITION 2.1 *Suppose* $-1 < \alpha < +\infty$ *and* φ *is a Möbius map of the disk. Then*

$$(\mathbf{B}_\alpha f) \circ \varphi = \mathbf{B}_\alpha (f \circ \varphi)$$

for every $f \in L^1(\mathbb{D}, dA_\alpha)$.

Proof. For every $z \in \mathbb{D}$, the Möbius map $\varphi_{\varphi(z)} \circ \varphi \circ \varphi_z$ fixes the origin. Thus, there exists a unimodular number ζ (depending on z) such that

$$\varphi_{\varphi(z)} \circ \varphi \circ \varphi_z(w) = \zeta w, \quad \text{that is,} \quad \varphi \circ \varphi_z(w) = \varphi_{\varphi(z)}(\zeta w),$$

for all $w \in \mathbb{D}$. It follows that

$$
\begin{aligned}
\mathbf{B}_\alpha (f \circ \varphi)(z) &= \int_{\mathbb{D}} f \circ \varphi \circ \varphi_z(w) \, dA_\alpha(w) \\
&= \int_{\mathbb{D}} f \circ \varphi_{\varphi(z)}(\zeta w) \, dA_\alpha(w) \\
&= (\mathbf{B}_\alpha f)(\varphi(z)).
\end{aligned}
$$

In the last equality above, we used the rotation invariance of dA_α. ∎

Since dA_α is a probability measure for $-1 < \alpha < +\infty$, the operator $\mathbf{B}_\alpha$ is clearly bounded on $L^\infty(\mathbb{D})$. Actually, $\|\mathbf{B}_\alpha f\|_\infty \le \|f\|_\infty$ for all $-1 < \alpha < +\infty$.

PROPOSITION 2.2 *Suppose* $-1 < \alpha < +\infty$, $1 \le p < +\infty$, *and that* $\beta \in \mathbb{R}$. *Then* $\mathbf{B}_\alpha$ *is bounded on* $L^p(\mathbb{D}, dA_\beta)$ *if and only if* $-(\alpha+2)p < \beta+1 < (\alpha+1)p$.

Proof. This is a direct consequence of Theorem 1.9. ∎

Fix an α, $-1 < \alpha < +\infty$. By Proposition 2.2, the operator $\mathbf{B}_\beta$ is bounded on $L^1(\mathbb{D}, dA_\alpha)$ if and only if $\beta > \alpha$. Actually, $\mathbf{B}_\beta$ is uniformly bounded on $L^1(\mathbb{D}, dA_\alpha)$ as $\beta \to +\infty$. To see this, first use Fubini's theorem to obtain

$$\int_{\mathbb{D}} |\mathbf{B}_\beta f(z)| \, dA_\alpha(z) \le (\beta+1) \int_{\mathbb{D}} |f(w)| \int_{\mathbb{D}} \frac{(1-|z|^2)^{2+\beta}}{|1-z\overline{w}|^{2\beta+4}} \, dA_\alpha(z) \, dA_\beta(w).$$

Making the change of variables $z \mapsto \varphi_w(z)$ in the inner integral, we get

$$\int_{\mathbb{D}} |\mathbf{B}_\beta f(z)| \, dA_\alpha(z) \le (\beta+1) \int_{\mathbb{D}} |f(w)| \int_{\mathbb{D}} \frac{(1-|z|^2)^{2+\alpha+\beta}}{|1-z\overline{w}|^{2\alpha+4}} \, dA(z) \, dA_\alpha(w).$$

Note that for all z, $w \in \mathbb{D}$, we have

$$\frac{1}{|1-z\overline{w}|} \le \frac{1}{1-|z|} = \frac{1+|z|}{1-|z|^2} \le \frac{2}{1-|z|^2}.$$

It follows that for $\beta > \alpha + 1$,

$$\int_{\mathbb{D}} |\mathbf{B}_\beta f(z)| \, dA_\alpha(z) \le C \int_{\mathbb{D}} |f(w)| \, dA_\alpha(w) \int_{\mathbb{D}} (1-|z|^2)^{\beta-(\alpha+2)} \, dA(z),$$

where $C = 4^{\alpha+2}(\beta + 1)$; that is,

$$\int_{\mathbb{D}} |\mathbf{B}_\beta f(z)|\, dA_\alpha(z) \le \frac{4^{\alpha+2}(\beta + 1)}{\beta - \alpha - 1} \int_{\mathbb{D}} |f(w)|\, dA_\alpha(w).$$

This clearly shows that $\mathbf{B}_\beta$ is uniformly bounded on $L^1(\mathbb{D}, dA_\alpha)$ when $\beta \to +\infty$.

PROPOSITION 2.3 *Suppose* $-1 < \alpha < +\infty$ *and* $f \in C(\overline{\mathbb{D}})$. *Then we have* $\mathbf{B}_\alpha f \in C(\overline{\mathbb{D}})$ *and* $f - \mathbf{B}_\alpha f \in C_0(\mathbb{D})$.

Proof. We use the formula

$$\mathbf{B}_\alpha f(z) = \int_{\mathbb{D}} f \circ \varphi_z(w)\, dA_\alpha(w), \qquad z \in \mathbb{D}.$$

Since $\varphi_z(w) \to z_0$ as $z \to z_0 \in \mathbb{T}$, the dominated convergence theorem shows that $\mathbf{B}_\alpha f(z) \to f(z_0)$ whenever $z \to z_0 \in \mathbb{T}$. This shows that $f - \mathbf{B}_\alpha f \in C_0(\mathbb{D})$. In particular, we have $\mathbf{B}_\alpha f \in C(\overline{\mathbb{D}})$.   ∎

PROPOSITION 2.4 *If* $-1 < \beta < \alpha < +\infty$, *then* $\mathbf{B}_\alpha \mathbf{B}_\beta = \mathbf{B}_\beta \mathbf{B}_\alpha$ *on* $L^1(\mathbb{D}, dA_\beta)$.

Proof. By Proposition 2.2, the operator $\mathbf{B}_\alpha$ is bounded on $L^1(\mathbb{D}, dA_\beta)$. Thus, $\mathbf{B}_\beta \mathbf{B}_\alpha f$ makes sense for every $f \in L^1(\mathbb{D}, dA_\beta)$. Also, the operator $\mathbf{B}_\beta$ maps $L^1(\mathbb{D}, dA_\beta)$ boundedly into $L^1(\mathbb{D}, dA_\alpha)$. Hence $\mathbf{B}_\alpha \mathbf{B}_\beta f$ is well defined for $f \in L^1(\mathbb{D}, dA_\beta)$.

Let $f \in L^1(\mathbb{D}, dA_\beta)$. To prove $\mathbf{B}_\alpha \mathbf{B}_\beta f = \mathbf{B}_\beta \mathbf{B}_\alpha f$ it suffices to show – according to Proposition 2.1 – that $\mathbf{B}_\alpha \mathbf{B}_\beta f(0) = \mathbf{B}_\beta \mathbf{B}_\alpha f(0)$. Now,

$$\begin{aligned} \mathbf{B}_\alpha \mathbf{B}_\beta f(0) &= \int_{\mathbb{D}} \mathbf{B}_\beta f(z)\, dA_\alpha(z) \\ &= C \int_{\mathbb{D}} f(w)\, dA(w) \int_{\mathbb{D}} \frac{(1 - |w|^2)^\beta (1 - |z|^2)^{\alpha+\beta+2}}{|1 - z\overline{w}|^{2\beta+4}}\, dA(z), \end{aligned}$$

where $C = (\alpha + 1)(\beta + 1)$. Making the change of variables $z \mapsto \varphi_w(z)$ in the inner integral, we find that α and β will switch positions, and hence $\mathbf{B}_\alpha \mathbf{B}_\beta f(0) = \mathbf{B}_\beta \mathbf{B}_\alpha f(0)$.   ∎

PROPOSITION 2.5 *Let* $-1 < \alpha < +\infty$ *and* $f \in L^1(\mathbb{D}, dA_\alpha)$. *Then* $\mathbf{B}_\beta f \to f$ *in* $L^1(\mathbb{D}, dA_\alpha)$ *as* $\beta \to +\infty$.

Proof. First, assume that f is continuous on the closed disk. Since dA_β is a probability measure, we have the formula

$$\mathbf{B}_\beta f(z) - f(z) = (\beta + 1) \int_{\mathbb{D}} (1 - |w|^2)^\beta \big(f \circ \varphi_z(w) - f(z)\big)\, dA(w).$$

Writing $\mathbb{D}$ as the union of a slightly smaller disk $\mathbb{D}_r$ of radius $r \in (0, 1)$ centered at 0 and an annulus, estimating the integral over $\mathbb{D}_r$ by the uniform continuity of

f on $\mathbb{D}$, and estimating the integral over $\mathbb{D} \setminus \mathbb{D}_r$ using the fact that f is bounded and that

$$(\beta + 1) \int_{\mathbb{D} \setminus \mathbb{D}_r} (1 - |z|^2)^\beta \, dA(z) \to 0, \quad \beta \to +\infty,$$

we easily find that

$$\mathbf{B}_\beta f(z) \to f(z), \quad z \in \mathbb{D},$$

as $\beta \to +\infty$. Since $\|\mathbf{B}_\beta f\|_\infty \leq \|f\|_\infty$ for every β, it follows from the dominated convergence theorem that $\mathbf{B}_\beta f \to f$ in $L^1(\mathbb{D}, dA_\alpha)$ as $\beta \to +\infty$. The general case then follows from a simple limit argument, using the density of $C(\overline{\mathbb{D}})$ in $L^1(\mathbb{D}, dA_\alpha)$ and the uniform boundedness of the operators $\mathbf{B}_\beta$ on $L^1(\mathbb{D}, dA_\alpha)$. ∎

PROPOSITION 2.6 *For each α with $-1 < \alpha < +\infty$, the operator $\mathbf{B}_\alpha$ is one-to-one on the space $L^1(\mathbb{D}, dA_\alpha)$.*

Proof. Suppose $f \in L^1(\mathbb{D}, dA_\alpha)$ and $\mathbf{B}_\alpha f = 0$. Let

$$F(z) = \int_{\mathbb{D}} \frac{f(w) \, dA_\alpha(w)}{(1 - z\overline{w})^{2+\alpha}(1 - \overline{z}w)^{2+\alpha}}, \quad z \in \mathbb{D}.$$

Since

$$F(z) = \frac{\mathbf{B}_\alpha f(z)}{(1 - |z|^2)^{2+\alpha}},$$

we have $F(z) = 0$ throughout $\mathbb{D}$, and hence

$$\frac{\partial^{n+m} F}{\partial z^n \partial \overline{z}^m}(0) = 0$$

for all nonnegative integers n and m. Differentiating under the integral sign, we find that

$$\int_{\mathbb{D}} \overline{w}^n \, w^m \, f(w) \, dA_\alpha(w) = 0$$

for all nonnegative integers n and m. This clearly implies that $f = 0$. ∎

2.2   Harmonic Functions

Recall that if f is a harmonic function in $L^1(\mathbb{D}, dA)$, then $\mathbf{B}f = f$. In this section we prove the converse, that is, the conditions $f \in L^1(\mathbb{D}, dA)$ and $\mathbf{B}f = f$ imply that f is harmonic.

In dealing with harmonic functions on the unit disk, we find it more convenient to use the invariant Laplacian $\boldsymbol{\Delta}$ instead of the usual Laplacian Δ. We shall use the operator

$$\Delta = \frac{\partial^2}{\partial z \partial \overline{z}} = \frac{1}{4}\left(\frac{\partial^2}{\partial x^2} + \frac{\partial^2}{\partial y^2}\right),$$

where $z = x + iy$, as the Laplacian (this is a quarter of the standard Laplacian). This renormalization has the advantage that certain formulæ assume a particularly attractive form; for instance, if f is a holomorphic function, then $\Delta |f|^2 = |f'|^2$. The *invariant Laplacian* is defined by

$$\boldsymbol{\Delta} f(z) = (1 - |z|^2)^2 \Delta f(z).$$

As its name suggests, the invariant Laplacian $\boldsymbol{\Delta}$ is Möbius invariant, namely,

$$\boldsymbol{\Delta}(f \circ \varphi)(z) = (\boldsymbol{\Delta} f)(\varphi(z))$$

for every Möbius map φ of the disk. We may interpret $\boldsymbol{\Delta}$ as the Laplace-Beltrami operator on $\mathbb{D}$, provided $\mathbb{D}$ is supplied with the Poincaré metric.

PROPOSITION 2.7 *For* $-1 < \alpha < +\infty$, *the identity*

$$\boldsymbol{\Delta} \mathbf{B}_\alpha f = (\alpha + 1)(\alpha + 2)\left(\mathbf{B}_\alpha f - \mathbf{B}_{\alpha+1} f\right)$$

holds for every $f \in L^1(\mathbb{D}, dA_\alpha)$.

Proof. By the Möbius invariance of both $\mathbf{B}_\alpha$ and $\boldsymbol{\Delta}$, it suffices to show that

$$\boldsymbol{\Delta} \mathbf{B}_\alpha f(0) = (\alpha + 1)(\alpha + 2)\left(\mathbf{B}_\alpha f(0) - \mathbf{B}_{\alpha+1} f(0)\right)$$

holds for every $f \in L^1(\mathbb{D}, dA_\alpha)$. This follows from differentiating under the integral sign and regrouping terms. ∎

In other words, for $-1 < \alpha < +\infty$, we have the operator identity

$$\mathbf{B}_{\alpha+1} = \left(1 - \frac{\boldsymbol{\Delta}}{(\alpha + 1)(\alpha + 2)}\right) \mathbf{B}_\alpha.$$

The following conclusion is immediate.

COROLLARY 2.8 *Suppose* n *is a positive integer, and set*

$$G_n(z) = \prod_{k=1}^{n} \left(1 - \frac{z}{k(k+1)}\right), \qquad z \in \mathbb{C}.$$

Then $\mathbf{B}_n = G_n(\boldsymbol{\Delta})\mathbf{B}$ *on* $L^1(\mathbb{D}, dA)$.

Let

$$G(z) = \prod_{k=1}^{+\infty} \left(1 - \frac{z}{k(k+1)}\right).$$

It is clear that G is an entire function and that $G_n(z) \to G(z)$ uniformly on compact sets of $\mathbb{C}$. It should not be surprising now to see that the function G plays an important role in our analysis of the Berezin transform. Throughout this section, we let

$$\Sigma = \{w \in \mathbb{C} : -1 < \operatorname{Re} w < 2\},$$

and

$$\Omega = \{z \in \mathbb{C} : z = -w(1 - w) \text{ for some } w \in \Sigma\}.$$

By the open mapping theorem for analytic functions, Ω is a connected open subset of $\mathbb{C}$.

PROPOSITION 2.9 *If* $z = -w(1 - w)$, *then*

$$G(z) = \frac{\sin(\pi w)}{\pi w(1 - w)}.$$

Furthermore, $G(z) \neq 1$ *for* $z \in \Omega \setminus \{0\}$.

Proof. The k-th factor in the product

$$G(z) = \prod_{k=1}^{+\infty} \left(1 + \frac{w(1 - w)}{k(k + 1)}\right)$$

equals

$$\left[\left(1 + \frac{w}{k}\right) e^{-w/k}\right]\left[\left(1 + \frac{1 - w}{k}\right) e^{-(1-w)/k}\right]\left[\left(1 + \frac{1}{k}\right)^{-1} e^{1/k}\right].$$

The desired formula for G then follows from the well-known identities

$$\frac{1}{\Gamma(z + 1)} = e^{-\gamma z} \prod_{k=1}^{+\infty} \left(1 + \frac{z}{k}\right) e^{-z/k}$$

and

$$\Gamma(z)\Gamma(1 - z) = \frac{\pi}{\sin \pi z}.$$

To show that $G(z) \neq 1$ for $z \in \Omega \setminus \{0\}$, it suffices to show that the function

$$\Phi(w) = \frac{\pi w(1 - w)}{\sin(\pi w)}$$

has $\Phi(w) \neq 1$ for $w \in \Sigma \setminus \{0, 1\}$.

Observe that Φ has the symmetry property

$$\Phi\left(\frac{1}{2} + w\right) = \Phi\left(\frac{1}{2} - w\right),$$

and that it has

$$\Phi\left(\frac{1}{2} + iy\right) = \frac{\pi\left(y^2 + \frac{1}{4}\right)}{\cosh(\pi y)} < 1$$

for all real y. Thus, it suffices to show that the only solution of $\Phi(w) = 1$ in the strip $-1 < \operatorname{Re} w < \frac{1}{2}$ is $w = 0$. We achieve this with the help of the Argument Principle.

By an easy estimate, we can choose a positive number A such that $|\Phi(w)| < 1$ for all $w = u + iv$, where $-1 \leq u \leq \frac{1}{2}$ and v is real with $|v| \geq A$. We now consider the positively oriented contour γ given by the following picture.

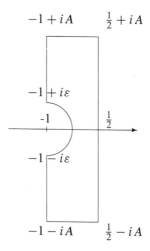

$-1 + iA$ $\frac{1}{2} + iA$

$-1 + i\varepsilon$

-1 $\frac{1}{2}$

$-1 - i\varepsilon$

$-1 - iA$ $\frac{1}{2} - iA$

We proceed to show that the image of the contour γ, $\Phi(\gamma)$, winds around the point 1 exactly once. We start from $w = \frac{1}{2}$ on γ and move upwards. The curve $\Phi(\gamma)$ will then start at $\pi/4$ and move toward 0 along the real axis. When w makes a left turn at $\frac{1}{2} + iA$ and moves horizontally to the left, the curve $\Phi(\gamma)$ oscillates in the half-plane to the left of the point 1. For w between $-1 + iA$ and $-1 + i\varepsilon$, we have

$$\Phi(-1 + iv) = \frac{\pi}{\sinh(\pi v)}\left[-3v + i(v^2 - 2)\right].$$

This part of $\Phi(\gamma)$ meets the real axis to the left of the point 1 when (and only when) $v = \sqrt{2}$. So far, the image of γ under Φ has not reached the real axis to the right of the point 1. Next, consider $\Phi(w)$ for w on the little semicircle near the point $w = -1$. An easy calculation shows that

$$\Phi(w) = \frac{2}{w + 1} + \Psi(w),$$

where $\Psi(w)$ is analytic near $w = -1$. It follows that

$$\Phi(-1 + \varepsilon e^{it}) = \frac{2}{\varepsilon}e^{-it} + O(\varepsilon).$$

This shows that if $\varepsilon > 0$ is small enough, then the curve

$$\Phi(-1 + \varepsilon e^{it}), \qquad -\frac{\pi}{2} \le t \le \frac{\pi}{2},$$

crosses the real axis near the point $2/\varepsilon$; the winding number of $\Phi(\gamma)$ around 1 will not depend on the exact number of times the above curve crosses the real axis. Finally, by the analysis above and the symmetry relation $\overline{\Phi(w)} = \Phi(\overline{w})$, when w moves downward from $-1 - \varepsilon i$ and comes back to the starting point $\frac{1}{2}$, the image $\Phi(w)$ will not cross the real axis from the right-hand side of the point 1. We conclude that the curve $\Phi(\gamma)$ winds around the point 1 exactly once. ∎

We need the following facts about eigenfunctions of the invariant Laplacian before we can prove our main result.

PROPOSITION 2.10 *Suppose α and λ are complex numbers related by $\lambda = -\alpha(1 - \alpha)$. Let X_λ be the eigenspace of Δ corresponding to the eigenvalue λ. Let*

$$g_\alpha(z) = \frac{(1 - |z|^2)^\alpha}{2\pi} \int_0^{2\pi} \frac{d\theta}{|1 - ze^{-i\theta}|^{2\alpha}}, \qquad z \in \mathbb{D}.$$

Then we have:

(1) *The function g_α belongs to X_λ.*

(2) *If $f \in X_\lambda$ and f is radial, then $f = f(0)g_\alpha$.*

(3) *The space X_λ contains a nonzero function in $L^1(\mathbb{D}, dA)$ if and only if $\alpha \in \Sigma$.*

Proof. Let $P(e^{i\theta}, z)$ be the Poisson kernel. Then the function g_α can be rewritten as

$$g_\alpha(z) = \frac{1}{2\pi} \int_0^{2\pi} \left[P(e^{i\theta}, z) \right]^\alpha d\theta, \qquad z \in \mathbb{D}.$$

Part (1) now follows from differentiating under the integral sign and the fact that $P(e^{i\theta}, z)$ is harmonic in z.

To prove (2), we let $f(z) = g(|z|^2)$ be a radial function in X_λ. It is easy to check that the function $g(x), 0 < x < 1$, is a solution to the following differential equation:

$$x(1 - x)^2 g''(x) + (1 - x)^2 g'(x) = \lambda g(x), \qquad 0 < x < 1.$$

The solution space of the above differential equation is two-dimensional, and we can exhibit a basis for it. In fact, by (1), the function $g_1(x) = g_\alpha(\sqrt{x})$ is a solution, and an easy calculation shows that the function

$$g_2(x) = g_1(x) \int_1^x \frac{dt}{t\, g_1(t)^2}, \qquad 0 < x < 1,$$

is also a solution. It is obvious that g_1 and g_2 are linearly independent. Thus, there exist constants a and b such that $g = a\,g_1 + b\,g_2$. Since the functions g and g_1 are bounded near $x = 0$ and g_2 is unbounded near $x = 0$, we must have $b = 0$ and hence $g = g(0)g_1$, so that $f = f(0)g_\alpha$.

To prove (3), let us assume that X_λ contains a nonzero function $f \in L^1(\mathbb{D}, dA)$. By invariance, we can also assume that $f(0) \neq 0$. It is easy to check that $f \in X_\lambda$ implies that its radialization

$$f^\#(z) = \frac{1}{2\pi} \int_0^{2\pi} f(ze^{it}) dt, \qquad z \in \mathbb{D},$$

also belongs to X_λ. By (2), we have $f^\# = f(0)g_\alpha$. This clearly implies that X_λ contains a nonzero function of $L^1(\mathbb{D}, dA)$ if and only if $g_\alpha \in L^1(\mathbb{D}, dA)$. By Theorem 1.7, the function g_α is in $L^1(\mathbb{D}, dA)$ if and only if $\alpha \in \Sigma$.   ∎

We now prove the main result of the section.

THEOREM 2.11 *Suppose* $f \in L^1(\mathbb{D}, dA)$. *Then* f *is harmonic if and only if* $\mathbf{B}f = f$.

Proof. Let M be the set of fixed points of $\mathbf{B}$ in $L^1(\mathbb{D}, dA)$. It is easy to see that M is a closed subspace of $L^1(\mathbb{D}, dA)$. We already know that every harmonic function in $L^1(\mathbb{D}, dA)$ belongs to M. We proceed to show that every function in M is harmonic.

By the integral formula for the operator $\mathbf{B}$, every function satisfying $f = \mathbf{B}f$ is real-analytic in $\mathbb{D}$. In particular, we can apply the Laplacian to every function in M. Let Δ_M be the restriction of Δ to M. By Proposition 2.7,

$$\Delta_M f = 2(f - \mathbf{B}_1 f), \qquad f \in M.$$

Since $\mathbf{B}_1$ is bounded on $L^1(\mathbb{D}, dA)$, we see that Δ_M maps M boundedly into $L^1(\mathbb{D}, dA)$. Moreover, since $\mathbf{B}\mathbf{B}_1 = \mathbf{B}_1\mathbf{B}$, we have

$$\mathbf{B}\Delta_M f = 2(\mathbf{B}f - \mathbf{B}\mathbf{B}_1 f) = 2(f - \mathbf{B}_1 f) = \Delta_M f, \qquad f \in M.$$

Thus Δ_M maps M into M, and hence Δ_M is a bounded linear operator on the Banach space M.

Recall from Corollary 2.8 that

$$\mathbf{B}_n f = G_n(\Delta)\mathbf{B}f = G_n(\Delta_M)f, \qquad f \in M.$$

Since $G_n \to G$ uniformly over compact subsets of $\mathbb{C}$, and Δ_M is a bounded linear operator on M, we have $G_n(\Delta_M) \to G(\Delta_M)$. This together with Proposition 2.5 shows that $G(\Delta_M)f = f$ for every $f \in M$, making $G(\Delta_M)$ the identity operator on M.

Suppose λ is an eigenvalue of Δ_M. By Proposition 2.10, we must have $\lambda \in \Omega$. Also, if f is a nonzero eigenfunction corresponding to λ, then

$$f = G(\Delta_M)f = G(\lambda)f.$$

It follows that $G(\lambda) = 1$. By Proposition 2.9, we must then have $\lambda = 0$. Thus, the only eigenvalue of the operator Δ_M is 0.

Recall that $G(z) - 1 = zH(z)$, where H is an entire function with $H(0) \neq 0$. By the holomorphic functional calculus (see, for instance Rudin's book [106]), we have

$$0 = G(\Delta_M) - I = H(\Delta_M)\Delta_M,$$

where I is the identity operator on M. Since the only eigenvalue of Δ_M is 0, the spectral mapping theorem (see, for instance, [106, Theorem 10.33]) implies that the only eigenvalue of $H(\Delta_M)$ is $H(0) \neq 0$. In particular, $H(\Delta_M)$ is one-to-one: after all, if for some $f \in M$, $H(\Delta_M)f = 0$ holds, then f is an eigenvector for (eigenvalue) 0, which is possible only for $f = 0$, as 0 fails to be an eigenvalue of $H(\Delta_M)$. Now if $f \in M$, then $H(\Delta_M)\Delta_M f = 0$. It follows that $\Delta_M f = 0$; in other words, f is harmonic. ∎

2.3   Carleson-Type Measures

Just as we can integrate the Poisson kernel against a measure on the circle, we can also integrate the kernel of the Berezin transform against a measure on the disk. More specifically, for a positive Borel measure μ on $\mathbb{D}$, we consider the function

$$\mathbf{B}\mu(z) = (1 - |z|^2)^2 \int_{\mathbb{D}} \frac{d\mu(w)}{|1 - z\overline{w}|^4}, \qquad z \in \mathbb{D}.$$

In this section we characterize those positive Borel measures μ on $\mathbb{D}$ such that $\mathbf{B}\mu$ is bounded. As a by-product we also characterize those measures μ such that $\mathbf{B}\mu(z) \to 0$ as $|z| \to 1^-$.

Recall that

$$\beta(z, w) = \frac{1}{2} \log \frac{|1 - z\overline{w}| + |z - w|}{|1 - z\overline{w}| - |z - w|}$$

is the Bergman metric on $\mathbb{D}$. Throughout this section, we fix some positive radius $0 < r < +\infty$ and *consider disks $D(z, r)$ in the Bergman metric*. The set

$$D(z, r) = \{w \in \mathbb{D} : \beta(z, w) < r\}, \qquad z \in \mathbb{D},$$

is called the *hyperbolic disk of radius r about z*. It is well known (see [49] or [135]) that $D(z, r)$ is a Euclidean disk with Euclidean center $(1 - s^2)z/(1 - s^2|z|^2)$ and Euclidean radius $(1 - |z|^2)s/(1 - s^2|z|^2)$, where $s = \tanh r \in (0, 1)$.

Let $|D(z, r)|_A$ denote the normalized area, or the dA-measure, of $D(z, r)$; the subscript indicates precisely that dA is used. Then $|D(z, r)|_A \sim (1 - |z|^2)^2$ as z approaches the unit circle. The following lemma lists some additional properties of the hyperbolic disks.

LEMMA 2.12 *Let r, s, and R be positive numbers. Then there exists a positive constant C such that for all z and w in $\mathbb{D}$, we have*

(1) $C^{-1}(1 - |z|^2) \le |1 - z\overline{w}| \le C (1 - |z|^2)$ when $\beta(z, w) \le r$.

(2) $C^{-1}|D(z, r)|_A \le |D(w, s)|_A \le C |D(z, r)|_A$ when $\beta(z, w) \le R$.

Proof. If $w \in D(z, r)$, then $w = \varphi_z(u)$ for some $|u| \le s$, where $s = \tanh r$. It follows that

$$1 - z\overline{w} = \frac{1 - |z|^2}{1 - z\overline{u}}.$$

This clearly implies (1). Since the condition $\beta(z, w) \le r$ is symmetric, (1) also holds with the positions of z and w interchanged. In particular, we have $1 - |z|^2 \sim 1 - |w|^2$ if $\beta(z, w) \le r$. Thus

$$|D(z, r)|_A \sim (1 - |z|^2)^2 \sim (1 - |w|^2)^2 \sim |D(w, s)|_A$$

for $\beta(z, w) \le R$.   ∎

LEMMA 2.13 *Fix* r, $0 < r < +\infty$. *There exists a positive integer* N *and a sequence* $\{a_n\}_n$ *in* $\mathbb{D}$ *such that:*

(1) The disk $\mathbb{D}$ *is covered by* $\{D(a_n, r)\}_n$.

(2) Every point in $\mathbb{D}$ *belongs to at most* N *sets in* $\{D(a_n, 2r)\}_n$.

(3) If $n \neq m$, *then* $\beta(a_n, a_m) \geq r/2$.

Proof. It is easy to construct a sequence $\{a_n\}_n$ in $\mathbb{D}$ satisfying conditions (1) and (3). We show that (2) has to hold, too. In fact, if we let N be the smallest integer such that $D(0, 2r)$ can be covered by N hyperbolic disks of radius $r/4$, then by Möbius invariance of the Bergman metric every hyperbolic disk of radius $2r$ can be covered by N hyperbolic disks of radius $r/4$. Now, if a point z in $\mathbb{D}$ belongs to $N + 1$ disks $D(a_{n_k}, 2r)$, $1 \leq k \leq N + 1$, then $a_{n_k} \in D(z, 2r)$ for $1 \leq k \leq N + 1$. Let $D(z_k, r/4)$, $1 \leq k \leq N$, be a cover of $D(z, 2r)$. Then at least one of the disks $D(z_k, r/4)$ contains two points from a_{n_k}, $1 \leq k \leq N + 1$. Two such points will have hyperbolic distance less than $r/2$, a contradiction to (3). $\blacksquare$

In connection with the above lemma, we mention that a sequence $\{a_j\}_j$ of points in $\mathbb{D}$ is said to be *separated* (or *uniformly discrete*) provided that

$$0 < \inf \left\{ \beta(a_j, a_k) : j \neq k \right\}.$$

LEMMA 2.14 *Fix an* r, $0 < r < +\infty$. *Then there exists a positive constant* $C = C(r)$ *such that*

$$|f(z)|^p \leq \frac{C}{|D(z, r)|_A} \int_{D(z,r)} |f(w)|^p \, dA(w), \qquad z \in \mathbb{D},$$

holds for all f *analytic in* $\mathbb{D}$ *and all* $0 < p < +\infty$.

Proof. Recall that $D(0, r)$ is a Euclidean disk centered at the origin. By the subharmonicity of $|f|^p$,

$$|f(0)|^p \leq \frac{1}{|D(0, r)|_A} \int_{D(0,r)} |f(w)|^p \, dA(w).$$

Replace f by $f \circ \varphi_z$ and make a change of variables. Then

$$|f(z)|^p \leq \frac{1}{|D(0, r)|_A} \int_{D(z,r)} |f(w)|^p \frac{(1 - |z|^2)^2}{|1 - z\overline{w}|^4} \, dA(w).$$

The desired result then follows from Lemma 2.12. $\blacksquare$

As a consequence of Lemmas 2.14 and 2.12, we obtain the following inequality:

$$(1 - |z|^2)^s |f(z)|^p \leq C \int_{D(z,r)} (1 - |w|^2)^{s-2} |f(w)|^p \, dA(w),$$

where f is analytic in $\mathbb{D}$, s is real, and $0 < p, r < +\infty$, and C is a constant depending on p, r, and s (but not on the function f and the point $z \in \mathbb{D}$).

We now prove the main result of this section.

THEOREM 2.15 *Fix* $0 < p, r < +\infty$, *and let* μ *be a positive Borel measure on* $\mathbb{D}$. *Then the following are equivalent:*

(1) The function $\mathbf{B}\mu$ *is bounded on* $\mathbb{D}$.

(2) The function $\widehat{\mu}_r(z) = \mu(D(z, r))/|D(z, r)|_A$ *is bounded on* $\mathbb{D}$.

(3) The Bergman space A^p *is boundedly contained in* $L^p(\mathbb{D}, d\mu)$.

Proof. Recall that

$$\mathbf{B}\mu(z) = \int_{\mathbb{D}} \frac{(1 - |z|^2)^2}{|1 - z\overline{w}|^4} \, d\mu(w) \geq \int_{D(z,r)} \frac{(1 - |z|^2)^2}{|1 - z\overline{w}|^4} \, d\mu(w).$$

That (1) implies (2) now follows from (1) of Lemma 2.12 and the fact that $|D(z, r)|_A$ is comparable to $(1 - |z|^2)^2$.

To see that (3) implies (1), assume that there exists a constant $C > 0$ such that

$$\int_{\mathbb{D}} |f(w)|^p \, d\mu(z) \leq C \int_{\mathbb{D}} |f(w)|^p \, dA(w)$$

for all $f \in A^p$. Fix $z \in \mathbb{D}$ and let

$$f(w) = \left[\frac{1 - |z|^2}{(1 - \overline{z}w)^2} \right]^{2/p}, \qquad w \in \mathbb{D}.$$

Then we obtain $\mathbf{B}\mu(z) \leq C$.

It remains to show that (2) implies (3). Thus, we assume there exists a positive constant C_1 such that $\mu(D(z, r)) \leq C_1 |D(z, r)|_A$ for all $z \in \mathbb{D}$. Pick a sequence $\{a_n\}_n$ in $\mathbb{D}$ satisfying the conditions in Lemma 2.13. For $f \in A^p$, we have

$$\int_{\mathbb{D}} |f(z)|^p \, d\mu(z) \leq \sum_{n=1}^{+\infty} \int_{D(a_n, r)} |f(z)|^p \, d\mu(z)$$

$$\leq \sum_{n=1}^{+\infty} \mu(D(a_n, r)) \sup\{|f(z)|^p : z \in D(a_n, r)\}.$$

By Lemmas 2.14 and 2.12, there exists a positive constant C_2 such that

$$\sup\{|f(z)|^p : z \in D(a_n, r)\} \leq \frac{C_2}{|D(a_n, r)|_A} \int_{D(a_n, 2r)} |f(z)|^p \, dA(z)$$

for all $n = 1, 2, 3, \ldots$. It follows that

$$\int_{\mathbb{D}} |f(z)|^p \, d\mu(z) \leq C_1 C_2 \sum_{n=1}^{+\infty} \int_{D(a_n, 2r)} |f(z)|^p \, dA(z).$$

Since every point in $\mathbb{D}$ belongs to at most N of the sets $D(a_n, 2r)$, we conclude that

$$\int_{\mathbb{D}} |f(z)|^p \, d\mu(z) \leq C_1 C_2 N \int_{\mathbb{D}} |f(z)|^p \, dA(z).$$

for every $f \in A^p$.   ∎

Note that if a positive measure μ satisfies any one of the three conditions in the theorem, then μ must be finite. The following is the "little oh" version of the above theorem.

THEOREM 2.16 *Fix* $1 < p < +\infty$ *and* $0 < r < +\infty$. *Let* μ *be a positive Borel measure on* $\mathbb{D}$. *Then the following conditions are equivalent:*

(1) The function $\mathbf{B}\mu$ *is in* $C_0(\mathbb{D})$.

(2) The function $\widehat{\mu}_r$ *is in* $C_0(\mathbb{D})$.

(3) $A^p \subset L^p(\mathbb{D}, d\mu)$ *and the inclusion map is compact.*

Proof. That (1) implies (2) follows from the estimate in the first paragraph of the proof of the previous theorem.

To prove (3) implies (1), recall that

$$\mathbf{B}\mu(z) = \int_{\mathbb{D}} |f_z(w)|^p \, d\mu(w), \qquad z \in \mathbb{D},$$

where

$$f_z(w) = \left[\frac{1 - |z|^2}{(1 - \bar{z}w)^2} \right]^{2/p}, \qquad z, w \in \mathbb{D}.$$

It is easy to check that $f_z \to 0$ weakly in A^p as $|z| \to 1^-$. Thus, the compactness of the inclusion map from A^p into $L^p(\mathbb{D}, d\mu)$ implies that $\mathbf{B}\mu(z) \to 0$ as $|z| \to 1^-$.

To prove (2) implies (3), let us assume that $\widehat{\mu}_r(z) \to 0$ as $|z| \to 1^-$ and $f_n \to 0$ weakly in A^p as $n \to +\infty$. We must show that $f_n \to 0$ in norm in $L^p(\mathbb{D}, d\mu)$ as $n \to +\infty$. Let $\{a_n\}_n$ be the sequence from Lemma 2.13. It is easy to see that $|a_n| \to 1^-$ as $n \to +\infty$. Given $\varepsilon > 0$, we can find a positive integer N_0 such that $\mu(D(a_n, r))/|D(a_n, r)|_A < \varepsilon$ for all $n \geq N_0$. Since $f_n \to 0$ weakly in A^p as $n \to +\infty$, we can find a positive constant C such that $\|f_n\|_p \leq C$ for all $n \geq 1$; see Exercise 1 of Chapter 1. The desired result now follows from the inequality

$$\int_{\mathbb{D}} |f_k(z)|^p \, d\mu(z) \leq \sum_{n=1}^{+\infty} \int_{D(a_n, r)} |f_k(z)|^p \, d\mu(z).$$

In fact, we can break the sum above into two parts; the first part is for $1 \leq n \leq N_0$ and the second part for $n > N_0$. The first part can be made arbitrarily small by choosing k sufficiently large, because $f_k \to 0$ weakly in A^p implies that $f_k(z) \to 0$ uniformly over compact sets. By the technique used in the proof of the previous theorem, the second part here can be shown to be less than a constant (independent of ε) times ε. We omit the details of this elementary ε-N argument.   ∎

2.4   BMO in the Bergman Metric

A well-known characterization of BMO on the unit circle is Garsia's lemma (see [49]), which says that a function f in L^2 of the circle belongs to BMO of the circle if and only if the function

$$z \mapsto \frac{1}{2\pi} \int_0^{2\pi} P(e^{it}, z)|f(e^{it})|^2 \, dt - \left| \frac{1}{2\pi} \int_0^{2\pi} P(e^{it}, z) f(e^{it}) \, dt \right|^2$$

is bounded, where $P(e^{it}, z)$ is the Poisson kernel at z. A similar result also holds for functions in VMO of the circle. The purpose of this section is to develop this theory in the Bergman metric.

Recall that for $0 < r < +\infty$ and $z \in \mathbb{D}$, the set $D(z, r)$ is the hyperbolic disk with hyperbolic center z and hyperbolic radius r. Also, $|D(z, r)|_A$ is the Euclidean area of $D(z, r)$ divided by π.

For a locally integrable function f on $\mathbb{D}$, we define the *averaging function* $\widehat{f}_r$ as follows:

$$\widehat{f}_r(z) = \frac{1}{|D(z, r)|_A} \int_{D(z,r)} f(w) \, dA(w), \qquad z \in \mathbb{D}.$$

If f is locally square-integrable, then we define the *mean oscillation* of f at z in the Bergman metric as

$$\mathrm{MO}_r(f)(z) = \left[\frac{1}{|D(z, r)|_A} \int_{D(z,r)} \left| f(w) - \widehat{f}_r(z) \right|^2 \, dA(w) \right]^{\frac{1}{2}}.$$

Let $\mathrm{BMO}_r = \mathrm{BMO}_r(\mathbb{D})$ denote the space of all locally square-integrable functions f such that

$$\|f\|_r = \sup \{\mathrm{MO}_r(f)(z) : z \in \mathbb{D}\} < +\infty.$$

The main result of this section is that the space BMO_r is independent of r and can be described in terms of the Berezin transform.

LEMMA 2.17 *Suppose r and s are positive numbers and β is the Bergman metric on $\mathbb{D}$. Then the following conditions on a function f defined on $\mathbb{D}$ are equivalent.*

(1) $M_r = \sup\{|f(z) - f(w)| : \beta(z, w) < r\} < +\infty.$

(2) $M_s = \sup\{|f(z) - f(w)| : \beta(z, w) < s\} < +\infty.$

(3) $|f(z) - f(w)| \leq C\,(\beta(z, w) + 1)$ *for some positive constant C and all $z, w \in \mathbb{D}$.*

Proof. Assume $r < s$. Then $M_r \leq M_s$, and hence (2) implies (1). It is clear that (3) implies (2). To prove the remaining implication, we fix two points z and w in $\mathbb{D}$ with $\beta(z, w) > r$; the desired inequality is obvious if $\beta(z, w) \leq r$. Let $\alpha(t)$, $0 \leq t \leq 1$, be the geodesic from z to w in the hyperbolic metric. Let N be the smallest integer greater than or equal to $\beta(z, w)/r$. For $t_k = k/N, 0 \le k \le N - 1$, we have

$$\beta(\alpha(t_k), \alpha(t_{k+1})) = \frac{\beta(z, w)}{N} \le r.$$

It follows that

$$|f(z) - f(w)| \le \sum_{k=1}^{N-1} \left| f(\alpha(t_k)) - f(\alpha(t_{k+1})) \right| \le N M_r.$$

By the choice of N, we have

$$N \le \frac{\beta(z, w)}{r} + 1 \le \frac{2}{r} \left(\beta(z, w) + 1 \right).$$

Thus,

$$|f(z) - f(w)| \le \frac{2M_r}{r} \left(\beta(z, w) + 1 \right)$$

for all $\beta(z, w) > r$.   ∎

The Bergman metric grows logarithmically:

$$\beta(z, 0) = \frac{1}{2} \log \frac{1 + |z|}{1 - |z|}, \qquad z \in \mathbb{D}.$$

It follows that a Borel measurable function f which satisfies any of the three equivalent conditions of Lemma 2.17 is in $L^p(\mathbb{D}, dA)$ for all finite positive exponents p.

We can now prove the main result of the section. For convenience, we introduce for $f \in L^2(\mathbb{D}, dA)$ the following notation:

$$\mathrm{MO}(f)(z) = \left[\mathbf{B}(|f|^2)(z) - |\mathbf{B}f(z)|^2 \right]^{\frac{1}{2}}.$$

It is easy to see that $\mathbf{B}(|f|^2)(z) \ge |\mathbf{B}f(z)|^2$, so that the above expression is well-defined. In fact, we can write

$$\mathrm{MO}(f)(z) = (1 - |z|^2)^2 \left[\int_{\mathbb{D}} \int_{\mathbb{D}} \left| \frac{f(u) - f(v)}{(1 - u\overline{z})^2(1 - v\overline{z})^2} \right|^2 dA(u) dA(v) \right]^{1/2}.$$

THEOREM 2.18 *Suppose $0 < r < +\infty$ and that the function f is locally square-integrable in $\mathbb{D}$. Then $f \in \mathrm{BMO}_r$ if and only if $f \in L^2(\mathbb{D}, dA)$ and the function $\mathrm{MO}(f)$ is bounded on $\mathbb{D}$.*

Proof. By Lemma 2.12, we can choose a small constant $\sigma > 0$ such that

$$|k_z(w)|^2 \ge \frac{\sigma}{|D(z, r)|_A}$$

for all $z \in \mathbb{D}$ and $w \in D(z, r)$, where

$$k_z(w) = \frac{1 - |z|^2}{(1 - w\overline{z})^2}$$

are the normalized reproducing kernels of A^2. In view of the above formula for $MO(f)$, we have

$$[MO(f)(z)]^2 = \frac{1}{2} \int_{\mathbb{D}} \int_{\mathbb{D}} |f(u) - f(v)|^2 |k_z(u)|^2 |k_z(v)|^2 \, dA(u) \, dA(v)$$

which we compare with

$$[MO_r(f)(z)]^2 = \frac{1}{2|D(z,r)|_A^2} \int_{D(z,r)} \int_{D(z,r)} |f(u) - f(v)|^2 \, dA(u) \, dA(v)$$

for $z \in \mathbb{D}$. By shrinking the domain of integration $\mathbb{D}$ to $D(z, r)$, we obtain

$$MO(f)(z) \geq \sigma \, MO_r(f)(z), \qquad z \in \mathbb{D}.$$

Thus, the boundedness of the function $MO(f)$ implies that $f \in BMO_r$.

Next, assume that f is in BMO_r. Let $r = 2s$, and recall that $\widehat{f}_s$ is the averaging function for f with parameter s. Write $f = f_1 + f_2$, where $f_1(z) = \widehat{f}_s(z)$ and $f_2(z) = f(z) - \widehat{f}_s(z)$. Since the space of functions f in $L^2(\mathbb{D}, dA)$ with bounded $MO(f)$ is linear, it suffices to show that both f_1 and f_2 have this property.

First, using the identity

$$\widehat{f}_s(z) - \widehat{f}_s(w) = \frac{1}{|D(z,s)|_A} \int_{D(z,s)} \left[f(u) - \widehat{f}_s(w) \right] \, dA(u)$$

and the Cauchy-Schwarz inequality we easily obtain

$$\left| \widehat{f}_s(z) - \widehat{f}_s(w) \right|^2$$
$$\leq \frac{1}{|D(z,s)|_A \, |D(w,s)|_A} \int_{D(z,s)} \int_{D(w,s)} |f(u) - f(v)|^2 \, dA(u) \, dA(v).$$

If $\beta(z, w) \leq s$, then

$$D(z, s) \subset D(z, r), \qquad D(w, s) \subset D(z, r),$$

and

$$|D(w, s)|_A \sim |D(z, s)|_A \sim |D(z, r)|_A;$$

see Lemma 2.12. Thus, there exists a positive constant C such that

$$\left| \widehat{f}_s(z) - \widehat{f}_s(w) \right|^2 \leq \frac{C}{2|D(z,r)|_A^2} \int_{D(z,r)} \int_{D(z,r)} |f(u) - f(v)|^2 \, dA(u) \, dA(v)$$
$$= C \, [MO_r(f)(z)]^2$$

for all $\beta(z, w) \leq s$. Since $MO_r(f)$ is bounded, it follows from Lemma 2.17 that there exists a positive constant C_1 such that

$$|\widehat{f}_s(z) - \widehat{f}_s(w)| \leq C_1 \, (\beta(z, w) + 1)$$

for all z and w in $\mathbb{D}$. In particular, $\widehat{f_s} \in L^2(\mathbb{D}, dA)$. Now,

$$
\begin{aligned}
2\left[\mathrm{MO}(\widehat{f_s})(z)\right]^2 \\
&= \int_{\mathbb{D}} \int_{\mathbb{D}} |f(u) - f(v)|^2 |k_z(u)|^2 |k_z(v)|^2 \, dA(u) dA(v) \\
&\leq C_1^2 \int_{\mathbb{D}} \int_{\mathbb{D}} (\beta(u, v) + 1)^2 |k_z(u)|^2 |k_z(v)|^2 \, dA(u) dA(v) \\
&= C_1^2 \int_{\mathbb{D}} \int_{\mathbb{D}} (\beta(u, v) + 1)^2 \, dA(u) \, dA(v).
\end{aligned}
$$

The last equality follows from a change of variables and the invariance of the hyperbolic metric. The last integral above can easily be checked to be finite. Hence the function $\mathrm{MO}(\widehat{f_s})$ is bounded.

Second, we look at $f_2 = f - f_s$. Then, by the triangle inequality,

$$
\begin{aligned}
\left[|\widehat{f_2}|^2_s(z)\right]^{\frac{1}{2}} &= \left[\frac{1}{|D(z, s)|_A} \int_{D(z,s)} |f(w) - \widehat{f_s}(w)|^2 \, dA(w)\right]^{\frac{1}{2}} \\
&\leq \left[\frac{1}{|D(z, s)|_A} \int_{D(z,s)} |f(w) - \widehat{f_s}(z)|^2 \, dA(w)\right]^{\frac{1}{2}} \\
&\quad + \left[\frac{1}{|D(z, s)|_A} \int_{D(z,s)} |\widehat{f_s}(z) - \widehat{f_s}(w)|^2 \, dA(w)\right]^{\frac{1}{2}}.
\end{aligned}
$$

The last term is bounded in z because of an earlier estimate on $\widehat{f_s}$. The term preceding it is bounded, too, because $f \in \mathrm{BMO}_r$ and

$$
\mathrm{MO}_s(f)(z) \leq C_2 \, \mathrm{MO}_r(f)(z), \qquad z \in \mathbb{D},
$$

which follows from Lemma 2.12 and the double-integral formula for $\mathrm{MO}_r(f)$ used earlier in the proof. By Theorem 2.15, the function $\mathbf{B}(|f_2|^2)$ is bounded, which obviously implies that $f_2 \in L^2(\mathbb{D}, dA)$ and that $\mathrm{MO}(f_2)$ is bounded. ∎

It follows from Theorem 2.18 that the space BMO_r does not depend on the parameter r, $0 < r < +\infty$ (but the norm changes with r, of course). Let us write $\mathrm{BMO}_\partial = \mathrm{BMO}_\partial(\mathbb{D})$ for the space BMO_r, for any $0 < r < +\infty$. The new notation signifies the independence of the parameter r; it also emphasizes the fact that whether or not a function from $L^2(\mathbb{D}, dA)$ belongs to BMO_∂ is a boundary property.

It is easy to check that BMO_∂ becomes a Banach space with the norm

$$
\|f\| = |\mathbf{B}f(0)| + \sup\{\mathrm{MO}(f)(z) : z \in \mathbb{D}\}.
$$

If the term involving

$$
\mathbf{B}f(0) = \int_{\mathbb{D}} f(z) \, dA(z)
$$

is removed, then what remains is only a seminorm. This seminorm is Möbius invariant, although the norm above is not.

Let VMO$_r$ be the space of locally square-integrable functions f in $\mathbb{D}$ such that $MO_r(f)(z) \to 0$ as $|z| \to 1^-$. It is clear that VMO$_r$ is contained in BMO$_r$.

THEOREM 2.19 *A locally square-integrable function f in $\mathbb{D}$ belongs to* VMO$_r$ *if and only if* $MO(f)(z) \to 0$ *as* $|z| \to 1^-$.

Proof. The proof is similar to that of the previous theorem; we leave the details to the interested reader. ∎

Again we let VMO$_\partial$ = VMO$_\partial(\mathbb{D})$ stand for the space VMO$_r$ for any r, $0 < r < +\infty$. It is easy to check that VMO$_\partial$ is a closed subspace of BMO$_\partial$ and that VMO$_\partial$ contains $C(\overline{\mathbb{D}})$.

THEOREM 2.20 *Let $H(\mathbb{D})$ be the space of analytic functions in $\mathbb{D}$. Then*

(1) BMO$_\partial \cap H(\mathbb{D}) = \mathcal{B}$.

(2) VMO$_\partial \cap H(\mathbb{D}) = \mathcal{B}_0$.

Proof. Since both BMO$_\partial$ and $\mathcal{B}$ are contained in $L^2(\mathbb{D}, dA)$, we may begin with a function f in A^2. By the symmetry of $\mathbb{D}$,

$$f'(0) = 2 \int_{\mathbb{D}} \overline{w} \left(f(w) - f(0) \right) dA(w).$$

Replacing f by $f \circ \varphi_z$ and performing an obvious estimate, we get

$$(1 - |z|^2)^2 |f'(z)|^2 \le 4 \int_{\mathbb{D}} |f \circ \varphi_z(w) - f(z)|^2 \, dA(w)$$

for every $z \in \mathbb{D}$. Since $\mathbf{B}f = f$ for analytic f, we easily verify that

$$\int_{\mathbb{D}} |f \circ \varphi_z(w) - f(z)|^2 \, dA(w) = \mathbf{B}(|f|^2)(z) - |\mathbf{B}f(z)|^2.$$

This shows that BMO$_\partial \cap H(\mathbb{D}) \subset \mathcal{B}$.

On the other hand, if $f \in \mathcal{B}$, then by Theorem 1.15, there exists a positive constant C such that $|f(z) - f(w)| \le C\beta(z, w)$ for all $z, w \in \mathbb{D}$. This, together with the integral formula for $\mathbf{B}(|f|^2) - |\mathbf{B}f|^2$ in the previous paragraph, then shows that $\mathcal{B} \subset$ BMO$_\partial \cap H(\mathbb{D})$.

The proof of the identity VMO$_\partial \cap H(\mathbb{D}) = \mathcal{B}_0$ is similar. ∎

2.5   A Lipschitz Estimate

Let $\alpha(t)$ be a smooth curve in $\mathbb{D}$. If $s(t)$ is the arc length of $\alpha(t)$ in the Bergman metric, then

$$\frac{ds}{dt} = \frac{|\alpha'(t)|}{1 - |\alpha(t)|^2}.$$

For a point $a \in \mathbb{D}$, we let Π_a denote the rank-one orthogonal projection from A^2 onto the one-dimensional subspace spanned by k_a, where

$$k_a(z) = \frac{1 - |a|^2}{(1 - \bar{a}z)^2}, \qquad z \in \mathbb{D},$$

which is a unit vector in A^2. In concrete terms,

$$\Pi_a f = \langle f, k_a \rangle k_a = (1 - |a|^2) f(a) k_a, \qquad a \in \mathbb{D}.$$

LEMMA 2.21 *Let $\alpha(t)$ be a smooth curve in $\mathbb{D}$, and let $s(t)$ be the arc length of $\alpha(t)$ in the Bergman metric. Then*

$$\frac{ds}{dt} = \frac{1}{\sqrt{2}} \left\| (I - \Pi_{\alpha(t)}) \left(\frac{d}{dt} k_{\alpha(t)} \right) \right\|,$$

where $\| \cdot \|$ is the norm in A^2 and I is the identity operator.

Proof. Since

$$\frac{d}{dt} k_{\alpha(t)}(z) = -\frac{\alpha'(t)\overline{\alpha(t)} + \alpha(t)\overline{\alpha'(t)}}{(1 - \overline{\alpha(t)}z)^2} + \frac{2z\overline{\alpha'(t)}(1 - |\alpha(t)|^2)}{(1 - \overline{\alpha(t)})^3},$$

a simple calculation gives

$$\Pi_{\alpha(t)} \left(\frac{d}{dt} k_{\alpha(t)} \right)(z) = \frac{-\alpha'(t)\overline{\alpha(t)} + \alpha(t)\overline{\alpha'(t)}}{(1 - \overline{\alpha(t)}z)^2},$$

and so

$$(I - \Pi_{\alpha(t)}) \left(\frac{d}{dt} k_{\alpha(t)} \right)(z) = \frac{2\overline{\alpha'(t)}\,(z - \alpha(t))}{(1 - \overline{\alpha(t)}z)^3}.$$

By a change of variables we then obtain

$$\left\| (I - \Pi_{\alpha(t)}) \left(\frac{d}{dt} k_{\alpha(t)} \right) \right\|^2 = \frac{2|\alpha'(t)|^2}{(1 - |\alpha(t)|^2)^2},$$

which clearly implies the desired result. ∎

THEOREM 2.22 *Let $\alpha(t)$ be a smooth curve in $\mathbb{D}$, and let $s(t)$ be the arc length of $\alpha(t)$ in the Bergman metric. Then, for any $f \in \mathrm{BMO}_\partial$, we have*

$$\left| \frac{d}{dt} \mathbf{B} f(\alpha(t)) \right| \leq 2\sqrt{2}\, \mathrm{MO}(f)(\alpha(t)) \frac{ds}{dt}.$$

Proof. Recall that

$$\mathbf{B} f(\alpha(t)) = \int_{\mathbb{D}} f(w) \left| k_{\alpha(t)}(w) \right|^2 dA(w).$$

Differentiation under the integral sign gives

$$\frac{d}{dt} \mathbf{B} f(\alpha(t)) = 2 \int_{\mathbb{D}} f(w) \operatorname{Re} \left[\left(\frac{d}{dt} k_{\alpha(t)}(w) \right) \overline{k_{\alpha(t)}(w)} \right] dA(w).$$

Also, differentiation of the identity $\langle k_{\alpha(t)}, k_{\alpha(t)} \rangle = 1$ gives

$$\operatorname{Re} \left\langle \frac{d}{dt} k_{\alpha(t)}, k_{\alpha(t)} \right\rangle = 0.$$

Using this and the formula

$$\Pi_{\alpha(t)} \left(\frac{d}{dt} k_{\alpha(t)} \right) = \left\langle \frac{d}{dt} k_{\alpha(t)}, k_{\alpha(t)} \right\rangle k_{\alpha(t)},$$

we then obtain

$$\operatorname{Re} \left[\Pi_{\alpha(t)} \left(\frac{d}{dt} k_{\alpha(t)} \right) (w) \overline{k_{\alpha(t)}(w)} \right] = 0.$$

It follows that

$$\frac{d}{dt} \mathbf{B} f(\alpha(t)) = 2 \int_{\mathbb{D}} f(w) \operatorname{Re} \left[(I - \Pi_{\alpha(t)}) \left(\frac{d}{dt} k_{\alpha(t)} \right) (w) \overline{k_{\alpha(t)}(w)} \right] dA(w).$$

On the other hand,

$$\int_{\mathbb{D}} (I - \Pi_{\alpha(t)}) \left(\frac{d}{dt} k_{\alpha(t)} \right) (w) \overline{k_{\alpha(t)}(w)} \, dA(w) = 0$$

by the definition of $\Pi_{\alpha(t)}$. Therefore, the derivative $d\mathbf{B} f(\alpha(t))/dt$ is equal to

$$2 \int_{\mathbb{D}} (f(w) - \mathbf{B} f(\alpha(t))) \operatorname{Re} \left[(I - \Pi_{\alpha(t)}) \left(\frac{d}{dt} k_{\alpha(t)} \right) (w) \overline{k_{\alpha(t)}(w)} \right] dA(w),$$

and hence $|d\mathbf{B} f(\alpha(t))/dt|$ is less than or equal to

$$2 \int_{\mathbb{D}} |f(w) - \mathbf{B} f(\alpha(t))| \, |k_{\alpha(t)}(w)| \left| (I - \Pi_{\alpha(t)}) \left(\frac{d}{dt} k_{\alpha(t)} \right) (w) \right| dA(w).$$

The desired result now follows from Lemma 2.21 and an application of the Cauchy-Schwarz inequality.   ∎

COROLLARY 2.23 *For $f \in \mathrm{BMO}_\partial$, we have*

$$|\mathbf{B} f(z) - \mathbf{B} f(w)| \le 2\sqrt{2} \, \|f\|_{\mathrm{BMO}} \, \beta(z, w)$$

for all z and w in $\mathbb{D}$, where

$$\|f\|_{\mathrm{BMO}} = \sup \{ \mathrm{MO}(f)(z) : z \in \mathbb{D} \}.$$

Proof. Fix z and w in $\mathbb{D}$ and let $\alpha(t)$, $0 \le t \le 1$, be the geodesic from z to w in the Bergman metric. Then, by the above theorem,

$$|\mathbf{B}f(z) - \mathbf{B}f(w)| = \left| \int_0^1 \frac{d}{dt} \mathbf{B}f(\alpha(t)) \, dt \right| \le 2\sqrt{2} \int_0^1 \mathrm{MO}(f)(\alpha(t)) \frac{ds}{dt} \, dt$$

$$\le 2\sqrt{2} \, \|f\|_{\mathrm{BMO}} \int_0^1 \frac{ds}{dt} \, dt = 2\sqrt{2} \, \|f\|_{\mathrm{BMO}} \, \beta(z, w),$$

as claimed.   ∎

2.6 Notes

The Berezin transform was introduced by Berezin in [17] and [18]. Most applications of the Berezin transform so far have been in the study of Hankel and Toeplitz operators; see [135]. Section 2.1 is elementary. All results in Section 2.2 are taken from the paper [4]. The results of Section 2.3, in various forms, are due to Hastings [54], Luecking [92], and Zhu [133]. The theory of BMO and VMO in the Bergman metric, as presented in Sections 2.4 and 2.5, was begun by Zhu in his thesis [132] and then developed by Békollé, Berger, Coburn, and Zhu in [15].

2.7 Exercises and Further Results

1. If $f \in L^1(\mathbb{D}, dA)$ is subharmonic, then $\mathbf{B}f$ is subharmonic and $f \le \mathbf{B}f$ on $\mathbb{D}$.

2. If $f \in L^\infty(\mathbb{D})$ and f has a nontangential limit L at some boundary point $\zeta \in \mathbb{T}$, then $\mathbf{B}f$ also has nontangential limit L at ζ.

3. Find a real-valued function $f \in L^1(\mathbb{D}, dA)$, strictly negative on a subset of positive area, such that $\mathbf{B}f$ is strictly positive on $\mathbb{D}$.

4. Show that there exist two functions f and g in A^2 such that $\mathbf{B}(|f|^2) \le \mathbf{B}(|g|^2)$ on $\mathbb{D}$, but nevertheless

$$\int_{\mathbb{D}} |f(z)p(z)|^2 \, dA(z) > \int_{\mathbb{D}} |g(z)p(z)|^2 \, dA(z)$$

holds for some polynomial p.

5. Show that the Berezin transform commutes with the invariant Laplacian on the space $C^2(\overline{\mathbb{D}})$.

6. If f is a bounded subharmonic function in $\mathbb{D}$, then $\{\mathbf{B}^n f\}_n$ converges to a harmonic function in $\mathbb{D}$.

7. If f is continuous on $\overline{\mathbb{D}}$, then $\{\mathbf{B}^n f\}_n$ converges uniformly in $\mathbb{D}$ to the harmonic extension of the boundary function f. See [42].

8. If f is bounded and radial, then $\mathbf{B}f \in C_0(\mathbb{D})$ if and only if

$$\frac{1}{1-r} \int_r^1 f(t)\,dt \to 0$$

as $r \to 1^-$. See [89].

9. If $f \in L^\infty(\mathbb{D})$, then $\mathbf{B}f \in C_0(\mathbb{D})$ if and only if

$$n \int_{\mathbb{D}} f(z)|z|^{2n}\,dA(z) \to 0$$

as $n \to +\infty$.

10. For $f(z) = -2\log|z|$ on $\mathbb{D}$, show that $\mathbf{B}f(z) = 1 - |z|^2$.

11. If $f \in C^2(\overline{\mathbb{D}})$, then

$$\mathbf{B}f(z) = F(z) - \int_{\mathbb{D}} \left[1 - |\varphi_z(z)|^2\right] \Delta_w f(w)\,dA(w), \qquad z \in \mathbb{D},$$

where F is the harmonic extension of the boundary function of f.

12. If $f \in C^2(\overline{\mathbb{D}})$, then

$$f(z) = F(z) + \int_{\mathbb{D}} \left[\log|\varphi_z(w)|^2\right] \Delta_w f(w)\,dA(w), \qquad z \in \mathbb{D},$$

where F is the harmonic extension of the boundary function of f.

13. For $f(z) = \log[1/(1 - |z|^2)]$ on $\mathbb{D}$, show that $\mathbf{B}f = f + 1$.

14. Let $0 < p < +\infty$. Characterize those functions $\varphi \in H^\infty$ such that

$$\sigma \int_{\mathbb{D}} |f(z)|^p\,dA(z) < \int_{\mathbb{D}} |\varphi(z)f(z)|^p\,dA(z)$$

for all $f \in A^p$ and some constant $\sigma > 0$ (depending on φ and p but not on f). See [29].

15. Suppose $2 \le p < +\infty$ and that f is an analytic function on $\mathbb{D}$. Show that $\mathrm{MO}(f) \in L^p(\mathbb{D}, d\lambda)$ if and only if $f \in B_p$ (the analytic Besov spaces). See Exercise 9 in Chapter 1 for the definition of $d\lambda$. See [135].

16. A bounded function φ on $\mathbb{D}$ is a pointwise multiplier of BMO_∂ if and only if $\mathrm{MO}(\varphi)\log(1 - |z|^2)$ is bounded in $\mathbb{D}$. See [134].

17. Fix a sequence $\{z_n\}_n$ in $\mathbb{D}$. For $t > 0$, let A_t be the operator on l^2 whose matrix under the standard basis has

$$\frac{(1 - |z_m|^2)^{t/2}(1 - |z_n|^2)^{t/2}}{(1 - z_m \bar{z}_n)^t}$$

as its (m, n) entry. For $t > 1$, A_t is bounded on l^2 if and only if $\{z_n\}_n$ is the union of finitely many separated sequences; for $t = 1$, A_t is bounded on l^2 if and only if $\{z_n\}_n$ is the union of finitely many (classical) interpolating sequences. See [142].

18. Show that the Bergman projection maps BMO_∂ onto the Bloch space. Similarly, the Bergman projection maps VMO_∂ onto the little Bloch space.

19. Fix $-1 < \alpha < +\infty$ and $0 < p < +\infty$. For a sequence $A = \{a_n\}_n$ in $\mathbb{D}$, let R_A be the operator that sends an analytic function f to the sequence

$$\left\{(1 - |a_n|^2)^{(2+\alpha)/p} f(a_n)\right\}_n.$$

Show that R_A is bounded from A_α^p to l^p if and only if A is the union of finitely many separated sequences. See [139].

20. If $f \in \text{BMO}_\partial$, then the function

$$(1 - |z|^2)|\nabla \mathbf{B} f(z)|$$

is bounded on $\mathbb{D}$. Here, ∇ stands for the gradient operator.

3

A^p-Inner Functions

In this chapter, we introduce the notion of A_α^p-inner functions and prove a growth estimate for them. The A_α^p-inner functions are analogous to the classical inner functions which play an important role in the factorization theory of the Hardy spaces. Each A_α^p-inner function is extremal for a z-invariant subspace, and the ones that arise from subspaces given by finitely many zeros are called finite zero extremal functions (for $\alpha = 0$, they are also called finite zero-divisors). In the unweighted case $\alpha = 0$, we will prove the expansive multiplier property of A^p-inner functions, and obtain an "inner-outer"-type factorization of functions in A^p. In the process, we find that all singly generated invariant subspaces are generated by its extremal function. In the special case of $p = 2$ and $\alpha = 0$, we find an analogue of the classical Carathéodory-Schur theorem: the closure of the finite zero-divisors in the topology of uniform convergence on compact subsets are the A^2-subinner functions. In particular, all A^2-inner functions are norm approximable by finite zero-divisors.

3.1 A_α^p-Inner Functions

Classical inner functions in $\mathbb{D}$ play an important role in the theory of Hardy spaces. Recall that a bounded analytic function φ in $\mathbb{D}$ is called inner if $|\varphi(z)| = 1$ for almost all $z \in \mathbb{T}$. This is clearly equivalent to

$$\frac{1}{2\pi} \int_0^{2\pi} (|\varphi(z)|^p - 1) z^n \, |dz| = 0$$

for all nonnegative integers n; and the condition above is independent of p, $0 < p < +\infty$. This motivates the following definition of inner functions for Bergman spaces.

DEFINITION 3.1 *A function φ in A_α^p is called an A_α^p-inner function if*

$$\int_{\mathbb{D}} (|\varphi(z)|^p - 1) \, z^n \, dA_\alpha(z) = 0$$

for all nonnegative integers n.

It follows easily from the above definition that a function φ in A_α^p is an A_α^p-inner function if and only if

$$\int_{\mathbb{D}} |\varphi(z)|^p \, q(z) \, dA_\alpha(z) = q(0)$$

for every polynomial q, and this condition is clearly equivalent to

$$\int_{\mathbb{D}} |\varphi(z)|^p \, h(z) \, dA_\alpha(z) = h(0),$$

where h is any bounded harmonic function in $\mathbb{D}$. In particular, every A_α^p-inner function is a unit vector in A_α^p.

An obvious example of an A_α^p-inner function is a constant times a monomial. In fact, for any $n = 0, 1, 2, \ldots$, the function

$$\varphi(z) = \left[\frac{\Gamma\left(\frac{np}{2} + \alpha + 2\right)}{\Gamma\left(\frac{np}{2} + 1\right) \Gamma(\alpha + 2)} \right]^{\frac{1}{p}} z^n$$

is A_α^p-inner. More examples of A_α^p-inner functions will be presented later when we study a certain extremal problem for invariant subspaces.

Our first goal is to show that A_α^p-inner functions grow much more slowly near the boundary than an arbitrary function from A_α^p does. The following lemma tells us how fast an arbitrary function from A_α^p grows near the boundary.

LEMMA 3.2 *If f is a unit vector in A_α^p, then*

$$|f(z)| \le \frac{1}{(1 - |z|^2)^{(2+\alpha)/p}}, \qquad z \in \mathbb{D}.$$

Proof. Let u be a positive subharmonic function in $\mathbb{D}$. Then by the sub-mean value property of subharmonic functions on circles and by using polar coordinates we have

$$u(0) \le \int_{\mathbb{D}} u(z) \, dA_\alpha(z).$$

Replace u by $u \circ \varphi_a$, where, for $a \in \mathbb{D}$,

$$\varphi_a(z) = \frac{a - z}{1 - \overline{a}z}, \qquad z \in \mathbb{D}.$$

We conclude that

$$u(a) \leq \int_{\mathbb{D}} u \circ \varphi_a(z) \, dA_\alpha(z)$$

for all $a \in \mathbb{D}$. Making an obvious change of variables, we obtain

$$u(a) \leq \int_{\mathbb{D}} u(z) \, |k_a^\alpha(z)|^2 \, dA_\alpha(z)$$

for all $a \in \mathbb{D}$, where

$$k_a^\alpha(z) = \frac{(1 - |a|^2)^{(2+\alpha)/2}}{(1 - \bar{a}z)^{2+\alpha}}$$

are the normalized reproducing kernels of A_α^2.

Now suppose f is a unit vector in A_α^p. Fix any $a \in \mathbb{D}$, and let

$$u(z) = \left| f(z) \left(k_a^\alpha(z) \right)^{-2/p} \right|^p, \qquad z \in \mathbb{D}.$$

Applying the estimate in the previous paragraph, we conclude that

$$\left| f(a) \left(k_a^\alpha(a) \right)^{-2/p} \right|^p \leq 1,$$

that is,

$$|f(a)| \leq \frac{1}{(1 - |a|^2)^{(2+\alpha)/p}}$$

for all $a \in \mathbb{D}$, completing the proof of the lemma.   ∎

Since the polynomials are dense in A_α^p, it is an immediate consequence of Lemma 3.2 that for $f \in A_\alpha^p$,

$$|f(z)| = o\left(\frac{1}{(1 - |z|^2)^{(2+\alpha)/p}} \right) \qquad \text{as } |z| \to 1,$$

which means that the boundary growth is actually not quite as fast as permitted by Lemma 3.2.

To obtain a better estimate for A_α^p-inner functions, we are going to show that every A_α^p-inner function is a contractive multiplier from the classical Hardy space H^p into A_α^p. Recall that H^p consists of analytic functions f in $\mathbb{D}$ such that

$$\|f\|_{H^p}^p = \sup_{0 < r < 1} \frac{1}{2\pi} \int_0^{2\pi} |f(re^{it})|^p \, dt < +\infty.$$

If $f \in H^p$, then the radial limits $f(e^{it})$ exist for almost all real t and

$$\|f\|_{H^p}^p = \frac{1}{2\pi} \int_0^{2\pi} |f(e^{it})|^p \, dt.$$

The books [37], [49], and [82] are excellent sources of information about Hardy spaces.

THEOREM 3.3 *If φ is A_α^p-inner, then φ is a contractive multiplier from H^p into A_α^p, and consequently,*

$$|\varphi(z)| \leq \frac{1}{(1 - |z|^2)^{(1+\alpha)/p}}, \qquad z \in \mathbb{D}.$$

Proof. Suppose $f \in H^p$ and let h be the least harmonic majorant of $|f(z)|^p$. More explicitly,

$$h(z) = \frac{1}{2\pi} \int_0^{2\pi} P(e^{it}, z)|f(e^{it})|^p \, dt, \qquad z \in \mathbb{D},$$

where $P(e^{it}, z)$ is the Poisson kernel at $z \in \mathbb{D}$. By Fatou's lemma and the definition of A_α^p-inner functions,

$$\int_\mathbb{D} |\varphi(z)|^p h(z) \, dA_\alpha(z) \leq \liminf_{r \to 1^-} \int_\mathbb{D} |\varphi(z)|^p h_r(z) \, dA_\alpha(z) = h(0),$$

where $h_r(z) = h(rz)$ for $z \in \mathbb{D}$. It follows that

$$\int_\mathbb{D} |\varphi(z)f(z)|^p \, dA_\alpha(z) \leq \int_\mathbb{D} |\varphi(z)|^p h(z) \, dA_\alpha(z) \leq h(0) = \|f\|_{H^p}^p,$$

so that φ is a contractive multiplier from H^p into A_α^p.

For any $z \in \mathbb{D}$, consider the function

$$f_z(w) = \left(\frac{1 - |z|^2}{(1 - \bar{z}w)^2} \right)^{1/p}, \qquad w \in \mathbb{D}.$$

Then f_z is a unit vector in H^p, and so φf_z has norm less than or equal to 1 in A_α^p. Applying Lemma 3.2 to the function φf_z, we conclude that

$$|\varphi(z)| \leq \frac{1}{(1 - |z|^2)^{(1+\alpha)/p}}, \qquad z \in \mathbb{D},$$

as claimed. ∎

3.2 An Extremal Problem

In this section, we exhibit the close relationship between A_α^p-inner functions and *invariant subspaces* of A_α^p. In particular, this will provide us with more examples of A_α^p-inner functions.

A closed subspace I of A_α^p is called *invariant* if $zf \in I$ whenever $f \in I$. Here, z denotes the identity function on $\mathbb{D}$. It is easy to see that a closed subspace I is invariant if and only if it is closed under multiplication by bounded analytic functions.

We give two examples of invariant subspaces in A_α^p. First, if $A = \{a_n\}_n$ is a sequence of points from $\mathbb{D}$, and if I_A consists of all functions in A^p whose zero sets contain A (counting multiplicities), then I_A is an invariant subspace of A^p. We call such spaces *zero-based invariant subspaces*.

Next, if f is any function in A_α^p, and if I_f is the closure in A_α^p of the set consisting of all polynomial multiples of f, then I_f is an invariant subspace (called the invariant subspace generated by f). We call such spaces singly generated invariant subspaces, or sometimes, cyclic invariant subspaces. Note that the notation $[f]$ is sometimes used instead of I_f.

For any invariant subspace I of A_α^p, we let $n = n_I$ denote the smallest nonnegative integer such that there exists a function $f \in I$ with $f^{(n)}(0) \neq 0$.

THEOREM 3.4 *Suppose I is an invariant subspace of A_α^p and G is any function that solves the extremal problem*

$$\sup \left\{ \operatorname{Re} f^{(n)}(0) : f \in I, \|f\|_{p,\alpha} \leq 1 \right\},$$

where $n = n_I$. Then G is an A_α^p-inner function.

Proof. It is obvious that G is a unit vector. We will prove the theorem by a variational argument.

Fix a positive integer k, and set

$$re^{i\theta} = \int_{\mathbb{D}} |G(z)|^p z^k \, dA_\alpha(z),$$

where $0 < r < 1$ and $-\pi < \theta \leq \pi$ (polar coordinates). For any complex number λ, we consider the function

$$f_\lambda(z) = \frac{G(z)(1 + \lambda z^k)}{\|G(1 + \lambda z^k)\|_{p,\alpha}}.$$

Since f_λ is a unit vector in I, the extremal property of G gives

$$\operatorname{Re} f_\lambda^{(n)}(0) \leq G^{(n)}(0).$$

This implies that

$$1 \leq \int_{\mathbb{D}} |G(z)|^p |1 + \lambda z^k|^p \, dA_\alpha(z)$$

for all $\lambda \in \mathbb{C}$, so that

$$1 \leq 1 + p \operatorname{Re} \left[\lambda \int_{\mathbb{D}} |G(z)|^p z^k \, dA_\alpha(z) \right] + O(|\lambda|^2).$$

Put $\lambda = -\varepsilon e^{-i\theta}$, where $\varepsilon > 0$ is small and θ is as above. We then obtain

$$0 \leq -r + O(\varepsilon).$$

Letting $\varepsilon \to 0$, we see that $r = 0$, and so G is A_α^p-inner. ∎

For any invariant subspace I, the extremal problem stated in the theorem above will be referred to as *the extremal problem for I*. It is now natural to ask when the extremal problem for I has a solution, and when the solution, if it exists, is unique.

PROPOSITION 3.5 *Suppose* $1 \leq p < +\infty$ *and* I *is an invariant subspace of* A_α^p. *Then the extremal problem for* I *has a unique solution.*

Proof. Let S be the supremum in the extremal problem for I. Choose a sequence $\{f_k\}_k$ of unit vectors in I such that

$$S = \lim_k f_k^{(n)}(0),$$

where $n = n_I$. By a normal family argument, we may assume $f_k(z) \to f(z)$ as $k \to +\infty$, uniformly on compact subsets of $\mathbb{D}$. By Fatou's lemma, $\|f\|_{p,\alpha} \leq 1$, and also, f is in the weak closure of I. Basic Functional Analysis tells us that the weak closure and norm closure of a subspace in A_α^p, for $1 \leq p < +\infty$, are the same. It follows that f belongs to I and solves the extremal problem for I.

To prove uniqueness, suppose f and g are two solutions to the extremal problem. Then f and g are unit vectors in I, and for every $t \in (0, 1)$, the function $tf+(1-t)g$ also solves the same extremal problem. It follows that

$$\|tf + (1 - t)g\|_p = 1 = \|tf\|_p + \|(1 - t)g\|_p,$$

for all $t \in (0, 1)$. From Real Analysis we know that

$$\|F + G\|_p = \|F\|_p + \|G\|_p$$

if and only if one of the two functions is a positive constant multiple of the other. From this we conclude that $f = g$. ∎

When $0 < p < 1$, the space A_α^p is no longer locally convex, and so we do not know automatically whether the weak and "norm" closures of I coincide. Neither the existence nor the uniqueness of solutions is known in general in the case $0 < p < 1$. However, if I is a zero-based invariant subspace in A_α^p, then the existence of a solution to the extremal problem for I, even when $0 < p < 1$, is easily established by a normal family argument; we are going to show later in the chapter that such a solution is also unique in the unweighted case.

If the extremal problem for I has a unique solution, we then denote it by G_I and call it the *extremal function* of I. In particular, if $I = I_A$ is a zero-based invariant subspace in A^p, then the corresponding extremal function $G_A = G_{I_A}$ will be called a *zero divisor*. The phrases *canonical divisor* or *contractive zero divisor* are sometimes used as well.

The extremal problem is explicitly solvable only in very special cases. We give several simple examples here.

First, if $p = 2$, then every invariant subspace I in A_α^2 has a reproducing kernel $K_I^\alpha(z, w)$. If in addition $n_I = 0$, then the extremal function G_I^α for I is simply

$$G_I^\alpha(z) = K_I^\alpha(z, 0)/\sqrt{K_I^\alpha(0, 0)}.$$

We now mention an iterative procedure for obtaining the reproducing kernel function for finite zero-based invariant subspaces, which by the above leads to explicit formulas for the corresponding extremal functions. Let $A = \{a_1, \ldots, a_N\}$ be a finite sequence of points in $\mathbb{D}$, and suppose $a \in \mathbb{D} \setminus A$. To simplify the notation, we write K_A^α for $K_{I_A}^\alpha$. Then the kernel function for an additional zero at a is given by

$$K_{A\cup\{a\}}^\alpha(z,w) = K_A^\alpha(z,w) - \frac{K_A^\alpha(z,a)K_A^\alpha(a,w)}{K_A^\alpha(a,a)}, \qquad (z,w)\in\mathbb{D}\times\mathbb{D}.$$

Iteratively this formula gives us the kernel function for finitely many distinct zeros. The first step is to apply the formula to the case of $A=\emptyset$, and get

$$K_a^\alpha(z,w) = \frac{1}{(1-z\overline{w})^{2+\alpha}} - \left(\frac{1-|a|^2}{(1-\overline{a}z)(1-a\overline{w})}\right)^{2+\alpha},$$

where we write a in place of $\{a\}$. As we insert this into the formula for the extremal function G_a for $I_a = \{f\in A_\alpha^2 : f(a) = 0\}$, we arrive at

$$G_a(z) = \left(1-(1-|a|^2)^{2+\alpha}\right)^{-\frac{1}{2}}\left[1-\left(\frac{1-|a|^2}{1-\overline{a}z}\right)^{2+\alpha}\right], \qquad z\in\mathbb{D}.$$

In general, for a finite zero sequence $A=\{a_1,\dots,a_N\}$ of distinct points in $\mathbb{D}$, the extremal function G_A^α is a linear combination of the functions

$$1,\quad \frac{1}{(1-\overline{a}_1 z)^{2+\alpha}},\quad \cdots\quad,\quad \frac{1}{(1-\overline{a}_N z)^{2+\alpha}},$$

which are the reproducing kernel evaluated at the zeros and at the origin. If multiple zeros are encountered, then derivatives of the kernel function,

$$\frac{\partial^j}{\partial\overline{w}^j}\frac{1}{(1-z\overline{w})^{2+\alpha}} = (2+\alpha)\cdots(j+1+\alpha)\frac{z^j}{(1-z\overline{w})^{2+j+\alpha}},$$

are needed for the construction of G_A^α.

We return to general p, $0<p<+\infty$, but put $\alpha=0$. For a point $a\in\mathbb{D}\setminus\{0\}$, let $I_{n\times a}$ be the invariant subspace of the unweighted Bergman space A^p consisting of functions having a zero at $z=a$ of order at least n. Then the extremal problem for $I_{n\times a}$ has a unique solution $G_{n\times a}$ (see Section 3.5 for uniqueness), which is given explicitly by

$$G_{n\times a}(z) = c\left(\frac{a-z}{1-\overline{a}z}\right)^n\left(1+\frac{np}{2}\frac{1-|a|^2}{1-\overline{a}z}\right)^{2/p},$$

where

$$c = \left(\frac{\overline{a}}{|a|}\right)^n\left(1+\frac{np}{2}(1-|a|^2)\right)^{-1/p}.$$

We leave the necessary verifications as an exercise to the reader. Hansbo has obtained an analogous explicit formula for the case of two different zeros of arbitrary multiplicities [52]; a Bessel-type function appears as a result of interaction between the zeros. Finally, if $0<\sigma<+\infty$, and

$$S_\sigma(z) = \exp\left(-\sigma\frac{1+z}{1-z}\right), \qquad z\in\mathbb{D},$$

is the *singular inner function* with atomic singularity at $z = 1$, then the invariant subspace $[S_\sigma]$ of the unweighted Bergman space A^p generated by S_σ gives rise to a unique extremal function G_σ (see Theorem 3.33 for uniqueness), where

$$G_\sigma(z) = (1 + p\sigma)^{-1/p} \left(1 + \frac{p\sigma}{1 - z} \right)^{2/p} \exp\left(-\sigma \frac{1 + z}{1 - z} \right), \qquad z \in \mathbb{D}.$$

This follows from the formula in the previous paragraph. In fact, if $a = a_n = 1 - \sigma/n$ in the extremal function in the previous paragraph, then $G_{n \times a_n} \to G_\sigma$ as $n \to +\infty$. Again, we leave the verification and justification of this limit as an exercise to the reader.

3.3   The Biharmonic Green Function

Our next goal in this chapter is to show that A^p-inner functions have the so-called expansive multiplier property, or equivalently, the contractive divisibility property.

We recall that we have normalized the Laplacian:

$$\Delta = \Delta_z = \frac{1}{4} \left(\frac{\partial^2}{\partial x^2} + \frac{\partial^2}{\partial y^2} \right), \qquad z = x + iy.$$

In terms of Wirtinger derivatives, we have the following alternative and often more convenient expression for the Laplacian:

$$\Delta_z = \frac{\partial^2}{\partial z \partial \bar{z}}.$$

We recall that a function f defined on a planar region is harmonic if $\Delta f = 0$. A real-valued twice differentiable function f on a planar domain is subharmonic if $\Delta f \geq 0$. We shall write ds for normalized length measure:

$$ds(z) = \frac{|dz|}{2\pi}.$$

The starting point for our proof of the expansive multiplier property is the classical Green formula below, which can be found in any book on multivariable calculus and whose proof will be omitted here.

THEOREM 3.6 *Suppose Ω is a domain in the complex plane whose boundary $\partial\Omega$ consists of a finite number of smooth curves. If f and g have continuous second derivatives on $\bar{\Omega}$, the closure of Ω, then*

$$\int_\Omega (f \Delta g - g \Delta f) \, dA = \frac{1}{2} \int_{\partial\Omega} \left(g \frac{\partial f}{\partial n} - f \frac{\partial g}{\partial n} \right) ds,$$

where $\partial/\partial n$ is the inward normal derivative.

COROLLARY 3.7 *Suppose* $0 < r < +\infty$ *and* f *has continuous second derivatives on the closed disk* $|w| \leq r$. *Then we have the identity*

$$\int_{|z|<r} (r^2 - |z|^2) \, \Delta f(z) \, dA(z) = r \int_{|z|=r} f(z) \, ds(z) - \int_{|z|<r} f(z) \, dA(z).$$

Proof. This is a straightforward application of Green's formula.   ∎

COROLLARY 3.8 *Suppose* $0 < r < +\infty$ *and* f *has continuous second derivatives on the closed disk* $|w| \leq r$. *Then for any fixed* z *with* $|z| < r$,

$$2 \int_{|w|<r} \log \left| \frac{r(z-w)}{r^2 - z\overline{w}} \right| \Delta f(w) \, dA(w) = f(z) - \frac{1}{r} \int_{|w|=r} \frac{r^2 - |z|^2}{|z-w|^2} f(w) \, ds(w).$$

Proof. For small positive ε, we remove a closed disk centered at z with radius ε from the disk $|w| < r$ and denote the remaining domain by Ω_ε. We note that the function

$$w \mapsto \log \left| \frac{r(z-w)}{r^2 - z\overline{w}} \right|$$

is harmonic in Ω_ε. The desired result then follows from applying Green's formula to the domain Ω_ε and then letting ε shrink to zero. We omit the routine details. ∎

The *Green function* for $\mathbb{D}$ is

$$G(z, w) = \log \left| \frac{z-w}{1-z\overline{w}} \right|^2 = 2 \log \left| \frac{z-w}{1-z\overline{w}} \right|, \qquad (z, w) \in \mathbb{D} \times \mathbb{D}.$$

The *Green potential* of a function f is then the function defined by

$$G[f](z) = \int_{\mathbb{D}} G(z, w) f(w) \, dA(w), \qquad z \in \mathbb{D}.$$

In what follows we let $C^k(X)$, where X is a planar set, denote the space of complex-valued functions on X whose k-th-order partial derivatives are all continuous on X. To make this precise, it is sometimes necessary to obtain extensions of the functions beyond X, and apply the differentiation to the extended function.

THEOREM 3.9 *Suppose* $f \in C(\overline{\mathbb{D}}) \cap C^2(\mathbb{D})$. *Then*

(1) $G[f] \in C(\overline{\mathbb{D}}) \cap C^2(\mathbb{D})$.

(2) $\Delta G[f] = f$ *in* $\mathbb{D}$.

(3) $G[f] = 0$ *on* $\mathbb{T}$.

Moreover, these conditions determine the Green potential $G[f]$ *uniquely.*

Proof. For any fixed $z \in \mathbb{D}$, we let φ_z denote the conformal mapping of $\mathbb{D}$ defined by

$$\varphi_z(w) = \frac{z - w}{1 - \bar{z}w}, \qquad w \in \mathbb{D}.$$

By a change of variable argument,

$$G[f](z) = \int_{\mathbb{D}} \log |w|^2 \, f \circ \varphi_z(w) \frac{(1 - |z|^2)^2}{|1 - \bar{z}w|^4} \, dA(w).$$

This clearly shows that $G[f] \in C^2(\mathbb{D})$.

Next we show that $G[f] \in C(\overline{\mathbb{D}})$ and $G[f] = 0$ on $\mathbb{T}$. Fix a point $a \in \mathbb{T}$, and write

$$\begin{aligned}
G[f](z) &= \int_{\mathbb{D}} \log |w|^2 \left[f \circ \varphi_z(w) - f(a) \right] \frac{(1 - |z|^2)^2}{|1 - \bar{z}w|^4} \, dA(w) \\
&\quad + f(a) \int_{\mathbb{D}} \log \left| \frac{z - w}{1 - \bar{z}w} \right|^2 dA(w).
\end{aligned}$$

If we use the function $|w|^2$ and $r = 1$ in Corollary 3.8, the result is

$$\int_{\mathbb{D}} \log \left| \frac{z - w}{1 - \bar{z}w} \right|^2 dA(w) = -(1 - |z|^2).$$

It follows that

$$G[f](z) = \int_{\mathbb{D}} \log |w|^2 \left[f \circ \varphi_z(w) - f(a) \right] \frac{(1 - |z|^2)^2}{|1 - \bar{z}w|^4} \, dA(w) - f(a) \, (1 - |z|^2).$$

The second term on the right-hand side above clearly tends to zero as $z \to a$. To see that the first term also goes to zero as $z \to a$, break the integral into two parts, one over the disk $|w| \leq \delta$, where $\delta \in (0, 1)$, and the other over the annulus $\delta < |w| < 1$. The integral over the annulus can be made arbitrarily small by choosing δ close enough to 1, because $\log |w| \to 0$ as $|w| \to 1$ and

$$\int_{\mathbb{D}} |f \circ \varphi_z(w) - f(a)| \frac{(1 - |z|^2)^2}{|1 - \bar{z}w|^4} \, dA(w) \leq 2\|f\|_\infty.$$

The integral over the disk $|w| \leq \delta$ tends to 0 as $z \to a$, because $f \circ \varphi_z(w) \to f(a)$ uniformly for $|w| \leq \delta$. This shows that $G[f](z) \to 0$ as $z \to a$ and hence completes the proof of (1) and (3).

To prove (2), take any C^∞ function g with compact support in $\mathbb{D}$. By Green's formula and Fubini's theorem,

$$\begin{aligned}
\int_{\mathbb{D}} \Delta G[f](z) g(z) \, dA(z) &= \int_{\mathbb{D}} G[f](z) \Delta g(z) \, dA(z) \\
&= \int_{\mathbb{D}} \Delta g(z) \, dA(z) \int_{\mathbb{D}} G(z, w) f(w) \, dA(w) \\
&= \int_{\mathbb{D}} f(w) \, dA(w) \int_{\mathbb{D}} G(z, w) \Delta g(z) \, dA(z).
\end{aligned}$$

By Corollary 3.8,

$$g(w) = \int_{\mathbb{D}} G(z, w) \Delta g(z) \, dA(z)$$

whenever g is C^{∞} with compact support in $\mathbb{D}$. It follows that

$$\int_{\mathbb{D}} \Delta G[f](z) \, g(z) \, dA(z) = \int_{\mathbb{D}} f(w) \, g(w) \, dA(w).$$

This clearly implies that $\Delta G[f] = f$.   ∎

The above theorem tells us how to solve the Laplace equation $\Delta f = g$. It also tells us that the solution is unique with the additional boundary condition $f = 0$ on $\mathbb{T}$.

It turns out that we can also solve the fourth-order partial differential equation $\Delta^2 f = g$ on $\mathbb{D}$ in a similar fashion. For that we introduce the *biharmonic Green function* for $\mathbb{D}$:

$$\Gamma(z, w) = |z - w|^2 \log \left| \frac{z - w}{1 - z\overline{w}} \right|^2 + (1 - |z|^2)(1 - |w|^2), \qquad (z, w) \in \mathbb{D} \times \mathbb{D}.$$

The *biharmonic Green potential* of a function f is then defined as the function

$$\Gamma[f](z) = \int_{\mathbb{D}} \Gamma(z, w) f(w) \, dA(w), \qquad z \in \mathbb{D}.$$

THEOREM 3.10 *Suppose* $f \in C^1(\overline{\mathbb{D}}) \cap C^4(\mathbb{D})$. *Then*

(1) $\Gamma[f] \in C^1(\overline{\mathbb{D}}) \cap C^4(\mathbb{D})$.

(2) $\Delta^2 \Gamma[f] = f$ *in* $\mathbb{D}$.

(3) $\Gamma[f] = \frac{\partial}{\partial n} \Gamma[f] = 0$ *on* $\mathbb{T}$.

Furthermore, these conditions uniquely determine the potential $\Gamma[f]$.

This theorem will not actually be used for the presentation of the material in this book, and its proof is rather tedious though analogous to that of Theorem 3.9, so we omit the proof here. However, the above properties are important for the general understanding of biharmonic Green potentials. We shall need several further properties of the biharmonic Green function. It is good to know approximately how big $\Gamma(z, w)$ is.

LEMMA 3.11 *For all* $z, w \in \mathbb{D}$, *we have*

$$\frac{(1 - |z|^2)^2(1 - |w|^2)^2}{2|1 - z\overline{w}|^2} \leq \Gamma(z, w) \leq \frac{(1 - |z|^2)^2(1 - |w|^2)^2}{|1 - z\overline{w}|^2}.$$

In particular, $\Gamma(z, w)$ *is strictly positive.*

Proof. Simple manipulations with the definition of $\Gamma(z, w)$ and the well-known identity

$$1 - \left| \frac{z - w}{1 - z\overline{w}} \right|^2 = \frac{(1 - |z|^2)(1 - |w|^2)}{|1 - z\overline{w}|^2}$$

yield the formula

$$\Gamma(z, w) = \frac{(1 - |z|^2)^2 (1 - |w|^2)^2}{|1 - z\overline{w}|^2} F\left(1 - \left| \frac{z - w}{1 - z\overline{w}} \right|^2 \right),$$

where

$$F(x) = \frac{(1 - x)\log(1 - x) + x}{x^2}.$$

It is easy to see that

$$F(x) = \sum_{k=0}^{+\infty} \frac{x^n}{(k + 1)(k + 2)}, \qquad |x| \leq 1.$$

This implies that

$$\frac{1}{2} \leq F(x) \leq 1, \qquad 0 \leq x \leq 1,$$

and the desired estimate for $\Gamma(z, w)$ follows. ∎

We emphasize three points that are easily seen from the asymptotic formula for $\Gamma(z, w)$ above. First, the biharmonic Green function is positive on $\mathbb{D} \times \mathbb{D}$. Second, if $z \in \mathbb{D}$ is fixed, then

$$\Gamma(z, w) \sim (1 - |w|^2)^2, \qquad |w| \to 1^-.$$

Finally, for fixed $z \in \mathbb{D}$, we have

$$\Gamma(z, w) = \frac{\partial}{\partial n(w)} \Gamma(z, w) = 0$$

for all $|w| = 1$ (this can be checked directly from the definition of Γ or from the power series expansion of F in the proof of the lemma above).

LEMMA 3.12 *Fix $w \in \mathbb{D}$. Then*

$$\Delta_z \Gamma(z, w) = G(z, w) + (1 - |w|^2) H(z, w), \qquad z \in \mathbb{D} \setminus \{w\},$$

where

$$H(z, w) = \frac{1 - |zw|^2}{|1 - z\overline{w}|^2}.$$

Proof. This is a straightforward calculation using Wirtinger derivatives. We omit the details. ∎

We need the following monotonicity property of the above function H.

LEMMA 3.13 *Fix $z \in \mathbb{T}$ and $w \in \mathbb{D}$. Then the function*

$$r \mapsto rH\left(z, \frac{w}{r}\right)$$

is increasing on the interval $(|w|, 1)$.

Proof. For $|w| < r < 1$, we have

$$H\left(z, \frac{w}{r}\right) = \frac{\left(r^2 - |w|^2\right)^2}{r^2 |r - \overline{w}z|^2}.$$

A computation shows that the derivative

$$\frac{d}{dr} H\left(z, \frac{w}{r}\right)$$

is equal to

$$4 \frac{r^2 - |w|^2}{r |r - \overline{w}z|^2} - \frac{\left(r^2 - |w|^2\right)^2}{|r - \overline{w}z|^2} \left(\frac{2}{r^3} + \frac{1}{r^2(r - \overline{w}z)} + \frac{1}{r^2(r - w\overline{z})} \right),$$

It then follows that the derivative

$$\frac{d}{dr}\left[rH\left(z, \frac{w}{r}\right) \right] = H\left(z, \frac{w}{r}\right) + r \frac{d}{dr} H\left(z, \frac{w}{r}\right)$$

is equal to

$$\frac{\left(r^2 - |w|^2\right)^2}{r |r - \overline{w}z|^2} \left(\frac{r - |w|}{r(r + |w|)} + \frac{2}{r - |w|} - \frac{1}{r - \overline{w}z} - \frac{1}{r - w\overline{z}} \right),$$

which is greater than or equal to

$$\frac{\left(r^2 - |w|^2\right)^2 (r - |w|)}{r^2 |r - \overline{w}z|^2 (r + |w|)},$$

because for $r > |w|$ we have

$$\frac{1}{r - w\overline{z}} + \frac{1}{r - \overline{w}z} = 2 \operatorname{Re}\left(\frac{1}{r - w\overline{z}} \right) \leq \frac{2}{|r - w\overline{z}|} \leq \frac{2}{r - |w|}.$$

The proof is complete. ∎

The monotonicity of H leads to a corresponding monotonicity property of the Green function Γ, as the two are related by the formula

$$\Gamma(z, w) = \frac{1}{\pi} \int_{\max\{|z|, |w|\}}^{1} \int_{-\pi}^{\pi} H\left(e^{i\theta}, \frac{z}{\xi}\right) H\left(e^{i\theta}, \frac{w}{\xi}\right) d\theta \, \xi \, d\xi.$$

This identity can be verified by explicit computation. However, it is more appropriate to view it as a special case of Hadamard's variational formula, which will be discussed in detail in Chapter 9.

LEMMA 3.14 *Fix* $z, w \in \mathbb{D}$. *Then the function*

$$r \mapsto r\Gamma\left(z, \frac{w}{r}\right)$$

is increasing on the interval $(|w|, 1)$.

Proof. By the above integral representation of the biharmonic Green function,

$$r\,\Gamma\left(z, \frac{w}{r}\right) = \frac{1}{\pi} \int_{\max\{|z|, |w|/r\}}^{1} \int_{-\pi}^{\pi} H\left(e^{i\theta}, \frac{z}{\xi}\right) r\xi\, H\left(e^{i\theta}, \frac{w}{r\xi}\right) d\theta\, d\xi$$

for $|w| < r < 1$. Note that the interval

$$\left[\max\left\{|z|, \frac{|w|}{r}\right\}, 1\right]$$

gets bigger as r increases. On the other hand, since $|w|/r < \xi$, we have $|w| < r\xi$, and hence

$$r\xi\, H\left(e^{i\theta}, \frac{w}{r\xi}\right)$$

is an increasing function of r, according to Lemma 3.13. The proof is complete. ∎

Recall that for a function f in $\mathbb{D}$ and $0 < r < 1$, the dilation f_r is defined by $f_r(z) = f(rz)$, $z \in \mathbb{D}$.

LEMMA 3.15 *If* f *is a positive locally summable function in* $\mathbb{D}$, *then*

$$r^3\Gamma[f_r](z) \leq \Gamma[f](z), \qquad z \in \mathbb{D},$$

for all $0 < r < 1$. *Moreover,* $r^3\Gamma[f_r]$ *increases monotonically to* $\Gamma[f]$ *pointwise on* $\mathbb{D}$ *as* $r \to 1^-$.

Proof. After a change of variables, the formula

$$r^3\Gamma[f_r](z) = r^3 \int_{\mathbb{D}} \Gamma(z, w)\, f_r(w)\, dA(w), \qquad z \in \mathbb{D},$$

becomes

$$r^3\Gamma[f_r](z) = \int_{|w|<r} r\Gamma\left(z, \frac{w}{r}\right) f(w)\, dA(w), \qquad z \in \mathbb{D}.$$

The assertion is now immediate from Lemma 3.14. ∎

LEMMA 3.16 *Suppose* $0 < p < +\infty$ *and* f *is an analytic function on* $\mathbb{D}$. *Then the potential* $r\,\Gamma[\Delta|f_r|^p]$ *increases monotonically to* $\Gamma[\Delta|f|^p]$, *pointwise in* $\mathbb{D}$, *as* $r \to 1^-$.

Proof. Since $|f|^p$ is subharmonic, the function $\Delta|f|^p$ is positive. Moreover,

$$\Delta|f_r(z)|^p = r^2\Delta|f|^p(rz), \qquad z \in \mathbb{D}.$$

The desired result then follows from Lemma 3.15. ∎

3.4   The Expansive Multiplier Property

In this section we prove an integral formula concerning A^p-inner functions. As a consequence we obtain the expansive multiplier property for such functions. Note that the results in this section are stated and proved in the unweighted case; they cannot be generalized to the weighted Bergman spaces A_α^p in the full range $-1 < \alpha < +\infty$.

We shall need the following lemma from integration theory.

LEMMA 3.17 *Suppose that μ is a finite positive measure on the measure space X and that $0 < p < +\infty$. If f_n and f are μ-measurable functions on X such that $f_n(x) \to f(x)$ μ-almost everywhere as $n \to +\infty$ and*

$$\limsup_{n \to +\infty} \int_X |f_n|^p \, d\mu \le \int_X |f|^p \, d\mu < +\infty,$$

then

$$\lim_{n \to +\infty} \int_X |f_n - f|^p \, d\mu = 0.$$

Proof. Let E be any measurable set in X. Then by Fatou's lemma,

$$
\begin{aligned}
\int_E |f|^p \, d\mu
&\le \liminf_n \int_E |f_n|^p \, d\mu \\
&\le \limsup_n \left[\int_X |f_n|^p \, d\mu - \int_{X \setminus E} |f_n|^p \, d\mu \right] \\
&\le \int_X |f|^p \, d\mu - \liminf_n \int_{X \setminus E} |f_n|^p \, d\mu \\
&\le \int_X |f|^p \, d\mu - \int_{X \setminus E} |f|^p \, d\mu = \int_E |f|^p \, d\mu.
\end{aligned}
$$

It follows that

$$\lim_n \int_E |f_n|^p \, d\mu = \int_E |f|^p \, d\mu.$$

Given any positive number ε, we can choose $\delta > 0$ such that

$$\int_E |f|^p \, d\mu < \varepsilon$$

whenever $\mu(E) < \delta$. By Egorov's theorem (see any book on Real Analysis), there exists $X_1 \subset X$ such that $\mu(X \setminus X_1) < \delta$ and $f_n \to f$ uniformly on X_1. It is elementary to check that

$$|z - w|^p \le 2^p \left(|z|^p + |w|^p \right)$$

for all $0 < p < +\infty$ and all complex numbers z and w. Thus,

$$\int_X |f_n - f|^p \, d\mu = \int_{X_1} |f_n - f|^p \, d\mu + \int_{X \setminus X_1} |f_n - f|^p \, d\mu$$

$$\leq \int_{X_1} |f_n - f|^p \, d\mu + 2^p \int_{X \setminus X_1} \left(|f_n|^p + |f|^p \right) d\mu.$$

Since the integral over X_1 above tends to zero by uniform convergence, we obtain

$$\limsup_n \int_X |f_n - f|^p \, d\mu \leq 2^{p+1} \int_E |f|^p \, d\mu < 2^{p+1} \varepsilon,$$

completing the proof of the lemma. ∎

Recall the expression of the kernel $\Delta_z \Gamma(z, w)$ from Lemma 3.12.

PROPOSITION 3.18 *If $0 < p < +\infty$ and f is analytic in $\mathbb{D}$, then*

$$\int_{\mathbb{D}} \Gamma(z, w) \Delta |f(z)|^p \, dA(z) = \int_{\mathbb{D}} \Delta_z \Gamma(z, w) |f(z)|^p \, dA(z)$$

for all $w \in \mathbb{D}$. Furthermore, either integral is finite if and only if $f \in A^p$.

Proof. We observe that for fixed $w \in \mathbb{D}$, $\Delta_z \Gamma(z, w)$ is positive except on a compact subset of $\mathbb{D}$, and that part makes a finite contribution to the right hand side. So, if the right-hand integral diverges, it is because of the contribution from points near the boundary, in which case the integral equals $+\infty$.

If f is analytic in $\overline{\mathbb{D}}$, then the desired identity follows directly from Green's formula and the fact that

$$\Gamma(z, w) = \frac{\partial}{\partial n(z)} \Gamma(z, w) = 0$$

for $z \in \mathbb{T}$; the zeros of f and the logarithmic singularity of $\Delta_z \Gamma(z, w)$ at $z = w$ can be taken care of by removing from $\mathbb{D}$ a finite number of disks with radius ε and then taking the limit as ε tends to zero. In particular, the identity holds if f is replaced by $f_r, 0 < r < 1$. The general case then follows from an obvious limit argument involving Lemma 3.16. ∎

COROLLARY 3.19 *Suppose $0 < p < +\infty$ and φ is an A^p-inner function. Then*

$$\int_{\mathbb{D}} \Gamma(z, w) \Delta |\varphi(z)|^p \, dA(z) = \int_{\mathbb{D}} G(z, w) |\varphi(z)|^p \, dA(z) + 1 - |w|^2$$

for all $w \in \mathbb{D}$.

Proof. Recall that by Lemma 3.12,

$$\Delta_z \Gamma(z, w) = G(z, w) + (1 - |w|^2) \frac{1 - |zw|^2}{|1 - z\overline{w}|^2}.$$

For any fixed $w \in \mathbb{D}$, the function

$$z \mapsto \frac{1 - |zw|^2}{|1 - z\overline{w}|^2}$$

is bounded and harmonic in $\mathbb{D}$. The assertion then follows from Proposition 3.18 and the definition of A^p-inner functions.   ∎

Using the identities

$$\int_{\mathbb{D}} G(z, w)\, dA(z) = \int_{\mathbb{D}} \log \left| \frac{z - w}{1 - z\overline{w}} \right|^2 dA(z) = -(1 - |w|^2), \qquad w \in \mathbb{D},$$

we can rewrite Corollary 3.19 as

$$\Gamma \left[\Delta |\varphi|^p \right] = G \left[|\varphi|^p - 1 \right].$$

Since the Green function $G(z, w)$ is negative and the biharmonic Green function $\Gamma(z, w)$ is positive, it follows that

$$0 \leq \Gamma \left[\Delta |\varphi|^p \right](z) = G \left[|\varphi|^p - 1 \right](z) \leq 1 - |z|^2, \qquad z \in \mathbb{D},$$

for all A^p-inner functions φ.

We can now prove the main result of the section.

THEOREM 3.20 *Suppose* $0 < p < +\infty$ *and* φ *is an* A^p*-inner function. Then*

$$\int_{\mathbb{D}} |\varphi|^p g\, dA = \int_{\mathbb{D}} g\, dA + \int_{\mathbb{D}} \int_{\mathbb{D}} \Gamma(z, w) \Delta g(w) \Delta |\varphi(z)|^p \, dA(z)\, dA(w)$$

for all $g \in C^2(\overline{\mathbb{D}})$. *In particular, if* g *is also subharmonic, then*

$$\int_{\mathbb{D}} g(z)\, dA(z) \leq \int_{\mathbb{D}} |\varphi(z)|^p g(z)\, dA(z).$$

Proof. By Theorem 3.9, there exists a bounded harmonic function h in $\mathbb{D}$ such that

$$g(z) = G\left[\Delta g \right](z) + h(z), \qquad z \in \mathbb{D}.$$

Using the definition of A^p-inner functions, Fubini's theorem, and the remark after Corollary 3.19, we obtain

$$
\begin{aligned}
\int_{\mathbb{D}} \left(|\varphi|^p - 1 \right) g \, dA &= \int_{\mathbb{D}} \left(|\varphi(z)|^p - 1 \right) \left[h(z) + G\,[\Delta g]\,(z) \right] dA(z) \\
&= \int_{\mathbb{D}} \left(|\varphi(z)|^p - 1 \right) \int_{\mathbb{D}} G(z, w) \Delta g(w) \, dA(w) \, dA(z) \\
&= \int_{\mathbb{D}} \Delta g(w) \int_{\mathbb{D}} G(z, w) \left(|\varphi(z)|^p - 1 \right) dA(z) \, dA(w) \\
&= \int_{\mathbb{D}} \Delta g(w) \int_{\mathbb{D}} \Gamma(z, w) \Delta |\varphi(z)|^p \, dA(z) \, dA(w) \\
&= \int_{\mathbb{D}} \int_{\mathbb{D}} \Gamma(z, w) \Delta g(w) \Delta |\varphi(z)|^p \, dA(z) \, dA(w).
\end{aligned}
$$

If g is also subharmonic, then $\Delta g \geq 0$ in $\mathbb{D}$; since $\Gamma(z, w)$ and $\Delta |\varphi(z)|^p$ are both positive, we conclude that

$$
\int_{\mathbb{D}} g(z) \, dA(z) \leq \int_{\mathbb{D}} |\varphi(z)|^p g(z) \, dA(z),
$$

as asserted.   ∎

The form of Theorem 3.20 that we will actually use runs as follows.

COROLLARY 3.21 *Suppose $0 < p, q < +\infty$ and φ is an A^p-inner function. Then*

$$
\int_{\mathbb{D}} |\varphi|^p |f|^q \, dA = \int_{\mathbb{D}} |f|^q \, dA + \int_{\mathbb{D}} \int_{\mathbb{D}} \Gamma(z, w) \Delta |f(w)|^q \, \Delta |\varphi(z)|^p \, dA(z) \, dA(w)
$$

provided f is analytic in $\overline{\mathbb{D}}$.

Proof. Although the function $g = |f|^p$ is not necessarily in $C^2(\overline{\mathbb{D}})$, the proof of Theorem 3.20 can easily be modified to work in this case; all one has to do is remove from $\mathbb{D}$ tiny disks centered at the finitely many zeros of f and then use a limit argument.   ∎

COROLLARY 3.22 *If $0 < p < +\infty$ and φ is A^p-inner, then*

$$
\int_{\mathbb{D}} |f(z)|^p \, dA(z) \leq \int_{\mathbb{D}} |\varphi(z) f(z)|^p \, dA(z)
$$

for all $f \in H^\infty$.

Proof. By an obvious approximation argument, we may assume that f is analytic in $\overline{\mathbb{D}}$. The assertion now follows from Corollary 3.21 with $q = p$, in view of the positivity of the biharmonic Green function and the subharmonicity of $|f|^p$.   ∎

COROLLARY 3.23 *Suppose* $0 < p < +\infty$ *and* φ *is a bounded* A^p-*inner function. Then*

$$\int_{\mathbb{D}} |f(z)|^p \, dA(z) \le \int_{\mathbb{D}} |\varphi(z) f(z)|^p \, dA(z)$$

holds for all $f \in A^p$.

Proof. This follows from Corollary 3.22 if we approximate f by its dilates f_r with $0 < r < 1$, given by $f_r(z) = f(rz)$ for $z \in \mathbb{D}$. ∎

The property exhibited in the two corollaries above will be called the *expansive multiplier property* of A^p-inner functions.

We observe from Corollary 3.21 that if $p = 2$ and φ is an A^2-inner function, then

$$\int_{\mathbb{D}} |\varphi h|^2 \, dA = \int_{\mathbb{D}} |h|^2 \, dA + \int_{\mathbb{D}} \int_{\mathbb{D}} \Gamma(z, w) |\varphi'(z)|^2 |h'(w)|^2 \, dA(z) \, dA(w),$$

where h is any function analytic in $\overline{\mathbb{D}}$.

We conclude the section with an integral estimate for dilated A^p-inner functions, which should be compared with Corollary 3.19.

PROPOSITION 3.24 *Suppose* $0 < p < +\infty$ *and* φ *is an* A^p-*inner function. Then, for* $0 < r < 1$, *we have*

$$\int_{\mathbb{D}} \Gamma(z, w) \Delta |\varphi_r(z)|^p \, dA(z) \le \int_{\mathbb{D}} G(z, w) |\varphi_r(z)|^p \, dA(z) + 1 - |w|^2$$

for all $w \in \mathbb{D}$.

Proof. By Lemma 3.12 and Proposition 3.18,

$$\int_{\mathbb{D}} \Gamma(z, w) \Delta |\varphi_r(z)|^p \, dA(z) = \int_{\mathbb{D}} G(z, w) |\varphi_r(z)|^p \, dA(z)$$
$$+ (1 - |w|^2) \int_{\mathbb{D}} \frac{1 - |zw|^2}{|1 - z\overline{w}|^2} |\varphi_r(z)|^p \, dA(z),$$

where we observe that the function

$$P(z, w) = \frac{1 - |zw|^2}{|1 - z\overline{w}|^2}$$

is the Poisson kernel extended harmonically to both variables. We are to show that

$$\int_{\mathbb{D}} P(z, w) |\varphi_r(z)|^p \, dA(z) \le 1.$$

Consider, for $\lambda \in \mathbb{D}$, the function

$$\int_{\mathbb{T}} P(\lambda, \xi) |\varphi(z\xi)|^p \, ds(\xi), \qquad z \in \mathbb{D},$$

where we recall the notation $ds(z) = |dz|/(2\pi)$ for normalized arc length measure. As a function of λ, it is harmonic with boundary values $|\varphi(z\lambda)|^p$, and in view of the fact that the function $|\varphi(z\lambda)|^p$ is subharmonic as a function of λ, we obtain from the sub-mean value property that

$$|\varphi(z\lambda)|^p \le \int_{\mathbb{T}} P(\lambda, \xi)\, |\varphi(z\xi)|^p \, ds(\xi), \qquad z \in \mathbb{D}.$$

Specializing to $\lambda = r$, we obtain

$$|\varphi_r(z)|^p \le \int_{\mathbb{T}} P(r, \xi)\, |\varphi(z\xi)|^p \, ds(\xi), \qquad z \in \mathbb{D}.$$

Since φ is A^p-inner, an application of Fubini's theorem gives

$$
\begin{aligned}
\int_{\mathbb{D}} P(z, w)\, |\varphi_r(z)|^p \, dA(z) \;&\le\; \int_{\mathbb{D}} P(z, w) \int_{\mathbb{T}} P(r, \xi)\, |\varphi(z\xi)|^p \, ds(\xi)\, dA(z) \\
&=\; \int_{\mathbb{T}} P(r, \xi) \int_{\mathbb{D}} P(z, w)\, |\varphi(z\xi)|^p \, dA(z)\, ds(\xi) \\
&=\; \int_{\mathbb{T}} P(r, \xi)\, P(0, w)\, ds(\xi) \\
&=\; \int_{\mathbb{T}} P(r, \xi)\, ds(\xi) = P(r, 0) = 1,
\end{aligned}
$$

as claimed.   ∎

The above proof actually shows that for an A^p-inner function φ and $0 < r < 1$, the dilation φ_r is a subinner function in the sense of Section 3.7, that is,

$$\int_{\mathbb{D}} h(z)\, |\varphi_r(z)|^p \, dA(z) \le h(0)$$

holds for all positive harmonic functions h on $\mathbb{D}$.

3.5   Contractive Zero Divisors in A^p

In this section we take a closer look at the extremal problem for invariant subspaces generated by zero sets of A^p. We show that a unique solution exists in this case, even when $0 < p < 1$. Recall that existence follows from a normal families argument (see Section 3.2). We will also prove that the extremal function, which we call a (contractive) zero divisor, is analytic across the unit circle when the zero set is finite.

We begin with the case of a single zero with multiplicity 1.

LEMMA 3.25 *Suppose $0 < p < +\infty$ and a is a nonzero point in $\mathbb{D}$. Let I_a be the invariant subspace of A^p consisting of functions $f \in A^p$ with $f(a) = 0$. Then the extremal problem for I_a has a unique solution, and the solution is given by*

$$G_a(z) = C_a\, \frac{z - a}{1 - \bar{a}z} \left[1 + \frac{p}{2}\left(1 + \bar{a}\,\frac{z - a}{1 - \bar{a}z} \right) \right]^{2/p},$$

where

$$C_a = -\frac{\bar{a}}{|a|}\left[1 + \frac{p}{2}(1 - |a|^2)\right]^{-1/p}.$$

Proof. Let q be any polynomial. Then a change of variables leads to

$$\int_{\mathbb{D}} |G_a(z)|^p q(z)\, dA(z) = (1 - |a|^2)^2 |C_a|^p \int_{\mathbb{D}} |k(w)|^2 |w|^p q\left[\frac{w + a}{1 + \bar{a}w}\right] dA(w),$$

where

$$k(w) = \frac{1 + (p/2)(1 + \bar{a}w)}{(1 + \bar{a}w)^2}.$$

Now, let

$$k(w) = \sum_{n=0}^{+\infty} b_n w^n \quad \text{and} \quad f(w) = k(w)\, q\left(\frac{w + a}{1 + \bar{a}w}\right) = \sum_{n=0}^{+\infty} c_n w^n,$$

and integrate by polar coordinates to get

$$\int_{\mathbb{D}} |G_a(z)|^p q(z)\, dA(z) = (1 - |a|^2)^2 |C_a|^p \sum_{n=0}^{+\infty} \frac{2}{2n + p + 2} \bar{b}_n c_n$$

$$= (1 - |a|^2)^2 |C_a|^p f(-a);$$

here, we used the fact that $b_n = (n + p/2 + 1)(-\bar{a}^n)$. The definitions of C_a and k easily reduce the last expression above to $q(0)$, so that G_a is an A^p-inner function that clearly belongs to I_a. Since G_a is analytic on $\bar{\mathbb{D}}$, the expansive multiplier property of G_a becomes $\|g\|_p \leq \|G_a g\|_p$ for all $g \in A^p$, or equivalently, $\|g/G_a\|_p \leq \|g\|_p$ for all $g \in I_a$, since G_a only vanishes at a in $\mathbb{D}$.

Suppose $g \in I_a$ and $\|g\|_p \leq 1$. Then

$$\left|\frac{g(0)}{G_a(0)}\right| \leq \left\|\frac{g}{G_a}\right\|_p \leq \|g\|_p \leq 1,$$

so that G_a is an extremal function for I_a. Since the first inequality above is strict unless g/G_a is constant, we see that G_a is the unique solution of the extremal problem for I_a; otherwise, taking g to be another extremal function would yield $g(0) < G_a(0)$, a contradiction. ∎

COROLLARY 3.26 *Suppose $0 < p < +\infty$, A is an A^p-zero set, and I_A is the corresponding invariant subspace in A^p. If G is any solution of the extremal problem for I_A, then G has no extraneous zeros.*

Proof. If A consists of a single zero of multiplicity 1, then an inspection of the formula for G_A in Lemma 3.25 reveals that G_A has no extraneous zeros.

In the general case let us assume, for the sake of arriving at a contradiction, that G has an extraneous zero at $z = a$; thus either a is a new zero, or $a \in A$ but G

has a zero at $z = a$ of order higher than prescribed in A. It is obvious that $a \neq 0$. Let G_a be the function from Lemma 3.25. Then G/G_a belongs to I_A, and the expansive multiplier property of G_a gives $\|G/G_a\|_p \leq 1$. Since $0 < G_a(0) < 1$, the function G/G_a solves the extremal problem for I_A better than G does, which is a contradiction.   ∎

Our next step is to show that for a finite zero set A the extremal problem for I_A in A^p has a unique solution and that the solution is analytic in a larger disk. Recall that the uniqueness for $1 \leq p < +\infty$ follows from the local convexity of the space A^p.

LEMMA 3.27 *Suppose $1 \leq q < +\infty$ and $p = mq$ for some integer $m \geq 2$. Let G be the zero divisor in A^p of a finite zero set $\{z_j\}_j$, and let H be the zero divisor in A^q of the zero set $\{w_j\}_j$ obtained from $\{z_j\}_j$ by including each z_j exactly m times. Then $H = G^m$.*

Proof. By an approximation argument, we may assume that $\{z_j\}_j$ does not contain 0. Since by Corollary 3.26, the zeros of H are exactly the z_j's, each of which is of order m, we see that $H^{1/m}$ (the branch with $H^{1/m}(0) > 0$) is analytic and has all the properties required in the extremal problem that determines G. Thus $H^{1/m} = G$, or $H = G^m$, by uniqueness.   ∎

We will need to use the reproducing kernel functions for a class of Hilbert spaces of analytic functions in $\mathbb{D}$. Thus, we consider a weight function

$$\omega(z) = |h(z)|^t, \qquad z \in \mathbb{D},$$

where $0 < t < +\infty$ and h is a function in the Bergman space A^t (not identically zero). For $0 < p < +\infty$ let $B^p(\omega)$ be the space of analytic functions f in $\mathbb{D}$ such that

$$\|f\|_{p,\omega} = \left(\int_{\mathbb{D}} |f(z)|^p \omega(z)\, dA(z) \right)^{1/p} < +\infty.$$

It is easy to show that each point evaluation in $\mathbb{D}$ is a bounded linear functional on the space $B^p(\omega)$. In fact, if f is any analytic function in $\mathbb{D}$, then the subharmonicity of the function $|f|^p \omega$ implies that any point evaluation at $z \in \mathbb{D}$, where $\omega(z) > 0$, is bounded on $B^p(\omega)$; actually, if K is any compact subset of $\mathbb{D}$ where ω is strictly positive, then point evaluations at z are uniformly bounded on $B^p(\omega)$ for z in K. If z is a point with $\omega(z) = 0$, then we can find a sufficiently small positive number r such that the circle $S = \{w \in \mathbb{C} : |w - z| = r\}$ is contained in $\mathbb{D}$ and ω is positive there. An application of Cauchy's formula, together with the earlier remark that point evaluations at $w \in S$ are uniformly bounded on $B^p(\omega)$, then shows that the point evaluation at z is also bounded on $B^p(\omega)$. Using the continuity of ω, we see that the argument above also works for z in a sufficiently small neighborhood of a zero of ω. We conclude that point evaluations at z are uniformly bounded on $B^p(\omega)$ if z is restricted to any compact subset of $\mathbb{D}$, and consequently, each space $B^p(\omega)$ is complete.

It is clear that each $B^p(\omega)$ contains H^∞. Let $A^p(\omega)$ be the closure of the set of polynomials in $B^p(\omega)$. Equivalently, $A^p(\omega)$ is the closure of H^∞ in $B^p(\omega)$. For $p = 2$, $B^2(\omega)$ is a Hilbert space, and then so is $A^2(\omega)$, with the inner product

$$\langle f, g \rangle_\omega = \int_{\mathbb{D}} f(z) \overline{g(z)} \, \omega(z) \, dA(z).$$

LEMMA 3.28 *Let $K_\omega(z, w)$ be the reproducing kernel for the Hilbert space $A^2(\omega)$. Then $K_\omega(z, w) \neq 0$ for all z and w in $\mathbb{D}$.*

Proof. We first show that $K_\omega(z, 0)$ is nonvanishing in $\mathbb{D}$. It is obvious that

$$K_\omega(0, 0) = \sup\{|f(0)|^2 : \|f\|_{\omega,2} \leq 1\} > 0; \qquad (3.1)$$

the maximizing function is $f(z) = K_\omega(z, 0)/\sqrt{K_\omega(0, 0)}$. If $K_\omega(\lambda, 0) = 0$ for some $\lambda \in \mathbb{D} \setminus \{0\}$, we consider the function

$$F(z) = K_\omega(z, 0)/G_\lambda(z), \qquad z \in \mathbb{D},$$

where

$$G_\lambda(z) = \frac{1}{|\lambda|\sqrt{2 - |\lambda|^2}} \left[1 - \left(\frac{1 - |\lambda|^2}{1 - \bar{\lambda}z} \right)^2 \right]$$

is the one-point zero divisor in A^2. The function G_λ is analytic on $\overline{\mathbb{D}}$ and vanishes only at λ in $\mathbb{D}$. By the subharmonicity of $|F|^2\omega$ and the expansive multiplier property of G_λ (see Theorem 3.20 and its corollaries; the lower degree of smoothness of ω at its zeros can easily be taken care of by a limit argument), we have

$$\int_{\mathbb{D}} |F(z)|^2 \omega(z) \, dA(z) \leq \int_{\mathbb{D}} |K_\omega(z, 0)|^2 \omega(z) \, dA(z) = K_\omega(0, 0).$$

Since $0 < G_\lambda(0) < 1$, the function $F(z)/\sqrt{K_\omega(0, 0)}$ solves the extremal problem in (3.1) better than the function $K_\omega(z, 0)/\sqrt{K_\omega(0, 0)}$ does. This contradiction shows that $K_\omega(z, 0) \neq 0$ for all $z \in \mathbb{D}$.

To show that $K_\omega(z, w) \neq 0$ for all z and w in $\mathbb{D}$, observe that for any Möbius map ϕ preserving the disk $\mathbb{D}$, we have

$$K_\omega(\phi(z), \phi(w)) = K_{\omega_\phi}(z, w)$$

for all z and w in $\mathbb{D}$, where

$$\omega_\phi(z) = |\phi'(z)|^2 \omega \circ \phi(z), \qquad z \in \mathbb{D},$$

is a weight of the same type as ω (since ϕ' is nonvanishing). Combining this with the result in the previous paragraph, we conclude that $K_\omega(z, w) \neq 0$ for all z and $w \in \mathbb{D}$. ∎

We return to the extremal problem for zero sets. For general p, $0 < p < +\infty$, and a finite zero set $A = \{a_j\}_{j=1}^n$, we are going to show that any solution of the extremal problem for I_A in A^p can be written as the corresponding finite Blaschke product

$$b_A(z) = \prod_{j=1}^n \frac{|a_j|}{a_j} \frac{a_j - z}{1 - \overline{a}_j z}$$

times the reproducing kernel of the weighted Bergman space $A^2(\omega)$, where $\omega = |b_A(z)|^p$, so the uniqueness of the extremal solution follows. Then we are going to show that the reproducing kernel of this $A^2(\omega)$ has an analytic continuation across the unit circle. Note that $A^2(\omega) = B^2(\omega) = A^2$ as spaces in the present situation.

LEMMA 3.29 *Suppose $0 < p < +\infty$, A is a finite zero set, and I_A is the corresponding invariant subspace in A^p. If $G = G_A$ solves the extremal problem for I_A, then*

$$G(z) = K_\omega(0, 0)^{-1/p} b(z) K_\omega(z, 0)^{2/p}, \qquad z \in \mathbb{D},$$

where $\omega = |b|^p$ and b is the Blaschke product corresponding to A. In particular, G is unique.

Proof. Since G has no extraneous zeros, we can write $G(z) = b(z) k(z)^{2/p}$, where k is a nonvanishing function in A^2. By the same variational argument as used in the proof of Theorem 3.4, with the variation $G^* = G + \lambda z h$ instead, where $h \in I_A$ and $\lambda \in \mathbb{C}$, we see that

$$\int_{\mathbb{D}} |G(z)|^{p-2} \overline{G(z)} z h(z) \, dA(z) = 0$$

for all $h \in I_A$. Since $I_A = b A^p$, the decomposition $f = (f - f(0)) + f(0)$ gives

$$\int_{\mathbb{D}} |G|^{p-2} \overline{G} b f \, dA = f(0) \int_{\mathbb{D}} |G|^{p-2} \overline{G} b \, dA,$$

where f is any function in A^p. Using the factorization of G, we obtain

$$\int_{\mathbb{D}} k^{1-2/p} \overline{k} f |b|^p \, dA = f(0) \int_{\mathbb{D}} k^{1-2/p} \overline{k} |b|^p \, dA, \qquad f \in A^p.$$

Combining this with the fact that $\|G\|_p = 1$, we get

$$1 = \int_{\mathbb{D}} k^{1-2/p} k^{2/p} \overline{k} |b|^p \, dA = k(0)^{2/p} \int_{\mathbb{D}} k^{1-2/p} \overline{k} |b|^p \, dA.$$

It follows that

$$\int_{\mathbb{D}} k^{1-2/p} f \overline{k} |b|^p \, dA = k(0)^{-2/p} f(0)$$

for all $f \in A^p$.

If $0 < p \le 2$, we can choose $f = h k^{(2/p)-1}$, where h is any polynomial. Then $f \in A^p$ (since $0 \le 2 - p < 2$) and

$$k(0) \int_{\mathbb{D}} h \overline{k} |b|^p \, dA = h(0).$$

A simple approximation argument then shows that the above also holds for all $h \in A^2(\omega)$; thus $k(0)\overline{k(z)} = K_\omega(0, z)$ by the uniqueness of the reproducing kernel in $A^2(\omega)$.

If $2 < p < +\infty$, we can choose an integer $m \geq 2$ such that $q = p/m \leq 2$. By Lemma 3.27, the function $G^m = b^m k^{2/q}$ is the zero divisor in A^q corresponding to the zero set of b^m. According to the previous paragraph, we must have $k(0)\overline{k(z)} = K_\omega(0, z)$, where $K_\omega(z, w)$ is the reproducing kernel of $A^2(|b^m|^q) = A^2(|b|^p)$. This shows that the desired representation of G holds for all p.   ∎

LEMMA 3.30 *Suppose* $0 < p < +\infty$ *and* I_A *is the invariant subspace of* A^p *corresponding to a finite zero set* A. *Then the zero divisor* G_A *(the unique solution of the extremal problem for* I_A*) has an analytic continuation across the unit circle.*

Proof. We assume that A consists of distinct points $a_1, \ldots, a_n$; the case of multiple zeros will then follow from the formulas in the following proof and an obvious limit argument.

Let $b = b_A$ be the Blaschke product corresponding to A, and for each $1 \leq j \leq n$, let b_j denote the Blaschke product corresponding to $A \setminus \{a_j\}$. For any $w \in \mathbb{D}$ and $h \in H^\infty$, an application of the residue theorem yields

$$\frac{1}{2\pi i} \int_{|z|=1} \frac{b(w)\overline{z}}{1 - \overline{z}w} \overline{b(z)}\, h(z)\, dz = h(w) - \sum_{k=1}^{n} \frac{b_k(w)}{1 - \overline{a}_k w} \frac{1 - |a_k|^2}{b_k(a_k)} h(a_k).$$

Rewriting the left-hand side as

$$\frac{1}{2\pi i} \int_{|z|=1} \frac{b(w)\overline{z}}{1 - \overline{z}w} \overline{b(z)}\, |b(z)|^p h(z)\, dz,$$

and applying the Cauchy-Green's formula to this integral, we obtain

$$\int_{\mathbb{D}} b(w) \left[\frac{\overline{b(z)}}{(1 - \overline{z}w)^2} + \left(\frac{p}{2} + 1\right) \frac{\overline{zb'(z)}}{1 - \overline{z}w} \right] h(z)|b(z)|^p\, dA(z)$$

$$= h(w) - \sum_{k=1}^{n} \frac{b_k(w)}{1 - \overline{a}_k w} \frac{1 - |a_k|^2}{b_k(a_k)} h(a_k).$$

Writing the right-hand side above as an integral involving the reproducing kernel $K_\omega(z, w)$ of $A^2(\omega)$, where $\omega = |b|^p$, and then using the uniqueness of the reproducing kernel, we conclude that

$$\overline{b(w)} \left[\frac{b(z)}{(1 - z\overline{w})^2} + \left(\frac{p}{2} + 1\right) \frac{zb'(z)}{1 - z\overline{w}} \right]$$

$$= K_\omega(z, w) - \sum_{k=1}^{n} \frac{\overline{b_k(w)}}{1 - a_k\overline{w}} \frac{1 - |a_k|^2}{\overline{b_k(a_k)}} K_\omega(z, a_k).$$

Setting $z = 0$ reveals that $K_\omega(0, w)$ has a conjugate analytic continuation across the unit circle. In view of Lemma 3.29, the proof is now complete. ∎

We can now prove the existence and uniqueness of the solution of the extremal problem for any zero-based invariant subspace in A^p.

THEOREM 3.31 *Suppose $0 < p < +\infty$, A is an A^p-zero set, and I_A is the corresponding invariant subspace in A^p. Then the extremal problem for I_A has a unique solution G_A. Furthermore, G_A has no extraneous zeros in $\mathbb{D}$, $\|G_A f\|_p \geq \|f\|_p$ for all $f \in A^p$, and $\|g/G_A\|_p \leq \|g\|_p$ for all $g \in I_A$.*

Proof. Recall that the existence of an extremal function follows from a simple normal family argument, and that any such extremal function has no extraneous zeros in $\mathbb{D}$.

Write $A = \{a_1, a_2, a_3, \ldots\}$, and for any positive integer n, write $A_n = \{a_1, \ldots, a_n\}$ for the corresponding cut-off sequence. Let G_n be the (unique) zero divisor for I_{A_n} in A^p. Let G be any solution of the extremal problem for I_A. By Corollary 3.26, the function G has no extraneous zeros in $\mathbb{D}$. We will show that $G_n \to G$ in norm, which clearly gives the desired uniqueness.

Since $\|G_n\|_p = 1$ for all n, the sequence $\{G_n\}_n$ is a normal family. Thus a subsequence $\{G_{n_k}\}_k$ converges uniformly on compact sets to an analytic function H in $\mathbb{D}$. Since each G_n has no extraneous zeros, Hurwitz's theorem tells us that either H is identically zero, or H has A as its zero set.

The function G belongs to each I_{A_n}. So the extremal property of G_n gives $G_n(0) \geq G(0) > 0$ (an obvious adjustment can be made if $0 \in A$). This implies that $H(0) \geq G(0)$, and hence H has A as its zero set. Also, Fatou's lemma tells us that $\|H\|_p \leq 1$. Combining this with the extremal property of G, we obtain $H(0) = G(0)$.

Another application of Fatou's lemma gives

$$1 = \left|\frac{G(0)}{H(0)}\right| \leq \left\|\frac{G}{H}\right\|_p \leq \|G\|_p = 1.$$

This implies that $G = H$. And using Fatou's lemma one more time, we see that $\|g/G\|_p \leq \|g\|_p$ for all $g \in I_A$.

The same arguments above show that each subsequence of $\{G_n\}_n$ has a subsequence that converges uniformly on compact sets to the function G. It follows that $G_n \to G$ uniformly on compact sets. Since $\|G_n\|_p = \|G\|_p = 1$, an application of Lemma 3.17 shows that $G_n \to G$ in norm. In particular, the extremal function G is unique.

Finally, given any $f \in A^p$, if $\|Gf\|_p = +\infty$, we automatically have $\|Gf\|_p \geq \|f\|_p$. Otherwise, using $g = Gf \in A^p$ we obtain

$$\|f\|_p = \|g/G\|_p \leq \|g\|_p = \|Gf\|_p.$$

This completes the proof of the theorem. ∎

We mention that even when A is infinite, the contractive zero divisor G_A in A^p has an analytic continuation across any open arc of the unit circle that does not contain any accumulation point of A. The rest of the book does not use this fact, so we have not included a proof here. The interested reader should consult [128] for details.

3.6 An Inner-Outer Factorization Theorem for A^p

Recall from the theory of Hardy spaces that every function $f \in H^p$ admits a factorization $f = GF$, where G is an inner function (a bounded analytic function whose boundary values have modulus 1 almost everywhere) and F is a cyclic vector in H^p (a function that has the whole H^p as its generated invariant subspace). In this section, we show that an analogue of this holds for the Bergman spaces A^p.

Also, recall that for a weight function of the form $w(z) = |g(z)|^t$, we defined the spaces $B^p(\omega)$ and $A^p(\omega)$ in the previous section. We introduce another space $X^p(\omega)$ here, which consists of all analytic functions f in $\mathbb{D}$ such that

$$\|f\|_{X^p(\omega)}^p = \|f\|_p^p + \int_{\mathbb{D}} \int_{\mathbb{D}} \Gamma(z, w) \Delta |f(z)|^p \Delta \omega(w) \, dA(z) \, dA(w) < +\infty.$$

We are going to use the dilations f_r ($0 < r < 1$), where $f_r(z) = f(rz)$ for $z \in \mathbb{D}$.

THEOREM 3.32 *Suppose $0 < p < +\infty$, φ is A^p-inner, and I_φ is the invariant subspace of A^p generated by φ. Then,*

(a) we have

$$I_\varphi = \varphi A^p(\omega) = \varphi X^p(\omega) \subset \varphi B^p(\omega) \subset A^p,$$

where $\omega = |\varphi|^p$;

(b) we have the norm relations

$$\|g/\varphi\|_{X^p(\omega)} = \|g\|_p$$

for $g \in I_\varphi$ and

$$\|f\varphi\|_p = \|f\|_{X^p(\omega)}$$

for $f \in X^p(\omega)$;

(c) for $g = f\varphi \in I_\varphi$, we have $\|f_r\varphi - f\varphi\|_p \to 0$ as $r \to 1^-$.

Proof. First, assume that $g \in I_\varphi$. Then there exists a sequence $\{p_n\}_n$ of polynomials such that $\|p_n\varphi - g\|_p \to 0$ as $n \to +\infty$. By the expansive multiplier property of φ, the sequence $\{p_n\}_n$ is Cauchy in A^p, so that $\{p_n\}_n$ converges in norm to some $f \in A^p$. In particular, $g = f\varphi$. Recall from Corollary 3.21 that

$$\|p_n\varphi\|_p^p = \|p_n\|_p^p + \int_{\mathbb{D}} \int_{\mathbb{D}} \Gamma(z, w) \Delta |p_n(z)|^p \Delta |\varphi(w)|^p \, dA(z) \, dA(w).$$

Since the biharmonic Green function $\Gamma(z, w)$ is positive, and since $\|p_n\varphi\|_p \to$ $\|g\|_p$ and $\|p_n\|_p \to \|f\|_p$, an application of Fatou's lemma shows that

$$\int_{\mathbb{D}}\int_{\mathbb{D}} \Gamma(z, w)\Delta|f(z)|^p \Delta|\varphi(w)|^p \, dA(z)\, dA(w) \le \|g\|_p^p - \|f\|_p^p < +\infty,$$

which we write as

$$\|f\|_{X^p(\omega)} \le \|f\varphi\|_p = \|g\|_p, \qquad g = f\varphi \in I_\varphi. \tag{3.2}$$

In particular, $f \in X^p(\omega)$, and hence $g \in \varphi X^p(\omega)$.

Next, assume that $g = \varphi f$, where $f \in X^p(\omega)$. Applying Corollary 3.21 to f_r with $q = p$ and then using Lemma 3.16, we obtain

$$
\begin{aligned}
r\|f_r\varphi\|_p^p &= r\|f_r\|_{X^p(\omega)}^p \\
&= r\|f_r\|_p^p + r\int_{\mathbb{D}} \Gamma[\Delta|f_r|^p](w)\, \Delta|\varphi(w)|^p\, dA(w) \\
&\le \|f\|_p^p + \int_{\mathbb{D}} \Gamma[\Delta|f|^p](w)\, \Delta|\varphi(w)|^p\, dA(w) \\
&= \|f\|_{X^p(\omega)}^p < +\infty,
\end{aligned}
$$

for all $0 < r < 1$. This together with Fatou's lemma shows that

$$\|f\varphi\|_p \le \|f\|_{X^p(\omega)} < +\infty, \tag{3.3}$$

so that $g = \varphi f \in A^p$. The function $f\varphi$ is then a weak limit of the functions $f_r\varphi \in I_\varphi$ as $r \to 1^-$, so that by basic Functional Analysis, we have $f\varphi \in I_\varphi$ in the case $1 \le p < +\infty$; the case $0 < p < 1$ will be handled a little later.

The space $\varphi B^p(\omega)$ coincides with the subspace of A^p of all functions that vanish on the zero set of φ, counting multiplicities, so that by the above, we have the inclusion $X^p(\omega) \subset B^p(\omega)$.

Combining the above two inequalities (3.2) and (3.3), we obtain the isometry

$$\|g/\varphi\|_{X^p(\omega)} = \|g\|_p, \qquad g \in I_\varphi.$$

Since M_φ, the operator of multiplication by φ, is an isometry from $A^p(\omega)$ to I_f, it follows from the above that $A^p(\omega) \subset X^p(\omega)$, and that the inclusion map is an isometry. Moreover, we have obtained the equality $A^p(\omega) = X^p(\omega)$ for $1 \le p < +\infty$.

We proceed to show that $\|f_r\varphi - f\varphi\|_p \to 0$ as $r \to 1^-$, provided that $g = f\varphi \in I_\varphi$. By Lemma 3.17, it suffices to show that $\|f_r\varphi\|_p \to \|f\varphi\|_p$ as $r \to 1^-$. In view of the isometry $\|h/\varphi\|_{X^p(\omega)} = \|h\|_p$ for $h \in I_\varphi$, this means that we should verify that $\|f_r\|_{X^p(\omega)} \to \|f\|_{X^p(\omega)}$ as $r \to 1^-$. Recall that

$$\|f_r\|_{X^p(\omega)}^p = \|f_r\varphi\|_p^p = \|f_r\|_p^p + \int_{\mathbb{D}} \Gamma[\Delta|f_r|^p](w)\, \Delta|\varphi(w)|^p\, dA(w).$$

From Lemma 3.16, we see that $r\Gamma[\Delta|f_r|^p]$ increases monotonically to $\Gamma[\Delta|f|^p]$ as $r \to 1^-$. So, by the Monotone Convergence Theorem,

$$\lim_{r\to 1^-} \int_{\mathbb{D}} \Gamma[\Delta|f_r|^p](w)\, \Delta|\varphi(w)|^p\, dA(w) = \int_{\mathbb{D}} \Gamma[\Delta|f|^p](w)\, \Delta|\varphi(w)|^p\, dA(w).$$

It follows that

$$\lim_{r \to 1^-} \|f_r \varphi\|_p = \|f\|_{X^p(\omega)}^p, \qquad f \in X^p(\omega).$$

It remains for us to prove that $I_\varphi = \varphi X^p(\omega)$ in the case $0 < p < 1$. To this end, observe that the above proof can be modified to produce the identity $X^2(\omega) = A^2(\omega)$ with equality of norms. Furthermore, if $f \in A^2(\omega)$, then $\|f_r - f\|_{2,\omega} \to 0$ as $r \to 1^-$.

For the remainder of this proof, we assume $0 < p < 1$. Let f be a function in $X^p(\omega)$. Then we know that $f \in B^p(\omega)$. We are going to show that $f\varphi \in I_\varphi$, or, in other words, $f \in A^p(\omega)$.

First, suppose that f is zero-free in $\mathbb{D}$. Then we may form the power $(f)^{p/2}$, which is an element of $B^2(\omega)$. Since

$$\left\| (f_r)^{p/2} \right\|_{2,\omega}^2 = \|f_r\|_{p,\omega}^p = \|f_r\|_{X^p(\omega)} \to \|f\|_{X^p(\omega)} \qquad \text{as } r \to 1^-,$$

the functions $(f_r)^{p/2}$ converges weakly to $(f)^{p/2}$ in $B^2(\omega)$ as $r \to 1^-$. But all the functions $(f_r)^{p/2}$ belong to $A^2(\omega)$, and we know that the weak closure of a subspace in $B^2(\omega)$ is the same as the norm closure, so it follows that $(f)^{p/2} \in A^2(\omega)$. By the observation in the previous paragraph, the functions $(f_r)^{p/2}$ converges to $(f)^{p/2}$ in norm in the space $A^2(\omega)$ as $r \to 1^-$. In view of Lemma 3.17, we get that $f_r \to f$ in norm in the space $B^p(\omega)$ as $r \to 1^-$, and hence $f \in A^p(\omega)$, as we wanted, because $f_r \in A^p(\omega)$ for each r, $0 < r < 1$.

Next, we consider the case when f has finitely many zeros. Let ψ be the extremal function in A^p for those zeros, which extends analytically to a neighborhood of $\overline{\mathbb{D}}$, by Lemma 3.30. Then $f = \psi g$, where the zero-free function g is in $X^p(\omega)$. In fact, from the early part of this proof, we have

$$\|g\|_{X^p(\omega)} = \lim_{r \to 1^-} \|g_r\|_{X^p(\omega)} = \lim_{r \to 1^-} \|g_r\|_{p,\omega} = \lim_{r \to 1^-} \|g_r \varphi\|_p,$$

because $r\|g_r\|_{X^p(\omega)}^p$ is monotonically increasing in r. If we apply this to both f and g, taking into account the properties of ψ, we find that $g \in X^p(\omega)$. We now apply the result proved in the previous paragraph to the zero-free function g to obtain that $g \in A^p(\omega)$ and that $g_r \to g$ in norm as $r \to 1^-$. Since ψ is bounded, we also have $f \in A^p(\omega)$ with $f_r \to f$ in norm as $r \to 1^-$.

Finally, we turn to the case when f has infinitely many zeros. Let $A = \{a_1, a_2, a_3, \dots\}$ denote the sequence of zeros of f in $\mathbb{D}$. Let N be a large positive integer, and split the zeros into two portions: $A_{(N)} = \{a_1, a_2, \dots, a_N\}$ and $A^{(N)} = \{a_{N+1}, a_{N+2}, \dots\}$. Let ϕ_N be the extremal function in A^p for the zero sequence $A_{(N)}$, and let ψ_N be the extremal function for $A^{(N)}$. For $0 < r < 1$, let $A_{(N)}(r) = (r^{-1} A_{(N)}) \cap \mathbb{D}$ and $A^{(N)}(r) = (r^{-1} A^{(N)}) \cap \mathbb{D}$ be the correspondingly dilated zero sequences, restricted to the unit disk. Observe that they are both finite sequences. We let $\phi_{N,r}$ be the extremal function in A^p for the zero sequence $A_{(N)}(r)$, and let $\psi_{N,r}$ be the extremal function for $A^{(N)}(r)$. It is easy to see that as $r \to 1^-$, $\phi_{N,r} \to \phi_N$ and $\psi_{N,r} \to \psi_N$ in the norm of A^p. Let $g_{N,r}$ be the function

$$g_{N,r}(z) = \frac{f_r(z)}{\phi_{N,r}(z)\psi_{N,r}(z)}, \qquad z \in \mathbb{D},$$

which is zero-free in the disk and extends analytically across the unit circle. From the contractive division property of the functions $\phi_{N,r}$ and $\psi_{N,r}$, together with the fact that $r\|f_r\|_{X^p(\omega)}^p$ is monotonically increasing in r, $0 < r \le 1$, we see that

$$
\begin{aligned}
\|g_{N,r}\|_{p,\omega} &= \|g_{N,r}\varphi\|_p \le \|f_r\varphi\|_p = \|f_r\|_{p,\omega} \\
&= \|f_r\|_{X^p(\omega)} \le r^{-1/p}\|f\|_{X^p(\omega)}
\end{aligned}
$$

for all $0 < r < 1$. Since each $g_{N,r}$ is zero-free, this is the same as

$$\left\|(g_{N,r})^{p/2}\right\|_{2,\omega} \le r^{-1/2}\|f\|_{X^p(\omega)}^{p/2}, \qquad 0 < r < 1.$$

For $0 < r < 1$, each function $g_{N,r}$ is holomorphic on $\overline{\mathbb{D}}$, so that in particular $(g_{N,r})^{p/2} \in A^2(\omega)$. As $r \to 1^-$, we have

$$\left(g_{N,r}(z)\right)^{p/2} \to \left(H_N(z)\right)^{p/2} = \left(\frac{f(z)}{\phi_N(z)\psi_N(z)}\right)^{p/2}, \qquad z \in \mathbb{D},$$

where the equality is used to define H_N, and the convergence is uniform compact subsets of $\mathbb{D}$. It follows that $(H_N)^{p/2}$ is a weak limit in $B^2(\omega)$ of functions in $A^2(\omega)$, and hence $(H_N)^{p/2} \in A^2(\omega)$. We now argue as we did for zero-free f, and obtain that $H_N \in A^p(\omega)$. The property of being in $A^p(\omega)$ is preserved under multiplication by an H^∞ function, and hence we have $\phi_N H_N = f/\psi_N \in A^p(\omega)$ as well. By the contractive divisor property of ψ_N,

$$\|f/\psi_N\|_{p,\omega} = \|f\varphi/\psi_N\|_p \le \|f\varphi\|_p,$$

and since $\psi_N(z) \to 1$ as $N \to +\infty$ uniformly on compact subsets of $\mathbb{D}$ (after all, the zero sequence $A^{(N)}$ for ψ_N gradually evaporates as $N \to +\infty$; see Exercise 3), it follows from Lemma 3.17 that $f/\psi_N \to f$ in the norm of $B^p(\omega)$. Since each f/ψ_N is in $A^p(\omega)$, we conclude that $f \in A^p(\omega)$ as well. ∎

Recall that if $0 < p < 1$, then neither the existence nor the uniqueness of the solution of the extremal problem for an arbitrary invariant subspace in A^p is known. In the previous section we proved the existence and uniqueness for zero-based invariant subspaces. The next result shows that we have both existence and uniqueness for singly generated (or cyclic) invariant subspaces in A^p. In addition, the extremal function generates the invariant subspace.

THEOREM 3.33 *If I is a cyclic invariant subspace of A^p, then there exists a unique solution φ to the extremal problem for I. Furthermore, $I = I_\varphi$.*

Proof. Suppose $I = I_f$ for some $f \in A^p$. We assume $f(0) \ne 0$; the remaining case is handled by the observation that for a function $g \in A^p$, we have $I_{zg} = zI_g$. Let the spaces $A^p(|f|^p)$ and $A^2(|f|^p)$ be the weighted Bergman spaces with weight $\omega = |f|^p$, with norms $\|\cdot\|_{p,|f|^p}$ and $\|\cdot\|_{2,|f|^p}$, respectively. Then multiplication by f, denoted by M_f, is an isometry from $A^p(|f|^p)$ to A^p; it is clear that the range is I_f, so that $I_f = fA^p(|f|^p)$. By Lemma 3.28, the reproducing kernel function $K_{|f|^p}$ for $A^2(|f|^p)$ never vanishes on the bidisk $\mathbb{D} \times \mathbb{D}$. We claim that the extremal function $\varphi = \varphi_f$ for I_f is unique and given by the formula

$$\varphi(z) = K_{|f|^p}(0,0)^{-1/p} f(z) K_{|f|^p}(z,0)^{2/p}, \qquad z \in \mathbb{D}, \tag{3.4}$$

which is analogous to the formula obtained for the invariant subspace associated with finitely many zeros, with the Blaschke product b in place of f. The function $K_{|f|^p}(\cdot, 0)$ maximizes the value at the origin among all unit vectors in $A^2(|f|^p)$. We claim that the function

$$H(z) = K_{|f|^p}(0,0)^{-1/p} K_{|f|^p}(z,0)^{2/p}$$

is the unique maximizer of the value at 0 among all unit vectors in $A^p(|f|^p)$. Clearly, $\|H\|_{p,|f|^p} = 1$. Let q_n be a (maximizing) sequence of polynomials such that $\|q_n\|_{p,|f|^p} = 1$ and

$$q_n(0) \to \sup \{|g(0)| : g \in A^p(|f|^p), \|g\|_{p,|f|^p} = 1\} \quad \text{as } n \to +\infty.$$

Similarly, let p_n be another (maximizing) sequence of polynomials such that $\|p_n\|_{2,|f|^p} = 1$ and

$$p_n(0) \to \sup \{|h(0)| : h \in A^2(|f|^p), \|h\|_{2,|f|^p} = 1\} \quad \text{as } n \to +\infty.$$

For each n, let φ_n be the contractive zero divisor in A^p corresponding to the zeros of q_n, and let ψ_n be the contractive divisor in A^2 corresponding to the zeros of p_n. Then, by Theorem 3.31,

$$\|q_n/\varphi_n\|_{p,|f|^p} = \|q_n f/\varphi_n\|_p \le \|q_n f\|_p = \|q_n\|_{p,|f|^p}.$$

Since $0 < \varphi_n(0) < 1$, we have $|q_n(0)| < |q_n(0)/\varphi_n(0)|$, so that by replacing q_n by q_n/φ_n if necessary, we may assume that each q_n has no zeros in $\mathbb{D}$. Now, of course, q_n need not be a polynomial any more, but at least it extends analytically to a neighborhood of the closed unit disk. Similarly, Theorem 3.31, for $p = 2$, has an analogue which states that

$$\|p_n/\psi_n\|_{2,|f|^p} \le \|p_n\|_{2,|f|^p};$$

the derivation is analogous to that of Theorem 3.31. Since $0 < \psi_n(0) < 1$, we have $|p_n(0)| < |p_n(0)/\psi_n(0)|$, so that by replacing p_n by p_n/ψ_n if necessary, we may assume that each p_n has no zeros in $\mathbb{D}$. Again, p_n need not be a polynomial any more, but it extends analytically to a neighborhood of the closed disk. We compare the sequences $\{(q_n)^{p/2}\}_n$ and $\{p_n\}_n$ in the one maximization problem, and $\{(p_n)^{2/p}\}_n$ and $\{q_n\}_n$ in the other. The conclusion is that the maximization problems are equivalent, and that since we know that $p_n \to (H)^{p/2}$ in the norm of $A^2(|f|^p)$, we must have that $q_n \to H$ in the norm of $A^p(|f|^p)$ as well (use Lemma 3.17). Since the maximization problem that q_n approximates is equivalent to the extremal problem for φ – in view of the fact that $M_f : A^p(|f|^p) \to I_f$ is an isometry – it follows that the extremal function φ exists, is unique, and is explicitly given by (3.4).

We proceed to show that $f \in I_\varphi$, which will complete the proof. In the following, we set

$$F(z) = (f/\varphi)^{p/2}(z) = K_{|f|^p}(0,0)^{1/2} K_{|f|^p}(z,0)^{-1}, \qquad z \in \mathbb{D}.$$

The zero-free function F is in $B^2(|\varphi|^p)$, because

$$\int_\mathbb{D} |F|^2 |\varphi|^p dA = \int_\mathbb{D} |f/\varphi|^p |\varphi|^p dA = \int_\mathbb{D} |f|^p dA < +\infty.$$

We recall the polynomials p_n from the maximization problem, and note that $p_n F \to 1$ as $n \to +\infty$, uniformly on compact subsets of $\mathbb{D}$. Moreover, since

$$\|p_n F\|_{2,|\varphi|^p}^2 = \int_\mathbb{D} |p_n F|^2 |\varphi|^p dA = \int_\mathbb{D} |p_n|^2 |f|^p dA = \|p_n\|_{2,|f|^p}^2 = 1$$

and $\|1\|_{2,|\varphi|^p}^2 = \|\varphi\|_p^p = 1$, Lemma 3.17 shows that $p_n F \to 1$ in the norm of $B^2(|\varphi|^p)$. For $j = 1, 2, 3, \ldots$, we find that

$$\int_\mathbb{D} z^j F(z) |\varphi(z)|^p dA(z) = K_{|f|^p}(0,0)^{-1/2} \int_\mathbb{D} z^j K_{|f|^p}(0,z) |f(z)|^p dA(z) = 0,$$

by the reproducing property of the kernel. Let J_F stand for the closure of the polynomial multiples of F in $B^2(|\varphi|^p)$; in other words, J_F is the invariant subspace generated by F in $B^2(|\varphi|^p)$. From the above, we know that $1 \in J_F$, and that $1 \perp z J_F$, which we compress to $1 \in J_F \ominus z J_F$. We can represent J_F as $F A^2(|F|^2|\varphi|^p) = F A^2(|f|^p)$. It follows that for fixed $\lambda \in \mathbb{D}$, $g \in (z - \lambda) J_F$ if and only if $g \in J_F$ and $g(\lambda) = 0$. Consequently, again for fixed $\lambda \in \mathbb{D}$, the function

$$z \mapsto \frac{F(z) - F(\lambda)}{z - \lambda}$$

is in J_F, and hence the fact that $1 \in J_F \ominus z J_F$ entails that

$$\int_\mathbb{D} z \frac{F(z) - F(\lambda)}{z - \lambda} |\varphi(z)|^p dA(z) = 0, \qquad \lambda \in \mathbb{D}. \tag{3.5}$$

Since

$$\left| \frac{zF(z) - \lambda F(\lambda)}{z - \lambda} \right|^2 = \left| F(\lambda) + z \frac{F(z) - F(\lambda)}{z - \lambda} \right|^2,$$

expanding the expression on the right-hand side and then integrating over the disk – with respect to the reproducing probability measure $|\varphi(z)|^p dA(z)$ – we obtain from the above identity (3.5) that

$$\int_\mathbb{D} \left[\left| \frac{zF(z) - \lambda F(\lambda)}{z - \lambda} \right|^2 - |\lambda|^2 \left| \frac{F(z) - F(\lambda)}{z - \lambda} \right|^2 \right] |\varphi(z)|^p dA(z)$$

$$= \left| \frac{f}{\varphi}(\lambda) \right|^p + \int_\mathbb{D} (|z|^2 - |\lambda|^2) \left| \frac{F(z) - F(\lambda)}{z - \lambda} \right|^2 |\varphi(z)|^p dA(z).$$

We are going to integrate this identity over the circle $|\lambda| = r$ for $0 < r < 1$, but first let us observe that we can write

$$\left| \frac{zF(z) - \lambda F(\lambda)}{z - \lambda} \right|^2 = \left| F(z) + \lambda \frac{F(z) - F(\lambda)}{z - \lambda} \right|^2$$

$$= |F(z)|^2 + 2\operatorname{Re}\left[\overline{F(z)} \lambda \frac{F(z) - F(\lambda)}{z - \lambda} \right] + |\lambda|^2 \left| \frac{F(z) - F(\lambda)}{z - \lambda} \right|^2,$$

and so, by the mean value property of harmonic functions,

$$\left| \frac{f}{\varphi}(z) \right|^p = \frac{1}{r} \int_{|\lambda|=r} \left[\left| \frac{zF(z) - \lambda F(\lambda)}{z - \lambda} \right|^2 - r^2 \left| \frac{F(z) - F(\lambda)}{z - \lambda} \right|^2 \right] ds(\lambda),$$

where we recall the notation $ds(z) = |dz|/(2\pi)$. It follows that

$$\int_{\mathbb{D}} |f(z)|^p \, dA(z) = \frac{1}{r} \int_{|\lambda|=r} \left| \frac{f}{\varphi}(\lambda) \right|^p ds(\lambda)$$

$$+ \frac{1}{r} \int_{\mathbb{D}} \int_{|\lambda|=r} (|z|^2 - r^2) \left| \frac{F(z) - F(\lambda)}{z - \lambda} \right|^2 ds(\lambda) \, |\varphi(z)|^p \, dA(z),$$

and so

$$\int_{\mathbb{D}} |f(z)|^p \, dA(z) \geq \frac{1}{r} \int_{|\lambda|=r} \left| \frac{f}{\varphi}(\lambda) \right|^p ds(\lambda)$$

$$- \frac{1}{r} \int_{|z|<r} \int_{|\lambda|=r} \frac{r^2 - |z|^2}{|z - \lambda|^2} |F(z) - F(\lambda)|^2 |\varphi(z)|^p \, dA(z) \, ds(\lambda). \qquad (3.6)$$

Write w in place of λ, and apply Corollary 3.7 to the first integral on the right-hand side, which gives

$$\int_{|w|=r} \left| \frac{f}{\varphi}(w) \right|^p ds(w) = \frac{1}{r} \int_{|w|<r} (r^2 - |w|^2) \Delta \left| \frac{f}{\varphi}(w) \right|^p dA(w)$$

$$+ \frac{1}{r} \int_{|w|<r} \left| \frac{f}{\varphi}(w) \right|^p dA(w).$$

Next, we apply Corollary 3.8 to the second integral on the right-hand side, which result in

$$\frac{1}{r} \int_{|w|=r} \frac{r^2 - |z|^2}{|z - w|^2} |F(z) - F(w)|^2 ds(w)$$

$$= -\int_{|w|<r} G\left(\frac{z}{r}, \frac{w}{r} \right) \Delta \left| \frac{f}{\varphi}(w) \right|^p dA(w).$$

Putting things together in (3.6), we then have

$$\int_{\mathbb{D}} |f(z)|^p \, dA(z) \geq \frac{1}{r^2} \int_{|w|<r} \left|\frac{f}{\varphi}(w)\right|^p \, dA(w)$$

$$+ \frac{1}{r^2} \int_{|w|<r} \left(r^2 - |w|^2\right) \Delta \left|\frac{f}{\varphi}(w)\right|^p \, dA(w)$$

$$+ \int_{|z|<r} \int_{|w|<r} G\left(\frac{z}{r}, \frac{w}{r}\right) \Delta \left|\frac{f}{\varphi}(w)\right|^p \, dA(w) \, |\varphi(z)|^p \, dA(z).$$

After an appropriate dilation in both variables, the inequality reads

$$\int_{\mathbb{D}} |f(z)|^p \, dA(z) \geq \int_{\mathbb{D}} \left|\frac{f}{\varphi}(rw)\right|^p \, dA(w)$$

$$+ \int_{\mathbb{D}} \left(1 - |w|^2\right) \Delta_w \left|\frac{f}{\varphi}(rw)\right|^p \, dA(w)$$

$$+ r^2 \int_{\mathbb{D}} \int_{\mathbb{D}} G(z, w) \Delta_w \left|\frac{f}{\varphi}(rw)\right|^p \, dA(w) \, |\varphi_r(z)|^p \, dA(z).$$

We now invoke Proposition 3.24 to handle the last integral expression on the right-hand side, and find that

$$\int_{\mathbb{D}} |f(z)|^p \, dA(z) \geq \int_{\mathbb{D}} \left|\frac{f}{\varphi}(rw)\right|^p \, dA(w)$$

$$+ (1 - r^2) \int_{\mathbb{D}} \left(1 - |w|^2\right) \Delta_w \left|\frac{f}{\varphi}(rw)\right|^p \, dA(w)$$

$$+ r^2 \int_{\mathbb{D}} \int_{\mathbb{D}} \Gamma(z, w) \Delta_w \left|\frac{f}{\varphi}(rw)\right|^p \, dA(w) \, \Delta_z |\varphi_r(z)|^p \, dA(z).$$

Letting $r \to 1^-$ and applying Fatou's lemma results in

$$\|f\|_p^p \geq \|f/\varphi\|_p^p + \int_{\mathbb{D}} \Gamma(z, w) \Delta \left|\frac{f}{\varphi}(w)\right|^p \Delta |\varphi(z)|^p \, dA(z) \, dA(w).$$

This shows that f/φ belongs to the space $X^p(|\varphi|^p)$, and hence according to Theorem 3.32, the function f is in I_φ, so that $I = I_f = I_\varphi$. ∎

We can now prove the "inner-outer" factorization for functions in A^p. First recall that a function $f \in A^p$ is called a cyclic vector if there exists a sequence $\{p_n\}_n$ of polynomials such that $\|p_n f - 1\|_p \to 0$ as $n \to +\infty$.

THEOREM 3.34 Suppose $0 < p < +\infty$ and $f \in A^p$. Then there exists an A^p-inner function G and a cyclic vector F in A^p such that $f = GF$. Furthermore, $\|F\|_p \leq \|f\|_p$.

Proof. Let I be the invariant subspace generated by f. According to Theorem 3.33, there is a unique solution G to the extremal problem for I, the quotient $F = f/G$ belongs to A^p, and there exists a sequence $\{p_n\}_n$ of polynomials such that $\|p_n f - G\|_p \to 0$ as $n \to +\infty$. The expansive multiplier property of G

together with the fact that $I = I_G$ implies that G is a contractive divisor on the whole space I. Thus $\{p_n F\}_n$ is a Cauchy sequence in A^p. Since $p_n(z)F(z) \to 1$ pointwise, we must have $\|p_n F - 1\|_p \to 0$, that is, F is cyclic in A^p.   ∎

In the classical theory of H^p spaces, the inner-outer factorization is unique (up to a unimodular constant multiple of the inner factor). Unfortunately, the factorization here in A^p does not have such a strong uniqueness property; counterexamples will be constructed in Chapter 8.

3.7   Approximation of Subinner Functions

A classical theorem of Carathéodory-Schur (see [49]) states that if F is an element of the closed unit ball of H^∞, then there exists a sequence of finite Blaschke products b_n such that $b_n \to F$ uniformly on compact subsets of $\mathbb{D}$ as $n \to +\infty$.

The purpose of this section is to show that a version of this result also holds for Bergman spaces. In other words, we are going to characterize the normal limits of finite zero divisors in Bergman spaces. Recall that G is called a finite zero divisor in A^p if there is a finite zero set Z such that G is the extremal function of the invariant subspace I_Z in A^p. We begin with the following simple necessary condition.

PROPOSITION 3.35 *If $\{G_n\}_n$ is a sequence of finite zero divisors in A^p and $G_n \to G$, as $n \to +\infty$, uniformly on compact sets, then*

$$\int_{\mathbb{D}} |G(z)|^p h(z)\, dA(z) \leq h(0)$$

holds for all positive harmonic functions h in $\mathbb{D}$.

Proof. Let h be any positive harmonic function in $\mathbb{D}$. For any $n = 1, 2, 3, \ldots$, define a bounded harmonic function h_n on $\mathbb{D}$ by

$$h_n(z) = h(nz/(n+1)), \qquad z \in \mathbb{D}.$$

Since each G_n is an A^p-inner function, we have

$$\int_{\mathbb{D}} |G_n(z)|^p h_n(z)\, dA(z) = h_n(0) = h(0), \qquad n = 1, 2, 3, \ldots.$$

The desired inequality now follows from this and Fatou's lemma.   ∎

The rest of this section is devoted to proving that the converse of the proposition above also holds, at least in the case $p = 2$. For convenience, we introduce the following.

DEFINITION 3.36 *A function $F \in A_\alpha^p$ called an A_α^p-subinner function if*

$$\int_{\mathbb{D}} |F(z)|^p h(z)\, dA_\alpha(z) \leq h(0)$$

holds for all (bounded) positive harmonic functions h on $\mathbb{D}$.

The Hardy space analogue of this definition, with the normalized measure dA_α on $\mathbb{D}$ replaced by normalized arc length measure ds on $\mathbb{T}$, requires that the function $F \in H^p$ have $|F(z)| \leq 1$ almost everywhere on $\mathbb{T}$, in which case F is in the closed unit ball of H^∞. The Carathéodory-Schur theorem then asserts that such functions are indeed normal limits of finite Blaschke products.

We now prove the following Bergman space analogue of the Carathéodory-Schur theorem.

THEOREM 3.37 *Let F be a subinner function in A^2. Then there exists a sequence of finite zero divisors φ_n in A^2 such that $\varphi_n \to F$ uniformly on compact subsets of $\mathbb{D}$ as $n \to +\infty$.*

Proof. For $N = 1, 2, 3, \ldots$, let Φ_N denote the Fejér kernel:

$$\Phi_N(e^{i\theta}) = \frac{1}{N+1} \frac{\sin^2 \frac{1}{2}(N+1)\theta}{\sin^2 \frac{1}{2}\theta}, \qquad \theta \in \mathbb{R}.$$

It is positive, and we use it to mollify the subinner function F,

$$F_N(z) = F * \Phi_N(z) = \int_\mathbb{T} F(z\bar{\zeta}) \, \Phi_N(\zeta) \, ds(\zeta), \qquad z \in \mathbb{D},$$

which constitutes a polynomial of degree N or less. Let h be a positive harmonic function in $\mathbb{D}$. Then, by the Cauchy-Schwarz inequality, Fubini's theorem, and the observation that the property of being a subinner function is rotation invariant,

$$
\begin{aligned}
\int_\mathbb{D} |F_N|^2 \, h \, dA &= \int_\mathbb{D} \left| \int_\mathbb{T} F(z\bar{\zeta}) \Phi_N(\zeta) \, ds(\zeta) \right|^2 h(z) \, dA(z) \\
&\leq \int_\mathbb{D} \int_\mathbb{T} |F(z\bar{\zeta})|^2 \, \Phi_N(\zeta) h(z) \, ds(\zeta) \, dA(z) \\
&\leq h(0) \int_\mathbb{T} \Phi_N(\zeta) \, ds(\zeta) = h(0).
\end{aligned}
$$

It follows that the polynomial F_N is a subinner function. As $N \to +\infty$, F_N approaches F, uniformly on compact subsets of $\mathbb{D}$. Consequently, if we can approximate each of the polynomials F_N by finite zero divisors, then F, too, is so approximable.

We may now, without loss of generality, assume that F itself is a polynomial; let N be the degree of F. Moreover, we can assume that the function is strictly subinner, in the sense that

$$\int_\mathbb{D} h(z) |F(z)|^2 \, dA(z) \leq (1 - \varepsilon) h(0) \tag{3.7}$$

holds for some small fixed ε, $0 < \varepsilon < 1$, and all bounded positive harmonic functions h on $\mathbb{D}$. For $f \in L^1(\mathbb{D})$, let $P^*[f]$ be the function

$$P^*[f](z) = \int_\mathbb{D} \frac{1 - |\zeta|^2}{|1 - z\bar{\zeta}|^2} f(\zeta) \, dA(\zeta), \qquad z \in \mathbb{T},$$

which is in $L^1(\mathbb{T})$ and has the property that

$$\int_{\mathbb{T}} h(z) P^*[f](z)\, ds(z) = \int_{\mathbb{D}} h(z) f(z)\, dA(z), \qquad (3.8)$$

for all bounded harmonic h. The function $P^*[f]$ is frequently called the *sweep* of f. Note that a function φ is A^2-inner if and only if

$$P^*[|\varphi|^2] = 1,$$

and a function F is A^2-subinner if and only if $0 \le P^*[|F|^2] \le 1$. As we apply the operation P^* to $|F|^2$, we obtain a trigonometric polynomial:

$$P^*[|F|^2](z) = \sum_{j=-\infty}^{+\infty} A_j z^j, \qquad z \in \mathbb{T}, \qquad (3.9)$$

where $A_j = \langle F, z^j F \rangle$ for $j \ge 0$ and $A_j = \langle z^{-j} F, F \rangle$ for $j < 0$. The inequality (3.7) together with the property (3.8) implies that

$$0 \le P^*[|F|^2] \le 1 - \delta$$

on $\mathbb{T}$. The function $1 - P^*[|F|^2]$ is a positive trigonometric polynomial of degree N. By a classical theorem of Fejér and Riesz (See Exercise 19), there exists an analytic polynomial G of degree N, zero-free in the closed disk $\overline{\mathbb{D}}$, such that $|G|^2 = 1 - P^*[|F|^2]$ on $\mathbb{T}$; then $\varepsilon \le |G|^2 \le 1$ on $\mathbb{T}$, and by the maximum principle, also on $\mathbb{D}$.

We now put, for $n = 1, 2, 3, \ldots,$

$$f_n(z) = F(z) + \sqrt{n+1}\, z^n G(z), \qquad z \in \mathbb{D}, \qquad (3.10)$$

and let M_n be the invariant subspace generated by f_n. Let φ_n be the extremal function for M_n. The functions φ_n are finite zero divisors, and we shall see that $\varphi_n \to F$, uniformly on compact subsets of $\mathbb{D}$. The assertion of the theorem is immediate once this has been achieved.

By the definition of f_n, we have

$$\begin{aligned} |f_n(z)|^2 &= |F(z)|^2 + (n+1)\,|z|^{2n}\,|G(z)|^2 \\ &\quad + 2\sqrt{n+1}\,\operatorname{Re}\left(z^n\, G(z)\, \bar{F}(z)\right) \end{aligned} \qquad (3.11)$$

for all $z \in \mathbb{D}$. As $n \to +\infty$,

$$\sqrt{n+1} \int_{\mathbb{D}} \left| z^n\, G(z)\, F(z) \right| dA(z) \to 0,$$

so that the third term on the right-hand side of (3.11) is eventually insignificant. As for the second term, we have

$$(n+1)\,|z|^{2n}\,|G(z)|^2\, dA(z) \to |G(z)|^2\, ds(z),$$

in the weak-star topology of Borel measures, where ds is the normalized arc-length measure on the unit circle $\mathbb{T}$, and

$$(n+1)\, P^*\!\left[|z^n\, G(z)|^2\right](z) \to |G(z)|^2$$

uniformly on $\mathbb{T}$; both limits are taken as $n \to +\infty$. From the construction of G we see that $P^*[|f_n|^2]$ is approximately 1 for large values of n, so that f_n is, in a sense, an approximate A^2-inner function. In other words, we expect φ_n not to differ much from f_n for large n. The rigorous demonstration of this requires some technical work.

LEMMA 3.38 *There exists a positive integer* $L = L(F, \varepsilon)$ *such that*

$$\frac{3}{8}\varepsilon \|g\|^2 \le \frac{3}{8}\varepsilon (n+1) \|z^n g\|^2 \le \|f_n g\|^2, \qquad g \in A^2,$$

for any n with $L \le n < +\infty$.

Proof. Since F is bounded and $\varepsilon \le |G|^2 \le 1$, there exists an $L = L(F, \varepsilon)$ such that for n with $L \le n < +\infty$,

$$|F(z)| \le \frac{1}{4}\sqrt{n+1}\,|z|^n\,|G(z)|, \qquad r_n < |z| < 1,$$

provided that $0 < r_n < 1$ and r_n is so close to 1 that its n-th power is bounded away from 0 as $n \to +\infty$. Then in the same annulus,

$$\frac{3}{4}\sqrt{(n+1)\varepsilon}\,|z|^n \le \frac{3}{4}\sqrt{n+1}\,|z|^n\,|G(z)| \le |f_n(z)|, \qquad r_n < |z| < 1.$$

It follows from the above estimate that for $g \in A^2$,

$$\frac{9}{16}(n+1)\varepsilon \int_{r_n < |z| < 1} |g(z)|^2\,|z|^{2n}\,dA(z)$$

$$\le \int_{r_n < |z| < 1} |f_n(z)\,g(z)|^2\,dA(z) \le \|f_n g\|^2. \tag{3.12}$$

Let M_g be the radial square-mean function

$$M_g(r) = \frac{1}{2\pi} \int_{-\pi}^{\pi} |g(re^{i\theta})|^2\,d\theta, \qquad 0 < r < 1,$$

which increases with r. Then

$$\int_{r_n < |z| < 1} |g(z)|^2\,|z|^{2n}\,dA(z) = 2\int_{r_n}^1 M_g(r)\,r^{2n+1}\,dr, \tag{3.13}$$

and using the monotonicity of M_g and redistributing masses along the interval $(0, 1)$, we have

$$\frac{1 - r_n^{2(n+1)}}{n+1} \int_0^1 M_g(r)\,r\,dr \le \left(1 - r_n^{2(n+1)}\right) \int_0^1 M_g(r)\,r^{2n+1}\,dr$$

$$\le \int_{r_n}^1 M_g(r)\,r^{2n+1}\,dr.$$

Combining this with (3.12) and taking into account (3.13), we arrive at

$$
\begin{aligned}
\frac{9}{16}\,\varepsilon \left(1 - r_n^{2(n+1)}\right) \|g\|^2 &= \frac{9}{8}\,\varepsilon \left(1 - r_n^{2(n+1)}\right) \int_0^1 M_g(r)\, r\, dr \\
&\leq \frac{9}{8}\,\varepsilon\, (n+1)\left(1 - r_n^{2(n+1)}\right) \int_0^1 M_g(r)\, r^{2n+1}\, dr \\
&= \frac{9}{16}\,\varepsilon\, (n+1)\left(1 - r_n^{2(n+1)}\right) \|z^n g\|^2 \\
&\leq \frac{9}{8}\,\varepsilon\, (n+1) \int_{r_n}^1 M_g(r)\, r^{2n+1}\, dr \\
&= \frac{9}{16}\,\varepsilon\, (n+1) \int_{r_n < |z| < 1} |g(z)|^2\, |z|^{2n}\, dA(z) \\
&\leq \|f_n\, g\|^2.
\end{aligned}
\tag{3.14}
$$

Choosing r_n such that $3 r_n^{2(n+1)} = 1$ then finishes the proof of the lemma. ∎

For large n, therefore, the polynomial f_n has no zeros on $\mathbb{T}$. The function φ_n is defined as the extremal function for the invariant subspace M_n in A^2 generated by f_n. We may assume, without loss of generality, that $f_n(0) \neq 0$, and after a rotation, that $f_n(0) > 0$. Then φ_n solves the extremal problem

$$
\max \left\{ \operatorname{Re} \varphi(0) : \varphi \in M_n,\ \|\varphi\| = 1 \right\},
$$

so that the function $q_n \in A^2$ defined by $\varphi_n = f_n q_n$ solves the related extremal problem

$$
\max \left\{ \operatorname{Re} q(0) : q \in A^2,\ \|f_n q\| = 1 \right\}.
$$

In a Hilbert space of holomorphic functions, the function maximizing the value at a point among the elements of the unit ball equals an appropriate constant multiple of the kernel function. In other words, if $A^2(|f_n|^2)$ stands for the space of holomorphic functions f on $\mathbb{D}$ with norm

$$
\|f\|_{A^2(|f_n|^2)} = \|f_n\, f\| < +\infty,
$$

and $K_{|f_n|^2}$ is the associated kernel function, then

$$
q_n(z) = K_{|f_n|^2}(0,0)^{-\frac{1}{2}}\, K_{|f_n|^2}(z,0), \qquad z \in \mathbb{D};
$$

compare with formula (3.4). We intend to show that φ_n converges to F as $n \to +\infty$, uniformly on compact subsets of $\mathbb{D}$. From the above identity, it follows that it suffices to show that $K_{|f_n|^2}(z,0) \to 1$ as $n \to +\infty$, uniformly on compact subsets of $\mathbb{D}$.

LEMMA 3.39 *If $L = L(F, \varepsilon)$ is the constant from Lemma 3.38, then for any n with $L \leq n < +\infty$ we have*

$$
0 < K_{|f_n|^2}(z,z) \leq \frac{8}{3\,\varepsilon}\, \frac{1}{(1 - |z|^2)^2}
$$

for $z \in \mathbb{D}$.

Proof. For $z \in \mathbb{D}$, the quantity $K_{|f_n|^2}(z, z)^{\frac{1}{2}}$ expresses the norm of the point evaluation functional at z in the space $A^2(\mathbb{D}, |f_n|^2)$. However, each element $A^2(|f_n|^2)$ is in A^2 (and vice versa), and for $f \in A^2$, we have the growth estimate

$$|f(z)| \leq \frac{\|f\|}{1 - |z|^2}, \qquad z \in \mathbb{D};$$

see Lemma 3.2. The assertion now follows from Lemma 3.38. ∎

To complete the proof of Theorem 3.37, we shall need to understand the behavior of the inner products $\langle z^j f_n, f_n \rangle$ for $j = 0, 1, 2, \ldots$. A computation based on (3.10) yields

$$\begin{aligned}
\langle z^j f_n, f_n \rangle &= \langle z^j F, F \rangle + \sqrt{n+1}\, \langle z^j F, z^n G \rangle \\
&\quad + \sqrt{n+1}\, \langle z^{n+j} G, F \rangle + (n+1)\, \langle z^{n+j} G, z^n G \rangle.
\end{aligned}$$

For n with $N < n < +\infty$ we have $\langle z^{n+j} G, F \rangle = 0$, as the functions G and F are polynomials of degree at most N. The above identity then simplifies to

$$\langle z^j f_n, f_n \rangle = \langle z^j F, F \rangle + \sqrt{n+1}\, \langle z^j F, z^n G \rangle + (n+1)\langle z^{n+j} G, z^n G \rangle. \tag{3.15}$$

Expanding the polynomial G in a power series

$$G(z) = \sum_{n=0}^{+\infty} \widehat{G}(n)\, z^n,$$

where $\widehat{G}(n) = 0$ for $n = N+1, N+2, \ldots$, we find that

$$\langle z^{n+j} G, z^n G \rangle = \sum_{k=0}^{+\infty} \frac{\widehat{G}(k)\, \overline{\widehat{G}}(j+k)}{j+k+n+1},$$

and hence

$$\langle z^j G, G \rangle_{H^2} - (n+1)\, \langle z^{n+j} G, z^n G \rangle = \sum_{k=0}^{+\infty} \frac{j+k}{j+k+n+1}\, \widehat{G}(k)\, \overline{\widehat{G}}(j+k).$$

Each term on the right-hand side vanishes for $N < j+k < +\infty$, and hence

$$\begin{aligned}
\left| \langle z^j G, G \rangle_{H^2} - (n+1)\, \langle z^{n+j} G, z^n G \rangle \right| \\
\leq \sum_{k=0}^{+\infty} \frac{j+k}{j+k+n+1}\, \left| \widehat{G}(k)\, \widehat{G}(j+k) \right| \\
\leq \frac{N}{N+n+1} \sum_{k=0}^{+\infty} \left| \widehat{G}(k)\, \widehat{G}(j+k) \right| \\
\leq \frac{N}{N+n+1}\, \|G\|_{H^2}^2 \leq \frac{N}{N+n+1},
\end{aligned} \tag{3.16}$$

where we used the fact that $|G| \leq 1$ on $\mathbb{T}$. The inner products $\langle z^j F, F \rangle$ appear as Fourier coefficients of the function $P^*[|F|^2]$, and the inner products $\langle z^j G, G \rangle_{H^2}$ are the Fourier coefficients of $|G|^2$. From the identity

$$P^*[|F|^2] + |G|^2 = 1$$

on $\mathbb{T}$ it then follows that

$$\langle z^j F, F \rangle + \langle z^j G, G \rangle_{H^2} = \delta_{j,0}, \qquad j = 0, 1, 2, 3, \ldots, \qquad (3.17)$$

where $\delta_{j,0}$ is the Kronecker delta symbol. The expanded expression (3.15) then assumes the form

$$
\begin{aligned}
\langle z^j f_n, f_n \rangle \quad = \quad & \delta_{j,0} + \sqrt{n+1}\, \langle z^j F, z^n G \rangle \\
& + (n+1)\, \langle z^{n+j} G, z^n G \rangle - \langle z^j G, G \rangle_{H^2}.
\end{aligned}
\qquad (3.18)
$$

Expanding the function F in a power series as well,

$$F(z) = \sum_{n=0}^{+\infty} \widehat{F}(n)\, z^n,$$

where $\widehat{F}(n) = 0$ for $n = N+1, N+2, \ldots$, we obtain

$$\langle z^j F, z^n G \rangle = \sum_{k=0}^{+\infty} \widehat{F}(k)\, \langle z^{j+k}, z^n G \rangle,$$

so that the Cauchy-Schwarz inequality yields

$$
\begin{aligned}
\left| \langle z^j F, z^n G \rangle \right| \quad \leq \quad & \sum_{k=0}^{+\infty} |\widehat{F}(k)|\, \left| \langle z^{j+k}, z^n G \rangle \right| \qquad (3.19) \\
\leq \quad & \left(\sum_{k=0}^{+\infty} \frac{|\widehat{F}(k)|^2}{k+1} \right)^{\frac{1}{2}} \left(\sum_{k=0}^{N} (k+1)\, \left| \langle z^{j+k}, z^n G \rangle \right|^2 \right)^{\frac{1}{2}}.
\end{aligned}
$$

A subinner function has norm at most 1, which shows that the first factor on the right-hand side of (3.19) is bounded by 1. Since $|G| \leq 1$ on $\mathbb{T}$, the maximum principle informs us that $|G| \leq 1$ on $\mathbb{D}$ as well. Consequently,

$$\left| \langle z^{j+k}, z^n G \rangle \right| \leq \int_{\mathbb{D}} |z|^{j+k+n}\, dA(z) = \frac{2}{j+k+n+2},$$

and hence (3.19) leads to the estimate

$$\left| \langle z^j F, z^n G \rangle \right| \leq 2 \left[\sum_{k=0}^{N} \frac{k+1}{(j+k+n+2)^2} \right]^{\frac{1}{2}} \leq \frac{2(N+1)}{N+n+j+2}. \qquad (3.20)$$

Summing up and using the identity (3.18) together with the estimates (3.16) and (3.20), we obtain

$$
\begin{aligned}
\left| \langle z^j f_n, f_n \rangle - \delta_{j,0} \right| &\leq 2 \frac{(N+1)\sqrt{n+1}}{N+n+j+2} + \frac{N}{N+n+1} \\
&\leq 3 \frac{N+1}{\sqrt{n+1}}, \qquad j = 0, 1, 2, \ldots. \qquad (3.21)
\end{aligned}
$$

Recall that we need to show that $K_{|f_n|^2}(z, 0) \to 1$ uniformly as long as z is confined to compact subsets of $\mathbb{D}$. We expand the kernel in a convergent power series,

$$
K_{|f_n|^2}(z, \zeta) = \sum_{j=0}^{+\infty} B_j(\zeta) z^j, \qquad (z, \zeta) \in \mathbb{D}^2,
$$

where the functions B_j are antiholomorphic. By the reproducing property of the kernel function applied to the constant function 1, we have

$$
1 = \sum_{j=0}^{+\infty} B_j(\zeta) \langle z^j f_n, f_n \rangle, \qquad \zeta \in \mathbb{D},
$$

which we may rewrite as

$$
1 - K_{|f_n|^2}(0, \zeta) = \left(\| f_n \|^2 - 1 \right) B_0(\zeta) + \sum_{j=1}^{+\infty} B_j(\zeta) \langle z^j f_n, f_n \rangle, \qquad (3.22)
$$

in view of the fact that $B_0(\zeta) = K_{|f_n|^2}(0, \zeta)$. We recover the Taylor coefficients $B_j(\zeta)$ via the integral formula

$$
B_j(\zeta) = (j+n+1) \int_{\mathbb{D}} \bar{z}^j \, K_{|f_n|^2}(z, \zeta) \, |z|^{2n} \, dA(z), \qquad \zeta \in \mathbb{D},
$$

which leads to the following estimate for $\zeta \in \mathbb{D}$ and large n:

$$
\begin{aligned}
|B_j(\zeta)| &\leq (j+n+1) \, \| z^{j+n} \| \, \| z^n K_{|f_n|^2}(\cdot, \zeta) \| \qquad (3.23) \\
&\leq \left(\frac{8}{3\varepsilon} \right)^{\frac{1}{2}} \frac{\sqrt{j+n+1}}{\sqrt{n+1}} \, \| f_n K_{|f_n|^2}(\cdot, \zeta) \| \\
&= \left(\frac{8}{3\varepsilon} \right)^{\frac{1}{2}} \sqrt{1 + \frac{j}{n+1}} \, K_{|f_n|^2}(\zeta, \zeta)^{\frac{1}{2}} \leq \frac{8}{3\varepsilon} \frac{\sqrt{1+j/n}}{1 - |\zeta|^2};
\end{aligned}
$$

here we used Lemma 3.38 as well as Lemma 3.39. Looking at the support set for the Taylor coefficients of f_n, we see that $\langle z^j f_n, f_n \rangle$ vanishes for j off the set $[0, N] \cup [n - N, n + N]$. For large n, the estimate (3.23) simplifies to (restricting it to the relevant interval)

$$
|B_j(\zeta)| \leq \frac{3}{\varepsilon} \frac{1}{1 - |\zeta|^2}, \qquad \zeta \in \mathbb{D}, \quad 0 \leq j \leq n + N. \qquad (3.24)
$$

It follows from (3.21) and (3.24) that for large n and $\zeta \in \mathbb{D}$,

$$\sum_{j=1}^{+\infty} |B_j(\zeta)| |\langle z^j f_n, f_n \rangle| = \sum_{j=1}^{N} |B_j(\zeta)| |\langle z^j f_n, f_n \rangle|$$

$$+ \sum_{j=n-N}^{n+N} |B_j(\zeta)| |\langle z^j f_n, f_n \rangle|$$

$$\leq \frac{27}{\varepsilon} \frac{(N+1)^2}{\sqrt{n+1}} \frac{1}{1-|\zeta|^2}, \tag{3.25}$$

by counting the number of terms. Similarly,

$$\left| \left(\|f_n\|^2 - 1 \right) B_0(\zeta) \right| \leq \frac{9}{\varepsilon} \frac{N+1}{\sqrt{n+1}} \frac{1}{1-|\zeta|^2}, \qquad \zeta \in \mathbb{D}. \tag{3.26}$$

We now see from the identity (3.22) and the estimates (3.25) and (3.26) that

$$K_{|f_n|^2}(0, \zeta) \to 1$$

as $n \to +\infty$, uniformly when ζ is confined to compact subsets of $\mathbb{D}$.

The proof of Theorem 3.37 is now complete, because $K_{|f_n|^2}(z, 0)$ equals the complex conjugate of $K_{|f_n|^2}(0, z)$.   ∎

3.8   Notes

The study of Bergman inner functions originated from Hedenmalm's paper [59], which marked the beginning of a very fruitful period for the study of Bergman spaces.

Lemma 3.2 is from Vukotić [131]. This estimate is almost "obvious", but it is critical in the proof of Theorem 3.3.

Theorem 3.3 is from Hedenmalm's paper [59] for $p = 2$ and $\alpha = 0$; the general case here was shown in Zhu [144]. This result will play an important role when we study zero sequences in Chapter 4 and interpolating sequences in Chapter 5.

The extremal problem of Section 3.2 is classical in the study of Hardy spaces. The importance of this extremal problem for the study of factorization and the structure of invariant subspaces in the Bergman spaces was first demonstrated by Hedenmalm in [59].

The connection between Bergman inner functions and the biharmonic Green function was found by Duren, Khavinson, Shapiro, and Sundberg in [38], and was further studied in [39], [40], and [7]. Lemma 3.13 is from Abkar's thesis [2].

The expansive multiplier property, or equivalently, the contractive divisibility property, of Bergman inner functions was first obtained by Hedenmalm [59] in A^2, and then by Duren, Khavinson, Shapiro, and Sundberg [38], [39], and [40], in A^p for general exponents p, $0 < p < +\infty$. Lemma 3.28 – even in the more general setting of logarithmically subharmonic weights – is due to Hedenmalm (see [40]).

Hedenmalm and Zhu [73] showed that the expansive multiplier property fails for the weighted Bergman spaces A_α^2 with $1 < \alpha < +\infty$. Shimorin [120, 121] later found that the expansive multiplier property remains valid for the spaces A_α^2 with $-1 < \alpha \leq 1$; Hedenmalm [60] had settled the case A_1^2 earlier (see also [65]). The case of $0 < \alpha < 1$ remains open for $p \neq 2$.

The uniqueness of the contractive zero divisors in A^p for $0 < p < +\infty$, $p \neq 2$, is due to Duren, Khavinson, Shapiro, and Sundberg. The proof of Lemma 3.30 here is taken from [94]. Sundberg [128] proves that the contractive zero divisor of I_A in A^p has an analytic continuation across each open arc of the unit circle that does not contain an accumulation point of A.

The material in Section 3.6 is from the fundamental paper [7] of Aleman, Richter, Sundberg. The final touch in the proof of Theorem 3.33, involving the dilation of the A^p-inner function, however, is new. It was inspired by the paper of Hedenmalm, Jakobsson, and Shimorin [69].

The material in Section 3.7 is from Shimorin's paper [125]. It is an open problem to do the same for A^p, $p \neq 2$.

3.9   Exercises and Further Results

1. Let $A = \{a_1, \ldots, a_n\}$ be a finite sequence of distinct points in $\mathbb{D}$. Show that G_A in A^2 is a linear combination of the functions

$$1, \quad \frac{1}{(1 - \bar{a}_1 z)^2}, \quad \cdots \quad, \quad \frac{1}{(1 - \bar{a}_n z)^2}.$$

2. If G is A_α^p-inner, then $1 \leq |G(\zeta)|$ whenever $\zeta \in \mathbb{T}$ is a point of continuity of G.

3. Let $A = \{a_1, a_2, a_3, \ldots\}$ be the zero sequence of a function in A^p. For positive integers N, let $A^{(N)}$ be the tail sequence

$$A^{(N)} = \{a_{N+1}, a_{N+2}, a_{N+3}, \ldots\}.$$

Let $G_{A^{(N)}}$ be the canonical divisor for the zero sequence $A^{(N)}$. Show that $G_{A^{(N)}}(z) \to 1$, uniformly on compact subsets of $\mathbb{D}$, as $N \to +\infty$.

4. Let us say that a sequence $A = \{a_1, a_2, a_3, \ldots\}$ is a sub-zero sequence for A^p if there exists a function $f \in A^p$ which vanishes along A without being identically zero. In other words, a sub-zero sequence is a subsequence of a zero sequence. Consider the extremal function G_A for A, and show by an argument which involves a competing function for the extremal problem that G_A vanishes precisely on A. Consequently, each sub-zero sequence is itself a zero sequence.

5. If G is A^p-inner and bounded, then G is a zero divisor whose zero set is the union of finitely many interpolating sequences. See [77].

6. Show that the contractive zero divisor G_A of an A^p-zero set A has an analytic continuation across any arc of $\mathbb{T}$ that does not contain an accumulation point of A. See [128].

7. Derive the formula for the zero divisor in A^p corresponding to a single point a repeated n times.

8. Derive the formula for the extremal function of the invariant subspace in A^p generated by the singular inner function with an atomic mass σ at $z = 1$.

9. Let G_σ be the extremal function of the invariant subspace in A^2 generated by the singular inner function S_σ with an atomic mass σ at $z = 1$. Show that

$$\int_{\mathbb{D}} |G_\sigma(z)|^2 \frac{1 - |z|^2}{|1 - z|^2} \, dA(z) = \frac{1}{2\sigma + 1}$$

for all $\sigma > 0$.

10. Let G be any A_α^p-inner function. Then

$$\int_{\mathbb{D}} |G(z)|^p \frac{1 - |z|^2}{|1 - \bar\zeta z|^2} \, dA_\alpha(z) = 1$$

for almost all $\zeta \in \mathbb{T}$.

11. Show that for certain α the zero divisor in A_α^2 with a single zero $a \in \mathbb{D}$ can have extraneous zeros. Then deduce that zero divisors in such cases fail to be contractive. See [73].

12. Show that if $-1 < \alpha \le 0$, then zero divisors in A_α^p are contractive. See [120], [122], as well as Chapter 9.

13. If G is the extremal function for an invariant subspace I of A^2, then

$$|G(z)|^2 \le \frac{1}{1 - |z|^2} - (1 - |z|^2)K_{I^\perp}(z, z)$$

for $z \in \mathbb{D}$, where $K_{I^\perp}(z, w)$ is the reproducing kernel for $I^\perp$.

14. Show that

$$\int_{\mathbb{D}} \left| f(z) \exp\left(-2\frac{1 - z}{1 + z}\right) \right|^2 dA(z) \le \int_{\mathbb{D}} |zf(z)|^2 \, dA(z)$$

for all $f \in A^2$. See [86].

15. Show that

$$\int_{\mathbb{D}} \left| \frac{f(z)}{2 - z} \right|^2 dA(z) \le \int_{\mathbb{D}} |zf(z)|^2 \, dA(z)$$

for all $f \in A^2$.

16. Show that

$$\int_{\mathbb{D}} |f(z)|^2 \, |\varphi_a(z)|^{b(a)} \, dA(z) \le \int_{\mathbb{D}} |zf(z)|^2 \, dA(z)$$

for all $a \in \mathbb{D}$ and $f \in A^2$, where $b(a) = 2(1 + |a|)/(1 - |a|)$, and $\varphi_a : \mathbb{D} \to \mathbb{D}$ is the usual Möbius involution associated with a.

17. Show that

$$\int_{\mathbb{D}} \left| \frac{\sqrt{2 - r^2}}{2 - rz} f(z) \right|^2 \, dA(z) \le \int_{\mathbb{D}} |zf(z)|^2 \, dA(z)$$

for all $0 < r < 1$ and $f \in A^2$. Moreover, equality holds above if and only if for some constant C,

$$f(z) = C \frac{2 - rz}{(1 - rz)^2}.$$

18. Suppose f is an analytic function f in $\mathbb{D}$ and $0 < p < +\infty$. Show that f belongs to A^p if and only if

$$\int_{\mathbb{D}} (1 - |z|^2)^2 \Delta |f(z)|^p \, dA(z) < +\infty.$$

19. Suppose f is a positive trigonometric polynomial on $\mathbb{T}$ of degree N. Then there exists an analytic polynomial $p(z)$ of degree N, zero-free on $\overline{\mathbb{D}}$, such that $|p|^2 = f$ on $\mathbb{T}$. This is usually referred to as a the Fejér-Riesz theorem.

20. If $\{I_n\}_n$ is a decreasing sequence of cyclic invariant subspaces of A^p, then $I = \cap_n I_n$ is cyclic (singly generated). Moreover, if $I \ne \{0\}$ and φ_n is the extremal function for I_n, then φ_n converges in A^p to the extremal function for I. In particular, if I_A is the invariant subspace of all functions that vanish on the zero set A, then I_A is generated by its extremal function G_A. For details, see [7].

4
Zero Sets

For an analytic function f in $\mathbb{D}$, not identically zero, we let Z_f denote the zero sequence of f, with multiple zeros repeated according to multiplicities. A sequence $A = \{a_n\}_n$ in $\mathbb{D}$ is called a zero set for A_α^p if there exists a nonzero function $f \in A_\alpha^p$ such that $A = Z_f$, counting multiplicities. Zero sets for other spaces of analytic functions are defined similarly.

In this chapter, we study the zero sets of functions in several Bergman-type spaces. It is well known that the zero sets cannot be described in terms of a simple Blaschke-type condition, because the angular distribution of the zeros plays a role. We shall obtain sharp necessary conditions for a sequence A to be a zero set for A_α^p, and sharp sufficient conditions as well. The gap between necessary and sufficient conditions is quite small. The characterizations are in terms of partial Blaschke sums on Stolz star domains and the Beurling-Carleson characteristic of the corresponding boundary set. In the case of the Bergman-Nevanlinna class A_α^0, however, we shall be able to characterize its zero sets by a simple Blaschke-type condition.

4.1 Some Consequences of Jensen's Formula

An effective tool for studying zeros of analytic functions is the classical Jensen formula, which gives us a relationship between the growth of the function and the growth of its zero set.

PROPOSITION 4.1 *Suppose the function f is analytic in $\mathbb{D}$ and $f(0) \neq 0$. For $0 < r < 1$, let $a_1, \ldots, a_n$ be the zeros of f in the disk $|z| < r$, repeated according* *to multiplicity. Then*

$$\log|f(0)| + \sum_{k=1}^{n} \log \frac{r}{|a_k|} = \frac{1}{2\pi} \int_0^{2\pi} \log|f(re^{i\theta})|\, d\theta.$$

Proof. First, assume that f is nonvanishing on the closed disk $|z| \le r$. Then

$$\log|f(0)| = \frac{1}{2\pi} \int_0^{2\pi} \log|f(re^{i\theta})|\, d\theta,$$

since the function $\log|f(z)|$ is harmonic on $|z| \le r$.

Next, assume that f is nonvanishing on $|z| < r$ but has a single zero $a = re^{it}$ on the circle $|z| = r$. Then the function $g(z) = f(z)/(z - a)$ is analytic and nonvanishing on $|z| \le r$, so that

$$\log|g(0)| = \frac{1}{2\pi} \int_0^{2\pi} \left[\log|f(re^{i\theta})| - \log|re^{i\theta} - re^{it}| \right] d\theta.$$

Since

$$\log|g(0)| = \log|f(0)| - \log r$$

and

$$\int_0^{2\pi} \log|1 - e^{i\theta}|\, d\theta = 0,$$

we conclude that

$$\log|f(0)| = \frac{1}{2\pi} \int_0^{2\pi} \log|f(re^{i\theta})|\, d\theta$$

whenever f is nonvanishing on $|z| < r$ and has a single zero on $|z| = r$. By induction, the above formula remains valid if f is nonvanishing on $|z| < r$ and has a finite number of zeros on $|z| = r$.

Finally, if $a_1, \cdots, a_n$ are the zeros of f in $|z| < r$, repeated according to multiplicity, then the function

$$F(z) = f(z) \prod_{k=1}^{n} \frac{r^2 - \bar{a}_k z}{r(z - a_k)}$$

is analytic in $\mathbb{D}$, nonvanishing on $|z| < r$, and has a finite number of zeros on $|z| = r$. Thus,

$$\log|F(0)| = \frac{1}{2\pi} \int_0^{2\pi} \log|F(re^{i\theta})|\, d\theta.$$

Since $|F(z)| = |f(z)|$ on $|z| = r$ and

$$F(0) = f(0) \prod_{k=1}^{n} \left(-\frac{r}{a_k} \right),$$

Jensen's formula results.   ■

Let f be an analytic function in $\mathbb{D}$, not identically zero, but with a zero of order m at $z = 0$, $m \geq 0$. Applying Jensen's formula to the function $g(z) = f(z)/z^m$, we obtain

$$\frac{1}{2\pi} \int_0^{2\pi} \log |f(re^{i\theta})| \, d\theta = m \log r + \log \left| \frac{f^{(m)}(0)}{m!} \right| + \sum_{k=1}^n \log \frac{r}{|a_k|},$$

where $0 < r < 1$ and $a_1, \ldots, a_n$ are the zeros of f in $0 < |z| < r$. It follows that for every $0 < \sigma < 1$, there exists a constant $C = C(\sigma)$ such that

$$C \leq \frac{1}{2\pi} \int_0^{2\pi} \log |f(re^{i\theta})| \, d\theta$$

for all r with $\sigma < r < 1$.

For f analytic in $\mathbb{D}$ and $0 < r < 1$, we let $n(r) = n_f(r)$ be the number of zeros of f in $|z| < r$, counting multiplicity. If $f(0) \neq 0$, we let

$$N(r) = N_f(r) = \int_0^r \frac{n(t)}{t} \, dt.$$

The counting functions $n(r)$ and $N(r)$ play important roles in the study of zeros of analytic functions.

PROPOSITION 4.2 *Suppose f is analytic in $\mathbb{D}$ with $f(0) \neq 0$. Then*

$$N(r) = \sum_{k=1}^n \log \frac{r}{|a_k|},$$

where $a_1, \ldots, a_n$ are the zeros of f in $|z| < r$, repeated according to multiplicity.

Proof. Since $n(t) = 0$ for $0 < t \leq |a_1|$, we have

$$N(r) = \sum_{k=1}^{n-1} \int_{|a_k|}^{|a_{k+1}|} \frac{n(t)}{t} \, dt + \int_{|a_n|}^r \frac{n(t)}{t} \, dt.$$

By definition, $n(t) = k$ for $|a_k| < t \leq |a_{k+1}|$, and $n(t) = n$ for $|a_n| < t \leq r$. It follows that

$$N(r) = \sum_{k=1}^{n-1} \left[k \left(\log |a_{k+1}| - \log |a_k| \right) \right] + n \left(\log r - \log |a_n| \right).$$

A little manipulation then shows that

$$N(r) = \sum_{k=1}^n \log \frac{r}{|a_k|},$$

as claimed.   ■

PROPOSITION 4.3 *Suppose μ is a probability measure on a measure space X and that g is a positive measurable function on the measure space X. Then*

$$\int_X \log g(x)\, d\mu(x) \leq \log\left[\int_X g(x)\, d\mu(x)\right].$$

Proof. This is a special case of a general result in Real Analysis, which is usually called the arithmetic-geometric mean inequality. ∎

We proceed to prove some necessary conditions for a sequence in $\mathbb{D}$ to be a zero set for A_α^p.

PROPOSITION 4.4 *Suppose $f \in A_\alpha^p$ with $f(0) \neq 0$. Then*

$$2\int_0^1 (1-r^2)^\alpha e^{pN(r)}\, r\, dr \leq \frac{1}{|f(0)|^p}\int_\mathbb{D} |f(z)|^p (1-|z|^2)^\alpha\, dA(z).$$

Proof. If f is analytic in $\mathbb{D}$ with $f(0) = 1$, then by Propositions 4.1 and 4.2,

$$N(r) = \frac{1}{2\pi}\int_0^{2\pi} \log|f(re^{i\theta})|\, d\theta$$

for all $0 < r < 1$. The desired result then follows from the arithmetic-geometric mean inequality. The general case follows by considering $g = f/f(0)$. ∎

COROLLARY 4.5 *Suppose $f \in A_\alpha^p$ with $f(0) \neq 0$. Let $a_1, a_2, a_3, \ldots$ be the zeros of f, repeated according to multiplicity and arranged so that $|a_1| \leq |a_2| \leq |a_3| \leq \ldots$. Then*

$$|f(0)|\prod_{k=1}^n \frac{1}{|a_k|} \leq Cn^{(\alpha+1)/p}\|f\|_{p,\alpha}$$

for all $n = 1, 2, 3, \ldots$, where C is a positive constant dependent only on p and α.

Proof. Recall from Proposition 4.2 that

$$e^{pN(r)} = \prod_{k=1}^{n(r)} \frac{r^p}{|a_k|^p},$$

where we recall that $n(r)$ counts the number of zeros in $|z| < r$. It follows that

$$e^{pN(r)} \geq \prod_{k=1}^n \frac{r^p}{|a_k|^p}$$

for every positive integer n. Combining this with Proposition 4.4, we obtain

$$|f(0)|^p \prod_{k=1}^n \frac{1}{|a_k|^p} \leq \frac{\Gamma(\alpha+2+np/2)}{\Gamma(\alpha+2)\Gamma(1+np/2)}\int_\mathbb{D} |f(z)|^p\, dA_\alpha(z).$$

The desired result now follows from Stirling's formula.   ■

It is easy to see that (see Exercise 15 of Chapter 1)

$$\lim_{\alpha \to -1^+} \int_{\mathbb{D}} |f(z)|^p \, dA_\alpha(z) = \frac{1}{2\pi} \int_0^{2\pi} |f(e^{it})|^p \, dt$$

for every function f in the Hardy space H^p. It follows from the proof of the above corollary that

$$|f(0)| \prod_{k=1}^{+\infty} \frac{1}{|a_k|} \leq \|f\|_{H^p}$$

for every $f \in H^p$ with $f(0) \neq 0$. This clearly implies the Blaschke condition

$$\sum_k (1 - |a_k|) < +\infty$$

for zero sets of functions in Hardy spaces.

We are going to show that zero sets for A_α^p satisfy a slightly weaker condition. To accomplish this, we need an estimate for the growth of $n(r)$ and $N(r)$ associated with functions in A_α^p.

PROPOSITION 4.6 *Suppose $f \in A_\alpha^p$ with $f(0) \neq 0$. Then there exists a positive constant C such that for all $r \in (0, 1)$,*

$$(1 - r)n(r) \leq C \log \frac{1}{1 - r}$$

and

$$N(r) \leq C + \frac{\alpha + 1}{p} \log \frac{1}{1 - r}.$$

Proof. By Proposition 4.4, the quantity

$$C = (\alpha + 1) \int_0^1 (1 - t)^\alpha e^{pN(t)} \, dt$$

is finite. Since $N(t)$ is increasing on $(0, 1)$, we have

$$
\begin{aligned}
C &\geq (\alpha + 1) \int_r^1 (1 - t)^\alpha e^{pN(t)} \, dt \\
&\geq (\alpha + 1) e^{pN(r)} \int_r^1 (1 - t)^\alpha \, dt = e^{pN(r)} (1 - r)^{\alpha+1}
\end{aligned}
$$

for every $r \in (0, 1)$. It follows that

$$N(r) \leq \frac{\alpha + 1}{p} \log \frac{1}{1 - r} + C_1$$

for all $r \in (0, 1)$ and some positive constant C_1.

Since $n(r)$ is monotone, we have

$$n(r^2)(r - r^2) \le \int_{r^2}^r n(t)\, dt \le N(r) \le \frac{\alpha + 1}{p} \log \frac{1}{1 - r} + C_1$$

for all $r \in (0, 1)$. This together with the assumption $f(0) \ne 0$ easily implies that

$$(1 - r)\, n(r) \le C_2 \log \frac{1}{1 - r}$$

for all $r \in (0, 1)$ and some positive constant C_2.   ∎

We can now prove the main result of the section.

THEOREM 4.7 *Suppose* $f \in A_\alpha^p$ *with* $f(0) \ne 0$. *If* $\{a_k\}_k$ *is the zero sequence of* f, *then for every positive* ε *we have*

$$\sum_{k=1}^{+\infty} \frac{1 - |a_k|}{\left[\log \frac{1}{1-|a_k|}\right]^{1+\varepsilon}} < +\infty.$$

Proof. It is clear that the desired result is equivalent to the convergence of the integral

$$I = \int_a^1 \frac{1 - t}{\left[\log \frac{1}{1-t}\right]^{1+\varepsilon}}\, dn(t),$$

where $a = |a_1| \in (0, 1)$. Integrating by parts and applying Proposition 4.6, we obtain

$$I = \int_a^1 \frac{1 + \varepsilon + \log \frac{1}{1-t}}{\left[\log \frac{1}{1-t}\right]^{2+\varepsilon}}\, n(t)\, dt = I_1 + I_2,$$

where

$$I_1 = (1 + \varepsilon) \int_a^1 \frac{n(t)\, dt}{\left[\log \frac{1}{1-t}\right]^{2+\varepsilon}},$$

whose convergence is also guaranteed by Proposition 4.6, and

$$I_2 = \int_a^1 \frac{n(t)\, dt}{\left[\log \frac{1}{1-t}\right]^{1+\varepsilon}} \le \int_a^1 \frac{dN(t)}{\left[\log \frac{1}{1-t}\right]^{1+\varepsilon}}.$$

We integrate by parts again to get

$$\int_a^1 \frac{dN(t)}{\left[\log \frac{1}{1-t}\right]^{1+\varepsilon}} = \frac{N(t)}{\left[\log \frac{1}{1-t}\right]^{1+\varepsilon}} \Bigg|_a^1 + \int_a^1 \frac{(1 + \varepsilon)N(t)\, dt}{(1 - t)\left[\log \frac{1}{1-t}\right]^{2+\varepsilon}}.$$

By Proposition 4.6, both terms above converge.   ∎

COROLLARY 4.8 *If* $\{a_n\}_n$ *is a zero set for some* A_α^p, *then for every* $\varepsilon > 0$,

$$\sum_n (1 - |a_n|^2)^{1+\varepsilon} < +\infty.$$

4.2   Notions of Density

In order to better understand the structure of zero sets for Bergman spaces, we need to introduce several notions of density for sequences in the unit disk. The reason is that it is well known that the zero sets cannot be captured by a simple Blaschke-type condition in terms of moduli: indeed, a spread-out zero set need not fulfill the Blaschke condition, whereas a concentrated one must do so – if, say, all the zeros are contained in a finite union of Stolz angles.

For a point $z \in \mathbb{T}$, we let $\mathfrak{s}_z$ denote the standard relatively closed *Stolz angle* in $\mathbb{D}$ with vertex at z and aperture $\pi/2$. Thus, $\mathfrak{s}_z$ is the convex hull of the set

$$\{z\} \cup \left\{ w \in \mathbb{C} : |w| \le 1/\sqrt{2} \right\},$$

with the vertex point z removed. The term Privalov "ice cream" cone is also used in the literature.

For an arc $I \subset \mathbb{T}$, let $|I|$ be its *arc length*, and $|I|_s = |I|/(2\pi)$ its *normalized arc length*. The subscript s refers to the measure $ds(z) = |dz|/(2\pi)$. For a closed and proper subset F of $\mathbb{T}$ with complementary arcs $\{I_n\}_n$, we define

$$\widehat{\kappa}(F) = \sum_n |I_n|_s \log \frac{e}{|I_n|_s},$$

where $e = 2.71828\ldots$ is the base for the natural logarithm. The quantity $\widehat{\kappa}(F)$ will be called the *Beurling-Carleson characteristic* of F. Sometimes the term *entropy* is also used for $\widehat{\kappa}(F)$. We define $\widehat{\kappa}(\emptyset) = 0$ for the empty set.

A closed subset F of $\mathbb{T}$ is called a *Beurling-Carleson set* if F is nonempty, has Lebesgue length measure zero, and $\widehat{\kappa}(F) < +\infty$. It is clear that $1 \le \widehat{\kappa}(F)$ for such sets, with equality occurring only for one-point sets F. Let $d_\mathbb{T}$ be the standard metric on the unit circle $\mathbb{T}$:

$$d_\mathbb{T}(z, w) = \left| \arg\left(\frac{z}{w} \right) \right|,$$

where the argument function is assumed to take values in the interval $(-\pi, \pi]$. The distance to a closed subset F of $\mathbb{T}$ is then

$$d_\mathbb{T}(z, F) = \inf \left\{ d_\mathbb{T}(z, w) : w \in F \right\},$$

and F is a Beurling-Carleson set if and only if

$$\widehat{\kappa}(F) = \int_\mathbb{T} \log \frac{\pi}{d_\mathbb{T}(z, F)} \, ds(z) < +\infty.$$

Actually, the left hand side only represents the integral over the complement of F, which does not matter as long as F has zero length. Let $d_{\mathbb{C}}$ stand for the Euclidean metric in $\mathbb{C}$: $d_{\mathbb{C}}(z, w) = |z - w|$. Then, for any closed subset F of $\mathbb{T}$,

$$\frac{2}{\pi} d_{\mathbb{T}}(z, F) \leq d_{\mathbb{C}}(z, F) \leq d_{\mathbb{T}}(z, F), \qquad z \in \mathbb{T},$$

where the distance to sets is defined in terms of an infimum as for $d_{\mathbb{T}}$, so that by the above,

$$\widehat{\kappa}(F) - \log \pi \leq \int_{\mathbb{T}} \log \frac{1}{d_{\mathbb{C}}(z, F)} \, ds(z) \leq \widehat{\kappa}(F) - \log 2,$$

provided F has zero length.

For most of our discussion, we assume that F is a finite set. In association with F, we define the Stolz star domain $\mathfrak{s}_F$ as follows:

$$\mathfrak{s}_F = \bigcup \{\mathfrak{s}_z : z \in F\}.$$

Let $A = \{a_n\}_n$ be a sequence of (not necessarily distinct) points from $\mathbb{D}$. We will not be primarily interested in the order that the the various points a_n appear, but rather think of A as a subset of $\mathbb{D}$, except that multiplicities are allowed. For an arbitrary subset E of $\mathbb{D}$, we form the partial Blaschke sum

$$\Sigma(A, E) = \frac{1}{2} \sum_n \{1 - |a_n|^2 : a_n \in E\}.$$

We note that for points $a \in \mathbb{D}$ close to $\mathbb{T}$, the quantitities $\frac{1}{2}(1 - |a|^2)$ and $1 - |a|$ are very close. Later on, we also need the related "logarithmic" sum

$$\Lambda(A, E) = \sum_n \left\{\log \frac{1}{|a_n|} : a_n \in E\right\};$$

again, for $a \in \mathbb{D}$ close to $\mathbb{T}$, the quantity we sum over, $\log[1/|a|]$, is very close to $1 - |a|$. Another thing to think of is that the above sums are in fact taken over the "sets" $A \cap E$, by which we mean that all the points of A are included *with multiplicities*, provided they are in E. We shall sum over the Stolz stars $\mathfrak{s}_F$, where $F \subset \mathbb{T}$ is finite: the κ-*density* of the sequence A in the Stolz star $\mathfrak{s}_F$ is

$$D(A, \mathfrak{s}_F) = \Sigma(A, \mathfrak{s}_F)/\widehat{\kappa}(F).$$

DEFINITION 4.9 *Let A be a sequence of points in $\mathbb{D}$ and F be finite subsets of $\mathbb{T}$. Then the quantities*

$$D^+(A) = \limsup_{\widehat{\kappa}(F) \to +\infty} D(A, \mathfrak{s}_F)$$

and

$$D^-(A) = \liminf_{\widehat{\kappa}(F) \to +\infty} D(A, \mathfrak{s}_F)$$

are called the upper and lower asymptotic κ-densities of A, respectively.

The upper asymptotic κ-density will be crucial for our description of zero sets for the Bergman spaces. It will also play an essential role later when we characterize sequences of interpolation for the Bergman spaces.

We proceed to give several equivalent definitions of the upper asymptotic κ-density. First, observe that replacing the standard Stolz angle $\mathfrak{s}_z$ by a general Stolz angle $\mathfrak{s}_{z,\alpha}$ with any fixed aperture $0 < \alpha < \pi$ and making the corresponding changes in the definitions of $\mathfrak{s}_F$, $\Sigma(A, \mathfrak{s}_F)$, and $D(A, \mathfrak{s}_F)$ will not alter the quantities $D^{\pm}(A)$. What is somewhat surprising is that the angle α can be reduced to 0 with no effect on $D^+(A)$. More specifically, for a finite set F and a sequence A of points in $\mathbb{D}$, we set

$$\mathfrak{r}_F = \{rz \in \mathbb{D} : 0 \leq r < 1, \, z \in F\}.$$

The set $\mathfrak{r}_F$ is the union of radii from 0 to the points of F. Then we have the following result.

PROPOSITION 4.10 *Let $A = \{a_n\}_n$ be any sequence of points from $\mathbb{D}$ and F be finite subsets of $\mathbb{T}$. Then*

$$D^+(A) = \limsup_{\widehat{\kappa}(F) \to +\infty} \frac{\Sigma(A, \mathfrak{r}_F)}{\widehat{\kappa}(F)}.$$

To prove the identity above, we need yet another notion of density based on *Carleson squares*. Recall that for an *open arc* $I \subset \mathbb{T}$, with $|I| = 2\pi |I|_s < 1$, the associated Carleson square is the set

$$Q(I) = \{w \in \mathbb{D} \setminus \{0\} : 1 - |I| < |w|, \, w/|w| \in I\};$$

for open arcs of bigger length, we let $Q(I)$ be the entire sector

$$Q(I) = \{w \in \mathbb{D} \setminus \{0\} : w/|w| \in I\}.$$

If $\{I_n\}_n$ are the complementary arcs of a finite set F in $\mathbb{T}$, we define

$$\mathfrak{q}_F = \mathbb{D} \setminus \bigcup_n Q(I_n).$$

We then arrive at another way of obtaining $D^+(A)$.

PROPOSITION 4.11 *Let $A = \{a_n\}_n$ be any sequence of points from $\mathbb{D}$ and F be finite subsets of $\mathbb{T}$. Then*

$$D^+(A) = \limsup_{\widehat{\kappa}(F) \to +\infty} \frac{\Sigma(A, \mathfrak{q}_F)}{\widehat{\kappa}(F)}.$$

We proceed now to prove the two above propositions, that is, the equivalence of all the three definitions of $D^+(A)$. Note that the lim sup in each of these definitions will not change if we allow closed countable sets F of finite entropy as well.

We start with the proof of Proposition 4.11.

Proof. Enlarge every finite set F by inserting on each complementary arc I of F additional points accumulating at the endpoints of I so that their distances from the nearest endpoint of I form a geometric progression with some fixed ratio q, $0 < q < 1$. An elementary computation then shows that the augmented set $F_1 \supset F$ will have the property

$$\widehat{\kappa}(F) < \widehat{\kappa}(F_1) < \widehat{\kappa}(F) + C,$$

where the constant C depends only on the ratio q. We can also choose q such that

$$\mathfrak{q}_F \subset \mathfrak{s}_F \subset \mathfrak{q}_{F_1},$$

so that

$$\mathbf{\Sigma}(A, \mathfrak{q}_F) \leq \mathbf{\Sigma}(A, \mathfrak{s}_F) \leq \mathbf{\Sigma}(A, \mathfrak{q}_{F_1}).$$

This proves Proposition 4.11.   ∎

We turn to the proof of Proposition 4.10.

Proof. Observe that $\mathfrak{r}_F \subset \mathfrak{s}_F$, and thus $\mathbf{\Sigma}(A, \mathfrak{r}_F) \leq \mathbf{\Sigma}(A, \mathfrak{s}_F)$, which implies

$$\limsup_{\widehat{\kappa}(F) \to +\infty} \frac{\mathbf{\Sigma}(A, \mathfrak{r}_F)}{\widehat{\kappa}(F)} \leq \limsup_{\widehat{\kappa}(F) \to +\infty} \frac{\mathbf{\Sigma}(A, \mathfrak{s}_F)}{\widehat{\kappa}(F)}.$$

By Proposition 4.11, the reverse inequality is equivalent to

$$\limsup_{\widehat{\kappa}(F) \to +\infty} \frac{\mathbf{\Sigma}(A, \mathfrak{q}_F)}{\widehat{\kappa}(F)} \leq \limsup_{\widehat{\kappa}(F) \to +\infty} \frac{\mathbf{\Sigma}(A, \mathfrak{r}_F)}{\widehat{\kappa}(F)}.$$

At first glance this looks highly improbable, since the sum defining $\mathbf{\Sigma}(A, \mathfrak{q}_F)$ involves all points from $\mathfrak{q}_F$, while the sum defining $\mathbf{\Sigma}(A, \mathfrak{r}_F)$ involves only those points lying on one of the radii from 0 to points of F. However, a more careful argument will prove the above inequality.

Without loss of generality, we may assume that the lim sup on the left-hand side of the desired inequality is positive. Let L be a positive number less than this lim sup. This implies that there are finite subsets F of $\mathbb{T}$ of arbitrarily large $\widehat{\kappa}(F)$ such that

$$\mathbf{\Sigma}(A, \mathfrak{q}_F) = \frac{1}{2} \sum_{a_k \in \mathfrak{q}_F} (1 - |a_k|^2) > L\widehat{\kappa}(F).$$

Until the end of the proof, we shall assume that F satisfies this inequality.

Let F_1 equal the set F plus the radial projections $z/|z|$ of points from the set $A \cap (\mathfrak{q}_F \setminus \mathfrak{r}_F)$, so that

$$\mathbf{\Sigma}(A, \mathfrak{q}_F) \leq \mathbf{\Sigma}(A, \mathfrak{r}_{F_1}).$$

Let k_n be the number of such radial projections (counting multiplicities) that lie on I_n, where I_n is a complementary arc to the finite set $F \subset \mathbb{T}$. Observe now that the contribution to $\widehat{\kappa}(F_1)$ from the complementary arcs of F_1 contained in I_n does not exceed the quantity

$$|I_n|_s \left[\log \frac{e}{|I_n|_s} + \log(k_n + 1) \right],$$

which corresponds to the case of k_n equidistant points of $(F_1 \setminus F) \cap I_n$. Therefore,

$$\widehat{\kappa}(F) \leq \widehat{\kappa}(F_1) \leq \widehat{\kappa}(F) + r(F),$$

where $r(F)$ is the "remainder" term

$$r(F) = \sum_n |I_n|_s \log(k_n + 1).$$

Suppose the point $a_j \in A \cap (\mathfrak{q}_F \setminus \mathfrak{r}_F)$ is such that its radial projection lies on I_n. Then $|I_n| = 2\pi |I_n|_s < 1$ by the construction of the Carleson squares forming the complement of $\mathfrak{q}_F$ in $\mathbb{D}$, and moreover, we have $|a_j| \leq 1 - |I_n|$. It follows that

$$|I_n|_s < \pi |I_n|_s = \frac{1}{2} |I_n| \leq \frac{1}{2} (1 + |a_j|)(1 - |a_j|) = \frac{1}{2} (1 - |a_j|^2).$$

This leads to the the conclusion

$$\sum_n k_n |I_n|_s \leq \Sigma(A, \mathfrak{q}_F \setminus \mathfrak{r}_F) \leq \Sigma(A, \mathfrak{q}_F).$$

We now show that the "remainder" term is small:

$$r(F) = o(\Sigma(A, \mathfrak{q}_F)) \qquad \text{as} \quad \widehat{\kappa}(F) \to +\infty.$$

To this end, we pick a positive integer N and split the sum defining $r(F)$ into two parts, keeping the above estimate in mind:

$$
\begin{aligned}
r(F) &= \left[\sum_{k_n \leq N} + \sum_{k_n > N} \right] |I_n|_s \log(k_n + 1) \\
&\leq \log(N + 1) + \frac{\log(N + 1)}{N} \sum_{k_n > N} k_n |I_n|_s \\
&\leq \log(N + 1) + \frac{\log(N + 1)}{N} \Sigma(A, \mathfrak{q}_F).
\end{aligned}
$$

Letting $\widehat{\kappa}(F) \to +\infty$, with $\Sigma(A, \mathfrak{q}_F) \to +\infty$, first holding N constant and then making $N \to +\infty$, we obtain $r(F) = o(\Sigma(A, \mathfrak{q}_F))$, as desired. Consequently,

$$\widehat{\kappa}(F_1) = \widehat{\kappa}(F) + o(\Sigma(A, \mathfrak{q}_F))$$

as $\widehat{\kappa}(F) \to +\infty$. Since by the above, $\Sigma(A, \mathfrak{q}_F) \leq \Sigma(A, \mathfrak{r}_{F_1})$, we get

$$\frac{\Sigma(A, \mathfrak{r}_{F_1})}{\widehat{\kappa}(F_1)} \geq \frac{\Sigma(A, \mathfrak{q}_F)}{\widehat{\kappa}(F) + o(\Sigma(A, \mathfrak{q}_F))}$$

as $\widehat{\kappa}(F) \to +\infty$. This implies that

$$\limsup_{\widehat{\kappa}(F) \to +\infty} \frac{\Sigma(A, \mathfrak{r}_{F_1})}{\widehat{\kappa}(F_1)} \geq L.$$

Since F_1 above can be substituted for F, and L can be chosen arbitrarily close to

$$\limsup_{\widehat{\kappa}(F) \to +\infty} \frac{\Sigma(A, \mathfrak{q}_F)}{\widehat{\kappa}(F)},$$

Proposition 4.10 is proved.   ∎

Let $A = \{a_n\}_n$ be a sequence in $\mathbb{D}$, and fix a real parameter $\varrho \in (0, +\infty)$. If, for every finite subset F of $\mathbb{T}$,

$$\Sigma(A, \mathfrak{s}_F) \leq \varrho \, \widehat{\kappa}(F) + C,$$

for some constant C independent of F, then by the inclusion $\mathfrak{r}_F \subset \mathfrak{s}_F$, we also have

$$\Sigma(A, \mathfrak{r}_F) \leq \varrho \, \widehat{\kappa}(F) + C.$$

Conversely, if for every finite subset F of $\mathbb{T}$,

$$\Sigma(A, \mathfrak{r}_F) \leq \varrho \, \widehat{\kappa}(F) + C,$$

then by Proposition 4.10, $D^+(A) \leq \varrho$, so that

$$\Sigma(A, \mathfrak{s}_F) \leq (\varrho + \varepsilon) \, \widehat{\kappa}(F) + C'(\varepsilon)$$

for every $\varepsilon > 0$, where $C'(\varepsilon)$ is a constant that is independent of the finite set $F \subset \mathbb{T}$, but may vary with ε.

We shall need a similar but more precise comparison between $\Sigma(A, \mathfrak{s}_F)$ and $\Sigma(A, \mathfrak{r}_F)$ for some slightly different asymptotic restrictions on the latter.

PROPOSITION 4.12 *Fix* $0 < \varrho, \eta < +\infty$. *Suppose that the sequence A in $\mathbb{D}$ is such that*

$$\Sigma(A, \mathfrak{r}_F) \leq \varrho \, \widehat{\kappa}(F) + \eta \log \widehat{\kappa}(F) + C,$$

for every finite nonempty subset F of $\mathbb{T}$, where C is a constant. Then

$$\Sigma(A, \mathfrak{s}_F) \leq \varrho \, \widehat{\kappa}(F) + (\eta + \varrho) \log \widehat{\kappa}(F) + C',$$

for every finite nonempty subset F of $\mathbb{T}$, for some other constant C'.

Proof. As in the proofs of Propositions 4.10 and 4.11, we can show that the second inequality here is equivalent to a similar estimate with summation over Stolz stars $\mathfrak{s}_F$ replaced by summation over the regions $\mathfrak{q}_F$ with omitted Carleson squares:

$$\Sigma(A, \mathfrak{q}_F) \leq \varrho \, \widehat{\kappa}(F) + (\eta + \varrho) \log \widehat{\kappa}(F) + O(1),$$

where $O(1)$ stands for a quantity that is bounded independently of the finite set F.

Let $F \subset \mathbb{T}$ be finite, and $\{I_n\}_n$ the collection of complementary arcs; $\{Q(I_n)\}_n$ are the associated Carleson squares. Project all points from $A \cap \mathfrak{q}_F$ (other than 0) radially to $\mathbb{T}$, and let

$$F' = \left\{ \frac{z}{|z|} \in \mathbb{T} : z \in A \cap \mathfrak{q}_F, \, z \neq 0 \right\}$$

be the resulting set, so that $\Sigma(A, \mathfrak{q}_F) \leq \Sigma(A, \mathfrak{r}_{F'})$. We put $k_n = \text{card}\,(I_n \cap F')$, and note that an elementary argument shows

$$\widehat{\kappa}(F) \leq \widehat{\kappa}(F') \leq \widehat{\kappa}(F) + \sum_n |I_n|_s \log(k_n + 1),$$

with equality occurring in the right hand side inequality if the k_n points from $F' \cap I_n$ divide I_n into $k_n + 1$ equal subarcs. On the other hand, as we saw in the proof of Proposition 4.10,

$$\sum_n k_n |I_n|_s \leq \Sigma(A, \mathfrak{q}_F) \leq \Sigma(A, \mathfrak{r}_{F'}).$$

Since $\sum_n |I_n|_s = 1$, the concavity of the function $\log t$ (that is, the geometric-arithmetic mean value inequality) gives

$$\sum_n |I_n|_s \log(k_n + 1) \leq \log\left(1 + \sum_n k_n |I_n|_s\right) \leq \log\left(1 + \Sigma(A, \mathfrak{q}_F)\right).$$

Now, replace F with F' in the assumption of the proposition and use two of the above inequalities to get

$$\Sigma(A, \mathfrak{q}_F) \leq \Sigma(A, \mathfrak{r}_{F'}) \leq \varrho\widehat{\kappa}(F') + \eta \log \widehat{\kappa}(F') + O(1)$$
$$\leq \varrho\widehat{\kappa}(F) + \varrho \log \Sigma(A, \mathfrak{q}_F) + \eta \log\left(\widehat{\kappa}(F) + \log \Sigma(A, \mathfrak{q}_F)\right) + O(1).$$

From the proofs of Propositions 4.10 and 4.11, we know that

$$\log \Sigma(A, \mathfrak{q}_F) \leq \log \widehat{\kappa}(F) + O(1),$$

and thus

$$\Sigma(A, \mathfrak{q}_F) \leq \varrho\widehat{\kappa}(F) + (\eta + \varrho) \log \widehat{\kappa}(F) + O(1),$$

which is equivalent to the inequality stated at the beginning of the proof.   ∎

4.3   The Growth Spaces $\mathcal{A}^{-\alpha}$ and $\mathcal{A}^{-\infty}$

In this section, we introduce a class of Bergman-type spaces, denoted by $\mathcal{A}^{-\alpha}$ and $\mathcal{A}^{-\infty}$, which are closely related to the Bergman spaces A_α^p and are sometimes called *growth spaces*, and begin the study of their zero sets.

DEFINITION 4.13 *For any $\alpha > 0$, the space $\mathcal{A}^{-\alpha}$ consists of analytic functions f in $\mathbb{D}$ such that*

$$\|f\|_{-\alpha} = \sup\left\{(1 - |z|^2)^\alpha |f(z)| : z \in \mathbb{D}\right\} < +\infty.$$

It is easy to verify that $\mathcal{A}^{-\alpha}$ is a (nonseparable) Banach space with the norm defined above. Each space $\mathcal{A}^{-\alpha}$ clearly contains all the bounded analytic functions. The closure in $\mathcal{A}^{-\alpha}$ of the set of polynomials will be denoted by $\mathcal{A}_0^{-\alpha}$, which is a separable Banach space and consists of exactly those functions f in $\mathcal{A}^{-\alpha}$ with

$$\lim_{|z| \to 1^-} (1 - |z|^2)^\alpha |f(z)| = 0.$$

We will also consider the space

$$\mathcal{A}^{-\infty} = \bigcup_{0 < \alpha < +\infty} \mathcal{A}^{-\alpha}.$$

It is clear that an analytic function f in $\mathbb{D}$ belongs to $\mathcal{A}^{-\infty}$ if and only if there exist positive constants C and N such that

$$|f(z)| \le \frac{C}{(1 - |z|^2)^N}, \qquad z \in \mathbb{D}.$$

It is also clear that

$$\mathcal{A}^{-\infty} = \bigcup_{0 < p < +\infty} A_\alpha^p$$

for any $\alpha \in (-1, +\infty)$. The space $\mathcal{A}^{-\infty}$ is a topological algebra when endowed with the inductive-limit topology (for a definition, see any book on Functional Analysis).

For an analytic function f in $\mathbb{D}$ that is not identically zero, we define its *hyperbolic exponential type*

$$t(f) = \limsup_{|z| \to 1^-} \frac{\log |f(z)|}{\log \frac{1}{1 - |z|}}.$$

The function f is said to be of *finite hyperbolic exponential type* if $t(f) < +\infty$. It is clear that

$$t(f) = \inf \{\alpha : f \in \mathcal{A}^{-\alpha}\}.$$

When $f \in \mathcal{A}^{-\alpha}$ for $\alpha = t(f)$, we say that f is of *exact type*. If $t(f) = 0$, we say that f is of *minimal type*. Clearly, $t(f) = 0$ if and only if $f \in \mathcal{A}^{-\alpha}$ for all $\alpha > 0$.

The space $\mathcal{A}^{-\infty}$ then consists of 0 and functions of finite hyperbolic exponential type.

In order to understand the complexity of the zero sets for Bergman spaces, we first show that zero sets for Bergman spaces cannot be characterized by any condition that involves only the modulus of the zeros.

Let z_0 be a point on the unit circle. Then, for $1 < a < +\infty$, the set

$$\Gamma_a(z_0) = \left\{z \in \mathbb{D} : \frac{|z - z_0|}{1 - |z|} \le a\right\}$$

behaves like a Stolz angle at z_0.

THEOREM 4.14 *Suppose f is in $\mathcal{A}^{-\infty}$. If the zeros of f, $A = \{a_n\}_n$, lie in some $\Gamma_a(z_0)$, with $z_0 \in \mathbb{T}$, then A satisfies the Blaschke condition*

$$\sum_n (1 - |a_n|) < +\infty.$$

Proof. Using a rotation if necessary, we may assume that $z_0 = 1$. By eliminating a finite number of zeros, our assumption then implies that the zeros of f all lie in the circle $|z - \frac{1}{2}| < \frac{1}{2}$.

Since f is in $\mathcal{A}^{-\infty}$, there exists a constant λ such that the function $(1 - |z|)^\lambda f(z)$ is bounded in $\mathbb{D}$. It follows easily that the function $g(z) = (1 - z)^{2\lambda} f(z)$ is bounded in the disk $|z - \frac{1}{2}| < \frac{1}{2}$. Since the disk $|z - \frac{1}{2}| < \frac{1}{2}$ is mapped to the unit disk $\mathbb{D}$ by the mapping $w = 2z - 1$, the function $h(w) = g((w + 1)/2)$ is bounded and analytic in $\mathbb{D}$.

The zeros of h are $w_k = 2a_k - 1$, $k = 1, 2, 3, \ldots$. Since the points $\{a_k\}_k$ lie in an angle at $z = 1$, it follows that the $\{w_k\}_k$ lie in an angle at $w = 1$. Thus, the Blaschke condition $\sum_k (1 - |w_k|) < +\infty$ implies that $\sum_k |1 - w_k| < +\infty$. Since $|1 - w_k| = 2|1 - a_k|$, we obtain $\sum_k |1 - a_k| < +\infty$. By the triangle inequality, this implies that $\sum_k (1 - |a_k|) < +\infty$. ∎

To better formulate the main results about zero sets for Bergman-type spaces, we introduce two additional types of spaces. Thus, we set

$$\mathcal{A}_+^{-\alpha} = \bigcap_{\beta : \beta > \alpha} \mathcal{A}^{-\beta} = \{0\} \cup \{f \in H(\mathbb{D}) : t(f) \leq \alpha\}$$

and

$$\mathcal{A}_-^{-\alpha} = \bigcup_{\beta : \beta < \alpha} \mathcal{A}^{-\beta} = \{0\} \cup \{f \in H(\mathbb{D}) : t(f) < \alpha\}.$$

It is clear that

$$\mathcal{A}_-^{-\alpha} \subset \mathcal{A}_0^{-\alpha} \subset \mathcal{A}^{-\alpha} \subset \mathcal{A}_+^{-\alpha}.$$

We can now state the main results of this chapter; the next two sections are devoted to their proofs.

THEOREM 4.15 *Let $A = \{a_n\}_n$ be a sequence in $\mathbb{D}$. Then A is a zero set for $\mathcal{A}_+^{-\alpha}$ if and only if $D^+(A) \leq \alpha$.*

In concrete terms, we prove that the condition $D^+(A) \leq \alpha$ is necessary and the condition $D^+(A) < \alpha$ is sufficient for A to be an $\mathcal{A}^{-\alpha}$ zero set. This clearly implies the following.

COROLLARY 4.16 *A sequence $A = \{a_n\}_n$ in $\mathbb{D}$ is an $\mathcal{A}_-^{-\alpha}$ zero set if and only if $D^+(A) < \alpha$.*

COROLLARY 4.17 *A sequence $A \subset \mathbb{D}$ is a zero set for $\mathcal{A}^{-\infty}$ if and only if $D^+(A) < +\infty$.*

4.4 $\mathcal{A}^{-\alpha}$ Zero Sets, Necessary Conditions

We begin the proof of the necessity of the condition $D^+(A) \leq \alpha$ for $\mathcal{A}^{-\alpha}$ zero sets with the following *balayage-type estimate*, which enables us to "sweep" zeros of an analytic function f radially to the circumference $\mathbb{T}$ and convert them into singular masses without increasing $|f|$ in a certain critical region.

LEMMA 4.18 *Let $\mathfrak{s}_1$ be the standard Stolz angle at $z = 1$. Then*

$$\left| \frac{a - z}{1 - az} \right| \geq \exp\left[(\log a) \frac{1 - |z|^2}{|1 - z|^2} \right]$$

for all $0 < a < 1$ and $z \in \mathbb{D} \setminus \mathfrak{s}_1$.

Proof. Using the *Cayley transform*

$$w = \phi(z) = \frac{1+z}{1-z}$$

from $\mathbb{D}$ onto the right half-plane $\mathbb{C}_+ = \{w \in \mathbb{C} : \operatorname{Re} w > 0\}$, we can rewrite the desired inequality as

$$\left| \frac{b-w}{b+w} \right| \geq \exp\left[\left(\log \frac{b-1}{b+1} \right) u \right],$$

where $w = u + iv \in \mathbb{C}_+ \setminus \phi(\mathfrak{s}_1)$ and

$$b = \frac{1+a}{1-a} > 1.$$

We are going to take the logarithm on both sides of this second inequality and show that it actually holds for w in the larger set $\mathbb{C}_+ \setminus \Omega$, where

$$\Omega = \{w = u + iv : u > 1, |v| < u\}.$$

To see that Ω is smaller than $\phi(\mathfrak{s}_1)$, observe that $\partial(K^{-1}(\Omega))$ consists of parts of two orthogonal circles through 1 and -1 and an arc of the circle through 0 and 1 tangent to $\mathbb{T}$ at 1. Then it is geometrically obvious that $K^{-1}(\Omega) \subset \mathfrak{s}_1$.

We now show that

$$\frac{1}{u} \log \frac{b^2 + u^2 + v^2 + 2bu}{b^2 + u^2 + v^2 - 2bu} \leq 2 \log \frac{b+1}{b-1},$$

where $b > 1$ and $w = u + iv \in \mathbb{C}_+ \setminus \Omega$. It is easy to check that the left-hand side above decreases, for any fixed u, as $|v|$ increases; and for $v = 0$, it is an increasing function of u. Thus, the inequality above holds in the strip $0 \leq u \leq 1$ with equality attained at $u = 1$ and $v = 0$. It remains to verify the case $|v| = u$:

$$\frac{1}{u} \log \frac{b^2 + 2u^2 + 2bu}{b^2 + 2u^2 - 2bu} \leq 2 \log \frac{b+1}{b-1}$$

for $u \geq 1$.

Let $u = bt$. It then suffices to show that

$$\frac{1}{t} \log \frac{1 + 2t^2 + 2t}{1 + 2t^2 - 2t} \leq 2b \log \frac{b+1}{b-1}$$

for $b > 1$ and $t > 0$. The right-hand side here is decreasing in b and tends to 4 as $b \to +\infty$. So it is enough to show that

$$\frac{1}{t} \log \frac{1 + 2t^2 + 2t}{1 + 2t^2 - 2t} \leq 4$$

for all $t > 0$. An easy computation shows that the function

$$f(t) = 4t + \log(1 + 2t^2 - 2t) - \log(1 + 2t^2 + 2t)$$

has a positive derivative on $(0, +\infty)$:

$$f'(t) = 8t^2(1 + 2t^2)/(1 + 4t^4).$$

Thus,

$$f(t) > f(0) = 0, \qquad t > 0,$$

and the proof of the proposition is complete.   ∎

Given a finite subset E of the punctured disk $\mathbb{D} \setminus \{0\}$, we define the *push-out measure* $d\Lambda_E$:

$$d\Lambda_E = \sum_{z \in E} \log \frac{1}{|z|} d\delta_{z^*},$$

where $z^* = z/|z| \in \mathbb{T}$ is the the pushed-out point and $d\delta_\zeta$ stands for the unit point mass at $\zeta \in \mathbb{T}$. This measure is related to the counting function $\Lambda(A, E)$ which we met back in Section 4.2. It can also be defined for more general subsets E of $\mathbb{D}$. For a finite Borel measure μ on $\mathbb{T}$, we recall the definition of the Poisson extension

$$P[\mu](z) = \int_{\mathbb{T}} P(z, w) \, d\mu(w), \qquad z \in \mathbb{D},$$

where

$$P(z, w) = \frac{1 - |zw|^2}{|1 - z\overline{w}|^2}$$

is the Poisson kernel. Lemma 4.18 states that the following assertion holds for a one-point set A; the general case follows by iteration.

COROLLARY 4.19 *Suppose $f \in A^{-\alpha}$ and $A = \{a_1, \dots, a_n\} \subset \mathbb{D} \setminus \{0\}$ are some of the zeros of f. Let B_A be the Blaschke product associated with A, and let $A^* = \{a_1/|a_1|, \dots a_n/|a_n|\}$ be the pushed-out sequence on $\mathbb{T}$. Then*

$$\left| \frac{f(z)}{B_A(z)} \right| \leq \frac{\|f\|_{-\alpha}}{(1 - |z|^2)^\alpha} \exp\left(P[\Lambda_A](z) \right), \qquad z \in \mathbb{D} \setminus \mathsf{s}_{A^*}.$$

We shall need some estimates for several auxiliary harmonic functions. Recall from Section 4.2 that for a closed set F in $\mathbb{T}$,

$$d_{\mathbb{C}}(z, F) = \inf \{|z - \zeta| : \zeta \in F\}, \qquad z \in \overline{\mathbb{D}},$$

is the Euclidean distance from z to F. Also, recall that ds is the normalized arc-length measure on $\mathbb{T}$.

LEMMA 4.20 *Suppose F is a finite set in $\mathbb{T}$ and its complementary arcs $I_1, \dots, I_n$ satisfy $|I_k| = 2\pi |I_k|_s < 1$, for all $k = 1, \dots, n$. Then the harmonic function*

$$U_F(z) = \int_{\mathbb{T}} \frac{1 - |z|^2}{|\zeta - z|^2} \log \frac{1}{d_{\mathbb{C}}(\zeta, F)} \, ds(\zeta), \qquad z \in \mathbb{D},$$

is positive and satisfies

$$\log \frac{1}{d_{\mathbb{C}}(z, F)} \le U_F(z), \qquad z \in \mathbb{D}.$$

Proof. We have

$$\log \frac{1}{d_{\mathbb{C}}(z, F)} = \max_{\zeta \in F} \log \frac{1}{|z - \zeta|}, \qquad z \in \mathbb{D},$$

so that the left hand side expresses a positive subharmonic function on $\mathbb{D}$ whose boundary values equal those of $U_F(z)$. Hence the desired inequality follows from the maximum principle. ∎

For $0 < p < 1$, consider the harmonic function

$$V_p(z, \zeta) = \left(\sec \frac{p\pi}{2} \right) \operatorname{Re} (1 - \bar{\zeta}z)^{-p}, \qquad z \in \mathbb{D},$$

where ζ is a point on $\mathbb{T}$. The choice of the constant factor involving the secant function ensures that

$$|1 - z\bar{\zeta}|^{-p} \le V_p(z, \zeta), \qquad (z, \zeta) \in \mathbb{D} \times \mathbb{T}.$$

Also, for $\zeta \in \mathbb{T}$, and $0 < c < \frac{1}{4}$, let $\gamma(\zeta, p, c)$ be the curve

$$\gamma(\zeta, p, c) = \left\{ z \in \overline{\mathbb{D}} : 1 - |z|^2 = c\,|\zeta - z|^{2-p} \right\},$$

which makes one loop around the origin and touches the unit circle exactly at ζ. More generally, for a finite subset F of $\mathbb{T}$, we define the curve

$$\gamma(F, p, c) = \left\{ z \in \overline{\mathbb{D}} : 1 - |z|^2 = c\,d_{\mathbb{C}}(z, F)^{2-p} \right\},$$

which encloses a star-shaped domain touching the unit circle exactly at the points of F (see Figure 4.1).

We now compare the kernel $V_p(z, \zeta)$ to the Poisson kernel $P(z, \zeta)$.

LEMMA 4.21 *Fix $0 < p < 1$ and $0 < c < \frac{1}{4}$. Then, for fixed $\zeta \in \mathbb{T}$,*

$$\frac{1 - |z|^2}{|\zeta - z|^2} = P(z, \zeta) < c\,V_p(z, \zeta)$$

for all z in the region between $\mathbb{T}$ and $\gamma(\zeta, p, c)$.

Proof. In the region between $\gamma(\zeta, p, c)$ and $\mathbb{T}$, we have

$$1 - |z|^2 < c\,|\zeta - z|^{2-p},$$

and there

$$\frac{1 - |z|^2}{|\zeta - z|^2} < c\,|1 - \bar{\zeta}z|^{-p} \le c\,V_p(z, \zeta),$$

by the inequality we derived before the statement of the lemma. ∎

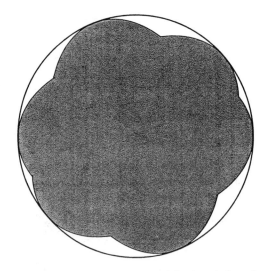

Figure 4.1. The curve $\gamma(F, p, c)$ and the domain it encloses

For a finite Borel measure μ on $\mathbb{T}$, let

$$V_p[\mu](z) = \int_{\mathbb{T}} V_p(z, \zeta)\, d\mu(\zeta), \qquad z \in \mathbb{D},$$

be corresponding potential, which represents a harmonic function on $\mathbb{D}$.

Nota bene: We restrict the parameters p and c to $0 < p < 1$ and $0 < c < \frac{1}{4}$, and assume that the finite set F has complementary arcs $\{I_k\}_k$ satisfying $|I_k| = 2\pi|I_k|_s < 1$ for all k.

LEMMA 4.22 *Let μ a finite positive Borel measure on $\mathbb{T}$, supported on a finite set F. Then the inequality*

$$P[\mu](z) \le c\, V_p[\mu](z)$$

holds for all z between $\mathbb{T}$ and the curve $\gamma(F, p, c)$.

Proof. The function $P[\mu]$ is a finite sum of Poisson kernels; apply Lemma 4.21 to each term. As the set of points between $\mathbb{T}$ and $\gamma(F, p, c)$ is the intersection of the domains described in Lemma 4.21 over $\zeta \in F$, the assertion is immediate. ∎

The key to our necessary conditions for $A^{-\alpha}$ zero sets is the following *Jensen-type inequality*. Recall the definition of the logarithmic sum

$$\Lambda(A, E) = \sum_j \left\{ \log \frac{1}{|a_j|} : a_j \in E \right\}$$

from Section 4.2, where $A = \{a_j\}_j$, counting multiplicities.

THEOREM 4.23 Let f be a nonzero function in $A^{-\alpha}$ having zeros (counting multiplicities) at $A = \{a_n\}_n$ with $0 \notin A$. Then, for any finite set F in $\mathbb{T}$,

$$\Lambda(A, \mathfrak{r}_F) - \alpha \log \Lambda(A, \mathfrak{r}_F)$$
$$\leq \alpha[\widehat{\kappa}(F) + \log \widehat{\kappa}(F)] - \alpha(\log \alpha - 2) + \log \|f\|_{-\alpha} - \log |f(0)|$$

whenever $4\alpha < \Lambda(A, \mathfrak{r}_F)\widehat{\kappa}(F)$.

Proof. We can assume A to be a finite sequence. By Corollary 4.19,

$$\log \left| \frac{f(z)}{B_{A \cap \mathfrak{r}_F}(z)} \right| \leq \log \|f\|_{-\alpha} + \alpha \log \frac{1}{1 - |z|^2} + P[\Lambda_{A \cap \mathfrak{r}_F}](z), \quad z \in \gamma(F, p, c),$$

where $B_{A \cap \mathfrak{r}_F}(z)$ is the Blaschke product for the zeros $A \cap \mathfrak{r}_F$ and the push-out measure $d\Lambda_{A \cap \mathfrak{r}_F}$ is as before. We now apply Lemmas 4.20 and 4.22, and use the geometric properties of the curve $\gamma(F, p, c)$, to obtain

$$\log \left| \frac{f(z)}{B_{A \cap \mathfrak{r}_F}(z)} \right| \leq \alpha(2 - p) U_F(z) + \alpha \log \frac{1}{c} + c V_p[\Lambda_{A \cap \mathfrak{r}_F}](z) + \log \|f\|_{-\alpha}$$

for $z \in \gamma(F, p, c)$; the function U_F is as in Lemma 4.20. The left-hand side here is a subharmonic function in the region enclosed by the curve $\gamma(F, p, c)$. Note that

$$\log |B_{A \cap \mathfrak{r}_F}(0)| = -\Lambda(A, \mathfrak{r}_F) \quad \text{and} \quad V_p[\Lambda_{A \cap \mathfrak{r}_F}](0) = \left(\sec \frac{p\pi}{2} \right) \Lambda(A, \mathfrak{r}_F).$$

Hence, by the maximum principle, we then have

$$\log \left| \frac{f(0)}{B_{A \cap \mathfrak{r}_F}(0)} \right| = \log |f(0)| + \Lambda(A, \mathfrak{r}_F)$$

$$\leq \alpha(2 - p) U_F(0) + \alpha \log \frac{1}{c} + c V_p[\Lambda_{A \cap \mathfrak{r}_F}](0) + \log \|f\|_{-\alpha}$$

$$= \alpha(2 - p) \int_{\mathbb{T}} \log \frac{1}{d_{\mathbb{C}}(\zeta, F)} ds(\zeta) + \alpha \log \frac{1}{c}$$
$$+ \left(c \sec \frac{p\pi}{2} \right) \Lambda(A, \mathfrak{r}_F) + \log \|f\|_{-\alpha},$$

By what we did in Section 4.2, the integral expression above is less than or equal to $\widehat{\kappa}(F)$, and it is elementary that

$$\sec \frac{p\pi}{2} < \frac{1}{1 - p}.$$

Thus,

$$\log |f(0)| \leq \alpha(2 - p)\widehat{\kappa}(F) + \left(\frac{c}{1 - p} - 1 \right) \Lambda(A, \mathfrak{r}_F) + \alpha \log \frac{1}{c} + \log \|f\|_{-\alpha}.$$

To minimize the right-hand side, we put

$$1 - p = \frac{1}{\widehat{\kappa}(F)}, \qquad c = \frac{\alpha}{\Lambda(A, \mathfrak{r}_F)\widehat{\kappa}(F)}.$$

The desired result then follows. ∎

Note that the result above implies that $\Lambda(A, \tau_F) < +\infty$ for every finite subset F of $\mathbb{T}$.

We now prove two necessary conditions for $\mathcal{A}^{-\alpha}$ zero sets.

THEOREM 4.24 *If $A = \{a_n\}_n$ is an $\mathcal{A}^{-\alpha}$ zero sequence, then*

$$\Sigma(A, \tau_F) \leq \alpha[\widehat{\kappa}(F) + 2\log\widehat{\kappa}(F)] + O(1),$$

where $O(1)$ stands for a quantity which is uniformly bounded independently of the finite nonempty subset F of $\mathbb{T}$.

Proof. Since

$$\frac{1}{2}(1 - t^2) < \log\frac{1}{t}, \qquad 0 < t < 1,$$

a comparison of the summation functions Σ and Λ shows that by Theorem 4.23,

$$\Sigma(A, \tau_F) - \alpha\log^+\Sigma(A, \tau_F) \leq \alpha[\widehat{\kappa}(F) + \log\widehat{\kappa}(F)] + O(1).$$

We readily find an "almost inverse" to the mapping $t \mapsto t - \alpha\log t$ for large positive t, which results in

$$\Sigma(A, \tau_F) \leq \alpha[\widehat{\kappa}(F) + 2\log\widehat{\kappa}(F)] + O(1),$$

as asserted. ∎

THEOREM 4.25 *If $A = \{a_n\}_n$ is an $\mathcal{A}^{-\alpha}$ zero sequence, then*

$$\Sigma(A, \mathfrak{s}_F) \leq \alpha[\widehat{\kappa}(F) + 3\log\widehat{\kappa}(F)] + O(1),$$

where $O(1)$ stands for a quantity which is uniformly bounded independently of the finite nonempty subset F of $\mathbb{T}$.

Proof. This is a direct consequence of the preceding theorem and Proposition 4.12. ∎

We derive two useful corollaries from the above necessary conditions.

COROLLARY 4.26 *Let $A = \{a_n\}_n$ be an $\mathcal{A}^{-\infty}$ zero sequence. Then*

$$S(r) = \sum_{|a_n| < r} (1 - |a_n|) = O\left(\log\frac{1}{1 - r}\right) \qquad \text{as } r \to 1^-,$$

and for each $\varepsilon > 0$, we have

$$\sum_n \frac{1 - |a_n|}{\left[\log\frac{e}{1 - |a_n|}\right]^{1+\varepsilon}} < +\infty.$$

Proof. Taking

$$F = \left\{\exp(2k\pi i/N) : 1 \leq k \leq N\right\}$$

in Theorem 4.25 and letting $N \to +\infty$ yields the first estimate, because the Stolz star $\mathfrak{s}_F$ will then cover a disk of radius $1 - \pi/N$, and a computation reveals that $\widehat{\kappa}(F) = 1 + \log N$. Since

$$\sum_n \frac{1 - |a_n|}{\left[\log \frac{e}{1-|a_n|}\right]^{1+\varepsilon}} = \int_0^1 \frac{dS(r)}{\left[\log \frac{e}{1-r}\right]^{1+\varepsilon}}$$

$$= S(0) + (1+\varepsilon) \int_0^1 \frac{S(r)\,dr}{(1-r)\left[\log \frac{e}{1-r}\right]^{2+\varepsilon}},$$

the second estimate then follows from the first one.   ∎

COROLLARY 4.27 *If A is an $\mathcal{A}^{-\alpha}$-zero sequence, then $D^+(A) \le \alpha$.*

4.5 $\mathcal{A}^{-\alpha}$ Zero Sets, a Sufficient Condition

In this section, we present a sufficient condition for a sequence A in $\mathbb{D}$ to be an $\mathcal{A}^{-\alpha}$ zero set. The proof of the main theorem consists of two key ideas: an "oblique" projection technique, and a technique from Linear Programming.

Throughout this section, we let $\mathfrak{s}_\zeta$ denote a Stolz angle with the vertex at $\zeta \in \mathbb{T}$ and an arbitrary but fixed aperture φ with $\pi/2 \le \varphi < \pi$. Thus, $\mathfrak{s}_\zeta$ is the convex hull of

$$\{\zeta\} \cup \{z \in \mathbb{C} : |z| \le \sin(\varphi/2)\},$$

with the vertex ζ removed. As before, for a finite subset F of $\mathbb{T}$,

$$\mathfrak{s}_F = \bigcup\{\mathfrak{s}_\zeta : \zeta \in F\}$$

is the corresponding Stolz star domain.

Given a point $\lambda \in \mathbb{D}$, contained in the annulus $\sin(\varphi/2) < |\lambda| < 1$, there are exactly two Stolz angles $\mathfrak{s}_\xi$ (with $\xi \in \mathbb{T}$) such that $\lambda \in \partial\mathfrak{s}_\xi$. Let ξ_1 and ξ_2 be the corresponding points of $\mathbb{T}$, which of course depend on λ. Given another point $\zeta \in \mathbb{T}$, we pick the one (out of ξ_1, ξ_2) which is the farthest away from ζ, and call it the *oblique projection* $\varpi_\zeta(\lambda)$ of λ. This can be done unless λ is on the straight line connecting ζ with $-\zeta$; however, we shall mainly be interested in $\lambda \in \mathbb{D} \setminus \mathfrak{s}_{\{\zeta,-\zeta\}}$. We also need the concept of a tent: for an open arc $I \subset \mathbb{T}$ with endpoints ω_1 and ω_2, we define the *tent* $\mathfrak{h}_I$ as the component of $\mathbb{D} \setminus \mathfrak{s}_{\{\omega_1,\omega_2\}}$ abutting on I. The geometric situation is illustrated in Figure 4.2.

LEMMA 4.28 *Fix the aperture of the Stolz angles $\varphi \in [3\pi/5, \pi)$. Then for all $z = t\zeta$, $0 < t < 1$, and $\lambda \in \mathbb{D} \setminus \mathfrak{s}_{\{\zeta,-\zeta\}}$, we have*

$$\log\left|\frac{\lambda - z}{1 - \bar{\lambda}z}\right| + \frac{(1 - |z|^2)(1 - |\lambda|^2)}{2|1 - \overline{\varpi}z|^2} \le 0,$$

where $\varpi = \varpi_\zeta(\lambda)$.

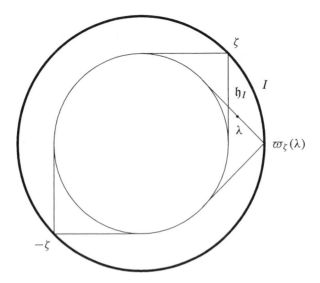

Figure 4.2. The oblique projection $\varpi_\zeta(\lambda)$

Note that the inequality above means that on the radius $\{z = t\zeta : 0 < t < 1\}$, the Blaschke factor $(\lambda - z)/(1 - \bar{\lambda}z)$ is dominated in modulus by the singular inner function $\exp[-\sigma(\varpi + z)/(\varpi - z)]$, where $\sigma = (1 - |\lambda|^2)/2$.

Proof. Using the identity

$$1 - \left| \frac{\lambda - z}{1 - \bar{\lambda}z} \right|^2 = \frac{(1 - |\lambda|^2)(1 - |z|^2)}{|1 - \bar{\lambda}z|^2}$$

we can rewrite the desired inequality as

$$\log(1 - 2\sigma a_2) + 2\sigma a_1 \le 0,$$

where

$$2\sigma = 1 - |\lambda|^2, \quad a_1 = \frac{1 - |z|^2}{|1 - \bar{z}\varpi|^2}, \quad a_2 = \frac{1 - |z|^2}{|1 - \bar{z}\lambda|^2}.$$

Since

$$
\begin{aligned}
\log(1 - 2\sigma a_2) + 2\sigma a_1 &= -\sum_{n=1}^{+\infty} \frac{(2\sigma a_2)^n}{n} + 2\sigma a_1 \\
&\le 2\sigma(a_1 - a_2) - \frac{(2\sigma a_2)^2}{2(1 - \sigma a_2)} \\
&= \frac{2\sigma(a_1 - a_2 - \sigma a_1 a_2)}{1 - \sigma a_2},
\end{aligned}
$$

it suffices for us to prove

$$a_1 - a_2 - \sigma a_1 a_2 \le 0,$$

which is equivalent to

$$\frac{1}{a_2} - \frac{1}{a_1} \le \sigma,$$

that is,

$$\left|\frac{1}{\bar{z}} - \lambda\right|^2 - \left|\frac{1}{\bar{z}} - \varpi\right|^2 \le \frac{1}{2}(1 - |\lambda|^2)\left(\frac{1}{|z|^2} - 1\right).$$

Let

$$\beta = |\arg(\zeta/\varpi)|, \qquad \gamma = |\arg(\zeta/\lambda)|,$$

where as usual the argument takes values in the interval $(-\pi, \pi]$. The definition of oblique projection implies that $0 < \beta/2 \le \gamma \le \beta < \pi$, and a geometric consideration reveals that

$$1 - |\lambda| \le (\beta - \gamma)\cos\frac{\varphi}{2} < \frac{1}{2}(\pi - \varphi)(\beta - \gamma).$$

Using the expansions

$$\left|\frac{1}{\bar{z}} - \lambda\right|^2 = \frac{1}{|z|^2} + |\lambda|^2 - 2\frac{|\lambda|}{|z|}\cos\gamma$$

and

$$\left|\frac{1}{\bar{z}} - \varpi\right|^2 = \frac{1}{|z|^2} + 1 - \frac{2}{|z|}\cos\beta,$$

we obtain the reformulation

$$\cos\beta - |\lambda|\cos\gamma \le \frac{1}{4}\left(|z| + \frac{1}{|z|}\right)(1 - |\lambda|^2).$$

Since

$$|z| + \frac{1}{|z|} > 2, \qquad z \in \mathbb{D} \setminus \{0\},$$

it is enough to prove

$$\cos\beta - |\lambda|\cos\gamma \le \frac{1}{2}(1 - |\lambda|^2), \qquad \lambda \in \mathbb{D} \setminus \mathfrak{s}_\zeta.$$

We can further assume $\beta < \pi/3$; otherwise, the above inequality holds for all $\lambda \in \mathbb{D}$. Solving the quadratic inequality, we are led to check that

$$0 \le 1 - |\lambda| \le (1 - \cos\gamma) + \left[(1 - \cos\gamma)^2 + 4\sin\frac{\beta - \gamma}{2}\sin\frac{\beta + \gamma}{2}\right]^{1/2}.$$

The right-hand side is actually greater than $(2/\pi)(\beta - \gamma)$. For $\frac{3}{5}\pi \le \varphi < \pi$, we have

$$1 - |\lambda| < \frac{1}{2}(\pi - \varphi)(\beta - \gamma) \le \frac{\pi}{5}(\beta - \gamma) < \frac{2}{\pi}(\beta - \gamma),$$

which completes the proof of the lemma.   ∎

In the remainder of this section, we assume that the aperture φ of the Stolz angles is chosen in the interval $[3\pi/5, \pi)$, so that the conclusion of Lemma 4.28 holds true. For instance, we can pick $\varphi = 3\pi/5$. Given an arc I of the circle $\mathbb{T}$, let $\kappa(I)$ be the quantity

$$\kappa(I) = |I|_s \log \frac{e}{|I|_s}.$$

DEFINITION 4.29 *Suppose $A = \{a_n\}_n$ is a finite sequence in $\mathbb{D}$, w_0 is a point in $\mathbb{T}$, and α is a positive number. A positive Borel measure μ on $\mathbb{T}$ is (A, α, w_0)-admissible if*

(i) $\mu(\{w_0\}) = 0$;

(ii) for each open arc $I \subset \mathbb{T}$, with $w_0 \notin I$, the following inequality holds:

$$0 \le \mu(I) \le \alpha \kappa(I) + \Sigma(A, \mathfrak{h}_I),$$

where $\mathfrak{h}_I$ is the tent associated with I.

The set of all (A, α, w_0)-admissible measures will be denoted by $\mathcal{M}(A, \alpha, w_0)$, or just $\mathcal{M}$. The second condition above, (ii), clearly implies that $\mu(\{\zeta\}) = 0$ for any $\zeta \in \mathbb{T}$, not just for $\zeta = w_0$.

LEMMA 4.30 *Suppose $0 < \alpha < +\infty$, $w_0 \in \mathbb{T}$, and $A = \{a_n\}_n$ is a finite sequence in $\mathbb{D}$. Then*

$$\sup\{\mu(\mathbb{T}) : \mu \in \mathcal{M}\} = \inf\{\alpha\widehat{\kappa}(F) + \Sigma(A, \mathbb{D} \setminus \mathfrak{s}_F) : F \subset \mathbb{T} \text{ finite}, w_0 \in F\},$$

where $\mathcal{M} = \mathcal{M}(A, \alpha, w_0)$ is the set of all (A, α, w_0)-admissible measures. Furthermore, there is at least one maximal admissible measure μ_0 for which

$$\mu_0(\mathbb{T}) = \inf\{\alpha\widehat{\kappa}(F) + \Sigma(A, \mathbb{D} \setminus \mathfrak{s}_F) : F \subset \mathbb{T} \text{ finite}, w_0 \in F\}.$$

Proof. The set $\mathbb{D} \setminus \mathfrak{s}_F$ is a disjoint union of tents of the kind $\mathfrak{h}_I$, with $w_0 \notin I$, so by the definition of the (A, α, w_0)-admissible measures, the "sup" on the left hand side is less than or equal to the "inf" on the right hand side.

Define a finite set F_0 consisting of w_0 and all those points on $\mathbb{T}$ which are "oblique projections" of points of A in the annulus $\sin(\varphi/2) < |z| < 1$. Here, that a point $\zeta \in \mathbb{T}$ is an "oblique projection" of a set $B \subset \mathbb{D}$ means that $(\partial \mathfrak{s}_\zeta) \cap B \ne \emptyset$. Let $\{I_k\}_1^N$ be the complementary arcs of F_0. The point w_0 acts as a divider; it permits us to order the arcs I_k according to their position relative w_0.

For every $\mu \in \mathcal{M}$, let $\tilde\mu$ denote the measure with constant density on each I_k and such that $\tilde\mu(I_k) = \mu(I_k)$ for all k. We claim that $\mu \in \mathcal{M}$ implies $\tilde\mu \in \mathcal{M}$. In fact, if the endpoints e^{it_1} and e^{it_2} of some arc I are not in F_0, then the right-hand side of the inequality in Definition 4.29 is a locally continuous and concave function of t_1 and t_2, and so replacing μ by $\tilde\mu$ will not invalidate that inequality. For the same reason, to ascertain that a measure $\mu = \tilde{\mu}$ is in $\mathcal{M}$, it is enough to check the inequality in Definition 4.29 only for arcs I with $w_0 \notin I$, whose endpoints are in F_0. Each such measure is described by a vector $x = (x_1, \ldots, x_N)$, where $N = \operatorname{card} F_0$ and $x_k = \mu(I_k)$. We are thus led to a standard optimization problem from *Linear Programming*: maximize the functional

$$L(x) = x_1 + \cdots + x_N,$$

where the positive vector $x = (x_1, \ldots, x_N)$ satisfies $N(N+1)/2$ restrictions of the type

$$x_k + x_{k+1} + \cdots + x_l \le b_{k,l}, \quad 1 \le k \le l \le N,$$

which correspond to the inequality in Definition 4.29 with arcs I whose endpoints are in F_0. Written out, the quantities $b_{k,l}$ are

$$b_{k,l} = \alpha \, \kappa(I_{k,l}) + \Sigma(A, \mathfrak{h}_{I_{k,l}}),$$

where $I_{k,l}$ is the arc obtained by filling in finitely many points in the union $I_k \cup I_{k+1} \cup \cdots \cup I_l$. We will refer to this as the *optimization problem*. Let $\mathcal{C}$ denote the closed convex polyhedron in $\mathbb{R}^N$ defined by the above-mentioned restrictions

$$x_k + x_{k+1} + \cdots + x_l \le b_{k,l}, \quad 1 \le k \le l \le N,$$

and denote by $\mathcal{C}_+$ its intersection with $\mathbb{R}_+^N$; $\mathbb{R}_+^N$ stands for the N-fold Cartesian product of the half-axis $\mathbb{R}_+ = [0, +\infty)$.

The "inf" over all finite subsets $F \subset \mathbb{T}$ appearing in the formulation of the lemma can only get bigger if we restrict F to be subsets of the "obliquely projected" set F_0, so it is clearly enough to prove the equality under the additional restriction $F \subset F_0$. In fact, one can argue that only subsets of F_0 have a chance of being extremal for the "inf". Thus, in terms of the optimization problem stated earlier, the assertion of the lemma can now be reformulated as follows:

$$\max\{L(x) : x \in \mathcal{C}_+\} = \min \sum_\nu b_{k_\nu, l_\nu},$$

where the minimum is taken over all simple coverings $\{[k_\nu, l_\nu]\}_\nu$ *of* $\mathbb{N}_N = \{1, 2, \ldots, N\}$. We will refer to this as the *min-max equation*. Note that we here deviate from standard notation and let $[k, l]$ stand for an interval consisting of integers and not of reals.

It is at least clear that on $\mathcal{C}_+$, $L(x)$ assumes its maximum somewhere. We claim that the maximum is in fact assumed at some point $x = (x_1, \ldots, x_N) \in \mathcal{C}_+$ with $x_j > 0$ for all $j = 1, \ldots, N$. To this end, take a point $x \in \mathcal{C}_+$, with $x_j = 0$ for some j. There may be a few zero slots clustering together, so say that $x_j = 0$ on the "interval" $k < j < l$, but that at the end points we have $x_k > 0$ and $x_l > 0$. For a small parameter $\varepsilon > 0$, consider the point

$$x' = (x_1, \cdots, x_{k-1}, x_k - \varepsilon(l - k - 1), \varepsilon, \cdots, \varepsilon, x_l, \cdots, x_N).$$

We now use a property of the given quantities $b_{k,l}$, namely that they are positive and strictly monotonically increasing in the interval $[k, l]$: $b_{k,l} < b_{k',l'}$ whenever

$[k, l]$ is strictly contained in $[k', l']$. It follows that the competing point x' is in C_+ for sufficiently small ε, and moreover, $L(x') = L(x)$. If x is the point where $L(x)$ assumes its maximum, we treat all clusters of zeros the same way, and find a (perhaps different) point $x' \in C_+$ with $L(x')$ maximal, and $x'_j > 0$ for all $j = 1, \ldots, N$.

We next claim that $L(x) \leq L(x')$ for all $x \in C$. By what we have done so far, it is so for all $x \in C_+$. Suppose for the moment that at some point $x^0 \in C$, the inequality $L(x^0) > L(x')$ holds. Then we consider points x close to x' along the line segment connecting x' with x^0. Such x will be in C by convexity, and they are in $\mathbb{R}_+^N$, and hence in C_+. The value of $L(x)$ must then be slightly bigger than $L(x')$, a contradiction.

We can now apply the standard *duality theorem of Linear Programming* due to Gale, Kuhn, and Tucker [48], [127, p. 28]. To formulate the result, we write the $N(N + 1)/2$ inequalities defining C as

$$\langle x, e^j \rangle \leq b_j, \qquad j = 1, 2, \ldots, N(N + 1)/2,$$

where b_j equals $b_{k,l}$ for the pair (k, l) numbered by j, and similarly, e^j stands for the vector $(0, \ldots, 0, 1, \ldots, 1, 0, \ldots, 0)$ in $\mathbb{R}^N$, with 1's precisely on the interval $[k, l]$ associated with the index j. Here, $\langle \cdot, \cdot \rangle$ is the usual inner product of $\mathbb{R}^N$:

$$\langle x, y \rangle = x_1 y_1 + \ldots + x_N y_N, \qquad x = (x_1, \ldots, x_N), \ y = (y_1, \ldots, y_N).$$

We also write $L(x) = \langle x, L \rangle$, where $L = (1, 1, \ldots, 1)$. The assertion of the duality theorem is

$$\max \left\{ \langle x, L \rangle : x \in C_+ \right\} = \max \left\{ \langle x, L \rangle : x \in C \right\}$$

$$= \min \left\{ \sum_j \theta_j b_j : \theta_j \in \mathbb{R}_+ \text{ for all } j, \ \sum_j \theta_j e^j = L \right\}.$$

The min-max equation we encountered earlier claims that the above minimum is achieved with coefficients $\theta_j \in \{0, 1\}$. The points $\theta = (\theta_1, \ldots, \theta_{N(N+1)/2}) \in \mathbb{R}_+^{N(N+1)/2}$ with

$$\sum_j \theta_j e^j = L$$

constitute – by inspection of the vectors involved (the e^j's and L) – a closed convex lower-dimensional polyhedron S contained in the cube $[0, 1]^{N(N+1)/2}$. We show that the polyhedron S is the (closed) convex hull of "edge points" $\theta \in S$ of the type that $\theta_j \in \{0, 1\}$ for every j. The min-max equation then follows easily. Points θ with positive rational coordinates are dense in S, and it suffices to obtain that they are in the convex hull of the "edge points". Multiplying by the least common denominator n of the positive rationals $\theta_1, \ldots, \theta_{N(N+1)/2}$, we have

$$\sum_j \vartheta_j e^j = nL, \tag{4.1}$$

where $\vartheta_j = n\theta_j \in \mathbb{Z}_+$. Here, $\mathbb{Z}_+ = \{1, 2, 3, \dots\}$ stands for the set of all positive integers. We interpret the above situation in terms of coverings. Let $\mathcal{J}$ stand for the set of all closed intervals $J = [k, l]$ in the integers $\mathbb{Z}$ whose endpoints are integers satisfying $1 \le k \le l \le N$. A system $\mathcal{P} = \{J_v\}_v = \{[k_v, l_v]\}_v$ of such intervals (repetitions are allowed) is called an n-*fold covering* (or n-covering, to shorten the notation) of $\mathbb{N}_N = \{1, 2, \dots, N\}$ if every $n \in \mathbb{N}_N$ belongs to exactly n intervals from $\mathcal{P}$ (if $n = 1$, we speak of a *simple* covering). In (4.1), we have an n-fold covering of $\mathbb{N}_N$ supplied by the various support intervals of the coordinates of the vectors e^j, with multiplicities as expressed by θ_j. We now claim:

Every n-covering $\mathcal{P}$ of $\mathbb{N}_N$ is the union of n simple coverings. In fact, every interval $J = [k, l] \in \mathcal{P}$ with $l < N$ has the property that $l + 1$ is covered n times by $\mathcal{P} \setminus \{J\}$ while l is covered only $n - 1$ times. This is possible only if there is an interval in $\mathcal{P} \setminus \{J\}$ whose left endpoint is $l + 1$. The rest is done by induction.

This means that the integer-valued vector $\vartheta = (\vartheta_1, \dots, \vartheta_{N(N+1)/2})$ can be written as a sum of n vectors of the type $\epsilon = (\epsilon_1, \dots, \epsilon_{N(N+1)/2})$, where $\epsilon_j \in \{0, 1\}$ for all j and

$$\sum_j \epsilon_j e^j = L;$$

each ϵ is then an "edge point" of S. That is, θ is a convex combination of "edge points", as claimed. The proof is complete. ∎

The reason why we introduced the splitting point $w_0 \in \mathbb{T}$ is that without it, we cannot assert that an n-covering is the union of n simple coverings, a technical point needed in the proof of the lemma. For example, there is a 2-covering of $\{1, 2, 3\}$ – made cyclic by declaring that after 3 comes again 1 – which cannot be decomposed as the union of two simple coverings.

We can now prove the main result of this section.

THEOREM 4.31 *Suppose $A = \{a_n\}_n$ is a sequence in $\mathbb{D}$. Suppose*

$$\Sigma(A, \mathfrak{s}_F) \le \alpha \widehat{\kappa}(F) + O(1)$$

holds for all finite subsets F of $\mathbb{T}$, where $O(1)$ is bounded independently of F. Then A is an $\mathcal{A}^{-\alpha}$ zero sequence.

Proof. Without loss of generality, we can assume that $0 \notin A$. Let A_0 be a *finite* subsequence of A. Now we choose an arbitrary $w_0 \in \mathbb{T}$, construct as in Lemma 4.30 a maximal (A_0, α, w_0)-admissible measure μ_0, and form the function

$$f_0(z) = B_{A_0}(z)\, \Phi(z),$$

where B_{A_0} is the Blaschke product for A_0 and Φ is the outer function

$$\Phi(z) = \exp\left\{ \int_{\mathbb{T}} \frac{\xi + z}{\xi - z} \, d\mu_0(\xi) \right\}.$$

We are going to obtain an upper estimate for $\|f_0\|_{-\alpha}$ and a lower estimate for $|f_0(0)|$, both independent of $A_0 \subset A$. To this end, we fix a point $\zeta \in \mathbb{T}$ and consider two subsequences of A_0: $A_0' = A_0 \cap \mathfrak{s}_{\{\zeta, -\zeta\}}$ and $A_0'' = A_0 \setminus A_0'$. Let $B_{A_0'}$ and $B_{A_0''}$ be the Blaschke products for A_0' and A_0'', respectively. For each $a_n \in A_0''$, let $\varpi_n = \varpi_\zeta(a_n)$ be its oblique projection. Form an atomic measure σ on $\mathbb{T}$ by placing at each ϖ_n a point mass of magnitude $\sigma_n = \frac{1}{2}(1 - |a_n|^2)$, and let $\Psi = \Phi S_\sigma$, where S_σ is the singular inner function

$$S_\sigma(z) = \exp\left\{-\int_\mathbb{T} \frac{\xi + z}{\xi - z} \, d\sigma(\xi)\right\}.$$

From its definition, we see that the measure σ has

$$\sigma(I) = \mathbf{\Sigma}(A_0'', \mathfrak{h}_I) = \mathbf{\Sigma}(A_0, \mathfrak{h}_I)$$

for each open arc I in the punctured circle $\mathbb{T} \setminus \{\zeta, -\zeta\}$. The (A_0, α, w_0)-admissibility of μ_0 means that

$$\mu_0(I) \le \alpha \kappa(I) + \mathbf{\Sigma}(A_0, \mathfrak{h}_I)$$

for any open arc I in $\mathbb{T} \setminus \{w_0\}$. We need this inequality for arcs that contain the point w_0, too. This is achieved by the following argument, if we pay a small price. If we partition an arc $I \subset \mathbb{T}$ into two arcs I_1 and I_2, then

$$|I|_s \log \frac{e}{|I|_s} \le |I_1|_s \log \frac{e}{|I_1|_s} + |I_2|_s \log \frac{e}{|I_2|_s} \le |I|_s \log \frac{e}{|I|_s} + (\log 2)|I|_s.$$

This implies that

$$\mu_0(I) \le \alpha \kappa(I) + \alpha(\log 2)|I|_s + \mathbf{\Sigma}(A_0, \mathfrak{h}_I)$$

holds for all arcs I, also those containing the point w_0.

The boundary measure for the zero-free function Ψ is $\mu_0 - \sigma$, and putting the above observations together, we have

$$(\mu_0 - \sigma)(I) \le \alpha \kappa(I) + \alpha(\log 2)|I|_s$$

for every arc I in $\mathbb{T} \setminus \{\zeta, -\zeta\}$. We apply this to arcs having ζ as one endpoint, Using integration by parts (see Exercise 1), we derive from this

$$|\Psi(z)| \le \frac{C}{(1 - |z|^2)^\alpha}, \qquad z = t\zeta,$$

for $0 \le t < 1$, where the constant $C = C(\alpha)$ only depends on α. At this point, we apply Lemma 4.28, to get

$$|B_{A_0''}(z)| \le |S_\sigma(z)|, \qquad z = t\zeta,$$

for $0 \le t < 1$. Since $|B_{A_0}(z)| \le |B_{A_0''}(z)|$, we obtain

$$\begin{aligned}
|f_0(z)| &= |B_{A_0}(z)\Phi(z)| \le |B_{A_0''}(z)\Phi(z)| \\
&\le |\Psi(z)| \le \frac{C}{(1 - |z|^2)^\alpha}, \qquad z = t\zeta,
\end{aligned}$$

where $0 \le t < 1$. The point $\zeta \in \mathbb{T}$ is arbitrary, and hence $\|f_0\|_{-\alpha} \le C$.

We note that

$$\log |B_{A_0}(0)| = -\mathbf{\Lambda}(A_0, \mathbb{D}) \qquad \text{and} \qquad \log |\Phi(0)| = \mu_0(\mathbb{T}),$$

where the logarithmic sum function $\mathbf{\Lambda}$ is as in Section 4.2, so that for the function $f_0 = B_{A_0}\Phi$, we have

$$\log |f_0(0)| = -\mathbf{\Lambda}(A_0, \mathbb{D}) + \mu_0(\mathbb{T}).$$

By Lemma 4.30 and the maximality of μ_0,

$$\mu_0(\mathbb{T}) = \inf \left\{ \alpha \, \widehat{\kappa}(F) + \mathbf{\Sigma}(A, \mathbb{D} \setminus \mathfrak{s}_F) : F \subset \mathbb{T} \text{ finite}, \, w_0 \in F \right\}.$$

We obtain

$$
\begin{aligned}
\log \frac{1}{|f_0(0)|} &= \mathbf{\Lambda}(A_0, \mathbb{D}) - \mu_0(\mathbb{T}) \\
&= -\inf \left\{ \alpha \, \widehat{\kappa}(F) + \mathbf{\Sigma}(A, \mathbb{D} \setminus \mathfrak{s}_F) : F \subset \mathbb{T} \text{ finite}, \, w_0 \in F \right\} \\
&\quad + \mathbf{\Sigma}(A_0, \mathbb{D}) + \left[\mathbf{\Lambda}(A_0, \mathbb{D}) - \mathbf{\Sigma}(A_0, \mathbb{D}) \right] \\
&= \sup \left\{ \mathbf{\Sigma}(A, \mathfrak{s}_F) - \alpha \, \widehat{\kappa}(F) : F \subset \mathbb{T} \text{ finite}, \, w_0 \in F \right\} \\
&\quad + \left[\mathbf{\Lambda}(A_0, \mathbb{D}) - \mathbf{\Sigma}(A_0, \mathbb{D}) \right].
\end{aligned}
$$

Since

$$0 \leq \log \frac{1}{t} - \frac{1}{2}(1 - t^2) = O[(1 - t)^2] \qquad \text{as } t \to 1,$$

and the assumption on the sequence $A = \{a_n\}_n$ easily implies (see the proof of Corollary 4.26)

$$\sum_n (1 - |a_n|)^2 < +\infty,$$

we have that

$$\mathbf{\Lambda}(A_0, \mathbb{D}) - \mathbf{\Sigma}(A_0, \mathbb{D}) = O(1),$$

with a bound that is independent of which particular finite subsequence A_0 we have picked. From the assumption of the theorem, we thus have

$$\log \frac{1}{|f_0(0)|} = O(1),$$

with a bound independent of $A_0 \subset A$.

Now, take a nested sequence of finite subsets of A, $A_1 \subset A_2 \subset A_3 \subset \cdots$, with $A = \cup_n A_n$, and construct as above functions f_n for each A_n. The functions $\{f_n\}_n$ form a normal family. Hence there is a subsequence $\{f_{n_k}\}_k$ converging to an analytic function f uniformly on compact subsets of $\mathbb{D}$; the function f is in $\mathcal{A}^{-\alpha}$ and its zero sequence is A. ∎

COROLLARY 4.32 *Suppose A is a sequence in $\mathbb{D}$ with $D^+(A) < \alpha$. Then A is a zero set for $\mathcal{A}^{-\alpha}$.*

4.6   Zero Sets for A_α^p

In this section, we consider zero sets for the standard weighted Bergman spaces A_α^p. The main work was done in the previous sections; we only have to take care of some minor technical points here.

We begin with a Blaschke-type product that can be used to divide out zeros of functions in weighted Bergman spaces. In the unweighted case, this product is not really necessary, since the contractive zero divisors will do the job even better.

Recall that the Blaschke factor induced by a single point a in $\mathbb{D}$ is defined as

$$B_a(z) = \frac{|a|}{a} \frac{a-z}{1 - \bar{a}z}, \qquad z \in \mathbb{D};$$

for $a = 0$ we set $B_0(z) = z$.

PROPOSITION 4.33 *Suppose $A = \{a_n\}_n$ is a sequence of points in $\mathbb{D}$ with*

$$\sum_{n=1}^{+\infty} (1 - |a_n|^2)^2 < +\infty.$$

Then the product

$$H_A(z) = \prod_{n=1}^{+\infty} B_{a_n}(z) \left[2 - B_{a_n}(z) \right]$$

converges uniformly on every compact subset of $\mathbb{D}$; the zero set of H_A is exactly A (counting multiplicities); and the function H_A is independent of the order of the factors.

Proof. Without loss of generality, we may assume $a_n \neq 0$ for every n. Then it is easy to check that

$$\left| 1 - B_{a_n}(z) \left(2 - B_{a_n}(z) \right) \right| = \left| 1 - B_{a_n}(z) \right|^2$$
$$= \left| \frac{a_n + z|a_n|}{a_n(1 - \bar{a}_n z)} \right|^2 (1 - |a_n|)^2.$$

The desired results now follow from standard facts about the convergence of infinite products of analytic functions. ∎

The function H_A is known as the *Horowitz product*.

LEMMA 4.34 *Suppose $0 < p < +\infty$ and $-1 < \alpha < +\infty$. Let $f \in A_\alpha^p$ be a function with $f(0) \neq 0$, and let $A = \{a_n\}_n$ be its sequence of zeros, counting multiplicities. Then there exists a positive constant $C = C(p, \alpha)$ such that*

$$\frac{|f(0)|}{\prod_{n=1}^{+\infty} |a_n|(2 - |a_n|)} \leq C \|f\|_{p,\alpha}.$$

Proof. It is clear that we may assume $f(0) = 1$. Let $n = n_f$ and $N = N_f$ be the usual counting functions associated with f. Consider the expression

$$S = \sum_{n=1}^{+\infty} \log \frac{1}{|a_n|(2 - |a_n|)} = \int_0^1 \log \frac{1}{r(2 - r)} \, dn(r).$$

By Proposition 4.6 and integration by parts (twice),

$$S = 2 \int_0^1 \frac{N(r) \, dr}{(2 - r)^2}.$$

Since $f(0) = 1$, Jensen's formula gives

$$N(r) = \frac{1}{2\pi} \int_0^{2\pi} \log |f(re^{i\theta})| \, d\theta.$$

It follows that

$$\begin{aligned} S p &= \int_{\mathbb{D}} \log |f(z)|^p \frac{dA(z)}{|z|(2 - |z|)^2} \\ &= C_1 + \int_{\mathbb{D}} \log \left(|f(z)|^p (1 - |z|^2)^\alpha \right) d\mu(z), \end{aligned}$$

where

$$C_1 = 2 \int_0^1 \frac{1}{r(2 - r)^2} \log \frac{1}{(1 - r^2)^\alpha} \, dr,$$

and

$$d\mu(z) = \frac{dA(z)}{|z|(2 - |z|)^2}$$

is a probability measure on $\mathbb{D}$. The desired result then follows from the arithmetic-geometric mean inequality; see Proposition 4.3. ∎

Since $|a|(2 - |a|) \leq 1$ for all $a \in \mathbb{D}$, we see that the lemma above remains true if we replace the zero sequence $\{a_n\}_n$ by any of its subsequences.

THEOREM 4.35 *Suppose $0 < p < +\infty$ and $-1 < \alpha < +\infty$. Then there exists a positive constant $C = C(p, \alpha)$ such that for every $f \in A_\alpha^p$ that has A as its zero set, we have $\|f / H_A\|_{\alpha, p} \leq C \|f\|_{\alpha, p}$.*

Proof. Let f be a function in A_α^p with zero set $A = \{a_n\}_n$. For every $w \in \mathbb{D} \setminus A$, let $f_w = f \circ \varphi_w$, where

$$\varphi_w(z) = \frac{w - z}{1 - \overline{w}z}, \qquad z \in \mathbb{D}.$$

Then f_w is in A_α^p and its zero set is $\{\varphi_w(a_n)\}_n$, which does not contain 0. Fix any $\beta > \alpha$, and apply Lemma 4.34 to the function f_w. Then

$$\frac{|f(w)|}{\prod_{n=1}^{+\infty} |\varphi_w(a_n)|(2 - |\varphi_w(a_n)|)} \leq C \|f_w\|_{p, \beta},$$

where C is a positive constant depending only on p, α, and β. Since

$$|\varphi_w(a)|(2 - |\varphi_w(a)|) \leq |B_a(w)(2 - B_a(w))|$$

for all $a, w \in \mathbb{D}$, we obtain

$$\left| \frac{f(w)}{H_A(w)} \right|^p \leq C^p(\beta + 1) \int_{\mathbb{D}} |f_w(z)|^p (1 - |z|^2)^\beta \, dA(z)$$

$$= C^p(\beta + 1) \int_{\mathbb{D}} |f(z)|^p \frac{(1 - |z|^2)^\beta (1 - |w|^2)^{\beta + 2}}{|1 - \overline{w}z|^{2\beta + 4}} \, dA(z)$$

for all w not in A. By continuity, the above also holds for other w's in $\mathbb{D}$. The desired norm estimate now follows from Fubini's theorem and Theorem 1.7. ∎

Again, the theorem above remains true if A is replaced by a subsequence of A. The next corollary is then obvious.

COROLLARY 4.36 *Suppose $0 < p < +\infty$ and $-1 < \alpha < +\infty$. Then any subset of an A_α^p-zero set is also an A_α^p-zero set.*

We now derive some very sharp conditions that are necessary or sufficient for a sequence to be an A_α^p-zero set. For $-1 < \alpha < +\infty$ and $0 < p < +\infty$, let

$$A_\alpha^{p+} = \bigcup_{q:p<q} A_\alpha^q$$

and

$$A_\alpha^{p-} = \bigcap_{q:q<p} A_\alpha^q.$$

THEOREM 4.37 *Suppose $0 < p < +\infty$, $-1 < \alpha < +\infty$, and that A is a sequence in $\mathbb{D}$. Then A is a zero set for A_α^{p-} if and only if $D^+(A) \leq (1 + \alpha)/p$.*

Proof. If $D^+(A) \leq (1 + \alpha)/p$, then, by Theorem 4.15, A is a zero set for $\mathcal{A}_+^{-\beta}$, where $\beta = (1 + \alpha)/p$. Since $\mathcal{A}_+^{-\beta} \subset A_\alpha^{p-}$, we conclude that A is a zero set for A_α^{p-}.

Conversely, if A is a zero set for A_α^{p-}, then A is a zero set for A_α^q for every $q < p$. Let G_q be an extremal function for the invariant subspace of A_α^q generated by the sequence A. Then $G_q \in \mathcal{A}^{-(1+\alpha)/q}$, by Theorem 3.3. Let A_q be the zero set of G_q. Then $A \subset A_q$, and hence by Theorem 4.15,

$$D^+(A) \leq D^+(A_q) \leq \frac{1 + \alpha}{q}.$$

Letting $q \to p^-$, we arrive at $D^+(A) \leq (1 + \alpha)/p$. ∎

COROLLARY 4.38 *Suppose $0 < p < +\infty$, $-1 < \alpha < +\infty$, and A is a sequence in $\mathbb{D}$. Then A is a zero set for A_α^{p+} if and only if $D^+(A) < (1 + \alpha)/p$.*

Note that the results above simply state that the condition $D^+(A) \leq (1+\alpha)/p$ is necessary and the condition $D^+(A) < (1+\alpha)/p$ is sufficient for A to be an A_α^p-zero set.

4.7 The Bergman-Nevanlinna Class

In this section, we consider the Bergman-Nevanlinna class A_α^0, $-1 < \alpha < +\infty$, consisting of analytic functions f in $\mathbb{D}$ with

$$\int_{\mathbb{D}} \log^+ |f(z)| \, (1 - |z|^2)^\alpha \, dA(z) < +\infty,$$

where $\log^+ x = \log x$ if $x \geq 1$ and $\log^+ x = 0$ if $0 < x < 1$. Our main purpose is to characterize the zero sets of A_α^0 by a Blaschke-type condition. The spaces A_α^0 appear in the limit as $p \to 0$ of the weighted Bergman spaces A_α^p, in the sense of

$$\lim_{p \to 0} \frac{t^p - 1}{p} = \log t, \qquad 0 < t < +\infty.$$

We begin with the elementary factors from the classical Weierstrass factorization theory for entire functions. For any nonnegative integer N, let E_N be the entire function defined by

$$E_N(z) = (1 - z) \exp\left(z + \frac{z^2}{2} + \cdots + \frac{z^N}{N} \right), \qquad z \in \mathbb{C}.$$

It is easy to see that its derivative is

$$E_N'(z) = -z^N \exp\left(z + \frac{z^2}{2} + \cdots + \frac{z^N}{N} \right).$$

Since $E_N(0) = 1$, we must have

$$E_N(z) = 1 + z^{N+1} F_N(z),$$

where F_N is an entire function. This proves the following lemma.

LEMMA 4.39 *For every nonnegative integer N, there exists a positive constant C such that*

$$|E_N(z)| \leq 1 + C\,|z|^{N+1}, \qquad \left|1 - E_N(z)\right| \leq C\,|z|^{N+1},$$

for all $|z| \leq 2$.

We can now prove the main result of this section.

THEOREM 4.40 *Fix $-1 < \alpha < +\infty$. A sequence $A = \{a_n\}_n$ in $\mathbb{D}$ is the zero set of a function in A_α^0 if and only if*

$$\sum_{n=1}^{+\infty} (1 - |a_n|)^{\alpha+2} < +\infty.$$

Proof. First, assume $f \in A^0_\alpha$ with zero set $A = \{a_n\}_n$. To simplify the use of Jensen's formula, we may as well assume $f(0) = 1$. In this case, Jensen's formula yields

$$\sum_n \log \frac{r}{|a_n|} \chi_{[|a_n|, 1)}(r) = \frac{1}{2\pi} \int_0^{2\pi} \log |f(re^{it})| \, dt \le \frac{1}{2\pi} \int_0^{2\pi} \log^+ |f(re^{it})| \, dt.$$

Integrating the inequality over $[0, 1)$ with respect to $2r(1 - r^2)^\alpha$ and applying Fubini's theorem to the left-hand side, we obtain

$$\sum_n \int_{|a_n|}^1 2r(1 - r^2)^\alpha \log \frac{r}{|a_n|} \, dr \le \int_{\mathbb{D}} \log^+ |f(z)| (1 - |z|^2)^\alpha \, dA(z).$$

By integration by parts, the integral on the left-hand side equals

$$\frac{1}{\alpha + 1} \int_{|a_n|}^1 (1 - r^2)^{\alpha+1} \frac{dr}{r},$$

which is obviously greater than

$$\frac{1}{\alpha + 1} \int_{|a_n|}^1 (1 - r)^{\alpha+1} \, dr = \frac{(1 - |a_n|)^{\alpha+2}}{(\alpha + 1)(\alpha + 2)}.$$

This shows that

$$\sum_n (1 - |a_n|^2)^{\alpha+2} \le (\alpha + 1)(\alpha + 2) \int_{\mathbb{D}} \log^+ |f(z)| (1 - |z|^2)^\alpha \, dA(z) < +\infty.$$

Next, assume that $A = \{a_n\}_n$ is a sequence in $\mathbb{D}$ satisfying

$$\sum_{n=1}^{+\infty} (1 - |a_n|)^{\alpha+2} < +\infty.$$

We shall construct a function in A^0_α whose zero set is exactly A, counting multiplicities. To do this, we may assume $a_n \ne 0$ for every n.

Fix a positive integer N such that $N > \alpha + 1$. Consider the product expression

$$f(z) = \prod_{n=1}^{+\infty} E_N \left(\frac{1 - |a_n|^2}{1 - \bar{a}_n z} \right).$$

It is clear that each factor in the above product is analytic in $\mathbb{D}$. Since

$$\left| \frac{1 - |a|^2}{1 - \bar{a} z} \right| \le 2, \qquad z \in \mathbb{D},$$

for each $a \in \mathbb{D}$, we have from Lemma 4.39

$$\left| 1 - E_N \left(\frac{1 - |a_n|^2}{1 - \bar{a}_n z} \right) \right| \le C \left| \frac{1 - |a_n|^2}{1 - \bar{a}_n z} \right|^{N+1}, \qquad z \in \mathbb{D},$$

for all $n = 1, 2, 3, \ldots$. This, along with

$$(1 - |a_n|^2)^{N+1} \le (1 - |a_n|^2)^{\alpha+2},$$

shows that the infinite product defining f converges, and that the function f is analytic in $\mathbb{D}$ with zero set $A = \{a_n\}_n$.

To show that the function f belongs to A_α^0, we apply Lemma 4.39 and the obvious inequality

$$\log(1 + x) \leq x, \qquad x > 0,$$

to obtain

$$\log |f(z)| \leq C \sum_{n=1}^{+\infty} \left(\frac{1 - |a_n|^2}{|1 - \overline{a}_n z|} \right)^{N+1}, \qquad z \in \mathbb{D}.$$

This implies that

$$\int_{\mathbb{D}} \log^+ |f(z)| \, (1 - |z|^2)^\alpha \, dA(z) \leq C \sum_{n=1}^{+\infty} (1 - |a_n|^2)^{N+1} \int_{\mathbb{D}} \frac{(1 - |z|^2)^\alpha \, dA(z)}{|1 - \overline{a}_n z|^{N+1}}.$$

Applying Theorem 1.7, we obtain another positive constant C' such that

$$\int_{\mathbb{D}} \log^+ |f(z)| \, (1 - |z|^2)^\alpha \, dA(z) \leq C' \sum_{n=1}^{+\infty} (1 - |a_n|^2)^{\alpha+2} < +\infty,$$

which completes the proof of the theorem.   ∎

4.8   Notes

It is still an open problem to characterize geometrically the zero sets for Bergman spaces. And it is well known that this problem is very difficult; for example, Theorem 4.14 clearly shows the subtlety of the problem. Only a handful of papers exist on zero sets for Bergman spaces.

Horowitz studied zero sets in his thesis and subsequent papers [76, 77]. His main tools were the classical Jensen's formula and lacunary series. In particular, Horowitz was able to show, using those elementary tools, that different Bergman spaces have different zero sets, that the union of two zero sets for a Bergman space can fail to be a zero set for the same space, but that any subset of a zero set for a Bergman space is still a zero set for the same space. Also, Theorem 4.35 is due to Horowitz.

Deeper properties of Bergman space zero sets were obtained by Korenblum in [83], and in a sharper form, by Seip in [112] and [113]. In particular, the results of Sections 4.2–4.5 are essentially Korenblum's, although the proofs here have been improved over those in the original paper. Theorem 3.3 allows us to obtain Korenblum's main results in the context of A_α^p; this is done in Section 4.6. The oblique projection technique is due to Seip [113]. The use of Linear Programming techniques first appeared in the context of $\mathcal{A}^{-\infty}$ zero sets in [83].

Luecking [93] has reformulated the condition on a sequence of points in $\mathbb{D}$ to be a zero sequences for the growth space $\mathcal{A}^{-\alpha}$ in terms of harmonic majorants; he also has a similar condition for the Bergman spaces A^p.

Probabilistic results on zero sets for the Bergman spaces were obtained by Shapiro [116], and later, in a different vein, by LeBlanc and Bomash [90, 24].

Theorem 4.40 is due to Heilper [74] in the case $\alpha = 0$; the proof here is basically from [36].

4.9   Exercises and Further Results

1. Carry out the integration by parts argument needed to show that $|\Psi(t\zeta)| \leq C(1 - t^2)^{-\alpha}$ for $0 < t < 1$ in the proof of Theorem 4.31. See [83].

2. If I is an invariant subspace of A^p and I contains a Blaschke product, then I is generated by a Blaschke product.

3. Fix a space A^p, $0 < p < +\infty$. For the singular inner function S_σ with a single point mass σ at $z = 1$, construct a sequence of Blaschke products $\{B_n\}_n$ such that $B_n \to S_\sigma$ uniformly on compact sets. If G_n is the extremal function of the invariant subspace generated by B_n in A^p, and G_σ is the extremal function of the invariant subspace generated by S_σ in A^p, show that $G_n \to G_\sigma$ in norm.

4. Explicitly construct a zero sequence for A^p_α that is not a Blaschke sequence.

5. Let I_A be the invariant subspace of A^2 generated by a zero set A. Show that the orthogonal complement of I_A is the closed linear span of the kernel functions $(1 - \bar{a}z)^{-2}$, with $a \in A$.

6. For a zero set A for the space A^p, let I_A be the corresponding zero-based invariant subspace, and G_A the associated extremal function. Suppose we have two zero sets A and B, with $A \subset B$. Show that the closure of $I_B/G_A = \{f/G_A : f \in I_B\}$ equals $I_{B\setminus A}$.

7. Suppose $\{A_n\}_n$ is a decreasing sequence of zero sets in A^p with

$$A = \bigcap_{n=1}^{+\infty} A_n.$$

Show that the closure of $\bigcup\{I_{A_n} : n \geq 1\}$ is I_A.

8. Fix $0 < \alpha < +\infty$. Construct a zero sequence for $\mathcal{A}^{-\alpha}$ that is not a zero sequence for $\mathcal{A}_0^{-\alpha}$. Hint: consider the regular sequences of Section 5.4.

9. Let A be a zero set for A^2 and let H_A be the product defined in Proposition 4.33. Show that $\|f/H_A\|_2 \leq \|f\|$ for all $f \in I_A$.

10. For any $0 < p < +\infty$ and $-1 < \alpha < +\infty$, there exist two zero sets A and B for A^p_α such that $A \cup B$ is no longer a zero set for A^p_α. See [76].

11. Decide whether for $(p, \alpha) \neq (q, \beta)$, the spaces A_α^p and A_β^q have different zero sets. This is probably still an open problem.

12. If $A = \{a_1, \dots, a_n\}$ is a finite sequence in $\mathbb{D}$, then $G_A(0) \geq |a_1 \cdots a_n|$.

13. Let $A = \{a_n\}_n$ be a zero set for A^2 with $a_n \neq a_m$ for $n \neq m$. If we apply the Gram-Schmidt process to the functions

$$K(z, a_1), K(z, a_2), \dots, K(z, a_n), \dots,$$

where K is the Bergman kernel, then the result is the following orthonormal system:

$$\frac{K(z, a_1)}{\sqrt{K(a_1, a_1)}}, \frac{K_{A_1}(z, a_2)}{\sqrt{K_{A_1}(a_2, a_2)}}, \dots, \frac{K_{A_n}(z, a_{n+1})}{\sqrt{K_{A_n}(a_{n+1}, a_{n+1})}}, \dots.$$

Here, $A_n = \{a_1, \dots, a_n\}$ and K_{A_n} is the reproducing kernel of I_{A_n}. In particular, the above system forms an orthonormal basis for $I_A^\perp$. See [140].

14. Let $A = \{a_n\}_n$ be a uniqueness sequence for A^2 (that is, A is not a zero set for A^2) with $a_n \neq a_m$ for $n \neq m$. For any $n \geq 1$ and $1 \leq k \leq n$, let $A_n^k = \{a_1, \dots, a_n\} \setminus \{a_k\}$ and let

$$\varphi_{n,k}(z) = K_{A_n^k}(z, a_k)/K_{A_n^k}(a_k, a_k), \qquad z \in \mathbb{D}.$$

Then

$$f = \lim_{n \to +\infty} \sum_{k=1}^{n} f(a_k)\, \varphi_{n,k}$$

for every $f \in A^2$. The convergence is in norm, and each $\varphi_{n,k}$ is a linear combination of the Bergman kernel functions $K(z, a_j)$, $1 \leq j \leq n$. See [140].

5

Interpolation and Sampling

In this chapter, we define and study sequences of interpolation and sampling for the Bergman spaces $\mathcal{A}^{-\alpha}$ and A_α^p. The main results include the characterization of interpolation sequences in terms of an upper density and the characterization of sampling sequences in terms of a lower density.

We will make use of the several notions of density introduced in Chapter 4. We will also introduce a new Möbius invariant density and show how it is related to the ones in Chapter 4.

As a final item, we show (in Section 5.4) how to compute the upper and lower densities of regular sequences.

5.1 Interpolation Sequences for $\mathcal{A}^{-\alpha}$

Recall that $\mathcal{A}^{-\alpha}$, with $0 < \alpha < +\infty$, is the Banach space of analytic functions f in $\mathbb{D}$ such that

$$\|f\|_{-\alpha} = \sup\{(1 - |z|^2)^\alpha |f(z)| : z \in \mathbb{D}\} < +\infty. \tag{5.1}$$

We say that a sequence $\Gamma = \{z_j\}_j \subset \mathbb{D}$ of distinct points is an *interpolation sequence* (or set) for $\mathcal{A}^{-\alpha}$ if the restriction operator R_Γ defined by

$$f \mapsto \{f(z_j)\}_j = f|_\Gamma$$

maps $\mathcal{A}^{-\alpha}$ onto $C^{-\alpha}(\Gamma)$. Here, $C^{-\alpha}(\Gamma)$ denotes the Banach space of sequences $a = \{a_j\}_j$ with the norm

$$\|a\|_{-\alpha,\Gamma} = \sup_j \left(1 - |z_j|^2\right)^\alpha |a_j|. \tag{5.2}$$

Write

$$K_\Gamma = \ker R_\Gamma = \{f \in \mathcal{A}^{-\alpha} : f|_\Gamma = 0\},$$

and observe that R_Γ induces the quotient map

$$R_\Gamma : \mathcal{A}^{-\alpha}/K_\Gamma \to C^{-\alpha}(\Gamma),$$

which has norm ≤ 1. Suppose Γ is an interpolating sequence for $\mathcal{A}^{-\alpha}$; this then means that R_Γ is onto, so that by Basic Functional Analysis, R_Γ is an invertible operator. Let $M(\Gamma) = M_\alpha(\Gamma)$ be the norm of the inverse of the above quotient map. Then it follows easily from a normal family argument that for every sequence $a \in C^{-\alpha}(\Gamma)$ there is a solution $f \in \mathcal{A}^{-\alpha}$ to the *interpolation problem*

$$f(z_j) = a_j \qquad \text{for all } j, \tag{5.3}$$

with $\|f\|_{-\alpha} \leq M_\alpha(\Gamma)\|a\|_{-\alpha,\Gamma}$. We write $M_\alpha(\Gamma) = +\infty$ if Γ fails to be an interpolation sequence for $\mathcal{A}^{-\alpha}$.

We shall also consider the separable subspace $\mathcal{A}_0^{-\alpha}$ of $\mathcal{A}^{-\alpha}$, whose elements satisfy

$$f(z) = o\left[(1 - |z|^2)^{-\alpha}\right], \qquad |z| \to 1^-,$$

as well as the corresponding sequence space $C_0^{-\alpha}(\Gamma)$. The definition of the concept of $\mathcal{A}_0^{-\alpha}$-interpolation sequence as well as the meaning of $M_0 = M_0(\Gamma) = M_{0,\alpha}(\Gamma)$ are then self-explanatory.

If Φ is a Möbius map of $\mathbb{D}$, then a straightforward computation shows that the transformation T_Φ defined by

$$(T_\Phi f)(z) = \Phi'(z)^\alpha f(\Phi(z)) \tag{5.4}$$

is a unitary operator on both $\mathcal{A}^{-\alpha}$ and $\mathcal{A}_0^{-\alpha}$. This implies that interpolation sets for $\mathcal{A}^{-\alpha}$ and $\mathcal{A}_0^{-\alpha}$ are Möbius invariant, and so are the interpolation constants:

$$M_\alpha(\Gamma) = M_\alpha(\Phi(\Gamma)), \qquad M_{0,\alpha}(\Gamma) = M_{0,\alpha}(\Phi(\Gamma)).$$

LEMMA 5.1 *Suppose $f \in \mathcal{A}^{-\alpha}$ and*

$$S_\alpha f(z) = (1 - |z|^2)^\alpha f(z), \qquad z \in \mathbb{D}.$$

Then there exists a constant C_α (depending only on α) such that

$$|S_\alpha f(a) - S_\alpha f(b)| \leq C_\alpha \|f\|_{-\alpha}\, \rho(a,b) \tag{5.5}$$

for all a and b in $\mathbb{D}$ with $\rho(a,b) \leq \frac{1}{2}$, where ρ is the pseudohyperbolic distance.

Proof. By Cauchy's formula,

$$f'(z) = \frac{1}{2\pi i} \int_{|\zeta - z| = (1 - |z|)/2} \frac{f(\zeta)\,d\zeta}{(\zeta - z)^2}, \qquad z \in \mathbb{D},$$

which yields the estimate

$$|f'(z)| \leq \gamma_\alpha \|f\|_{-\alpha} (1 - |z|^2)^{-\alpha - 1}, \qquad z \in \mathbb{D}.$$

If $\rho(a, b) \leq \frac{1}{2}$, then $1 - |z|^2 \sim 1 - |a|^2$ for all z lying on the line segment joining a and b; also,

$$|a - b| \sim \rho(a, b)(1 - |a|^2).$$

Therefore, if $dS_\alpha(z)$ is the total differential of S_α at z, then

$$
\begin{aligned}
|dS_\alpha f(z)| &\leq \left[2(1 - |z|^2)^{\alpha - 1} |f(z)| + (1 - |z|^2)^\alpha |f'(z)| \right] |dz| \\
&\leq \left[(2 + \gamma_\alpha)(1 - |z|^2)^{-1} \|f\|_{-\alpha} \right] |dz|.
\end{aligned}
$$

It follows that

$$|S_\alpha f(a) - S_\alpha f(b)| \leq \gamma'_\alpha \|f\|_{-\alpha} (1 - |a|^2)^{-1} |a - b| \sim \gamma'_\alpha \|f\|_{-\alpha}\, \rho(a, b),$$

as claimed. ∎

Recall that a sequence $\Gamma = \{z_j\}_j$ of points in $\mathbb{D}$ is called *separated* (in the hyperbolic or pseudohyperbolic metric) if

$$0 < \inf\{\rho(z_k, z_l) : k \neq l\}.$$

COROLLARY 5.2 *Every $\mathcal{A}^{-\alpha}$-interpolation sequence $\Gamma = \{z_j\}_j$ is separated.*

Proof. Fix k, and define a sequence $\{a_j\}_j$ by

$$a_j = (1 - |z_k|^2)^{-\alpha} \delta_{j,k},$$

where $\delta_{j,k}$ is the Kronecker delta symbol. Choose $f \in \mathcal{A}^{-\alpha}$ such that $\|f\|_{-\alpha} \leq M_\alpha(\Gamma)$ and $f(z_j) = a_j$ for all j. If $j \neq k$, then according to Lemma 5.1, either $\rho(z_j, z_k) > \frac{1}{2}$ or

$$1 = |S(z_j) - S(z_k)| \leq C_\alpha M_\alpha(\Gamma) \rho(z_j, z_k).$$

Thus, it follows that

$$\rho(z_j, z_k) \geq \min\left\{ \frac{1}{2}, \frac{1}{C_\alpha M_\alpha(\Gamma)} \right\}.$$

∎

If $\Gamma = \{z_j\}_j$ is a separated sequence in $\mathbb{D}$, then by the estimate following Lemma 2.14 we can find a positive constant C (independent of f) such that

$$\sum_j (1 - |z_j|^2)^s |f(z_j)|^p \leq C \int_{\mathbb{D}} |f(z)|^p (1 - |z|^2)^{s-2}\, dA(z) \qquad (5.6)$$

for all analytic f in $\mathbb{D}$. Obviously, the above estimate is also possible if Γ is the union of finitely many separated sequences.

We already mentioned that $\mathcal{A}^{-\alpha}$-interpolation sequences are Möbius invariant. Next, we show that $\mathcal{A}^{-\alpha}$-interpolation sequences are invariant under small perturbations with respect to the hyperbolic metric.

PROPOSITION 5.3 *Let* $\Gamma = \{z_j\}_j$ *be an* $\mathcal{A}^{-\alpha}$*-interpolation sequence. Then there is some constant* $\delta \in (0, \frac{1}{2}]$, *depending only on* $M_\alpha(\Gamma)$ *and* α, *such that each sequence* $\Gamma' = \{z'_j\}_j$ *satisfying* $\rho(z_j, z'_j) \leq \delta$ *is also an* $\mathcal{A}^{-\alpha}$*-interpolation sequence.*

Proof. We fix $\delta > 0$, whose value will be determined later, and assume that $\{z'_j\}_j$ is a sequence satisfying $\rho(z_j, z'_j) \leq \delta$ for all j. Given a sequence $\{w_j\}_j$ with $|w_j| \leq 1$ for all j, we wish to solve the interpolation problem

$$(1 - |z'_j|^2)^\alpha f(z'_j) = w_j = w_j^{(0)}, \qquad j = 1, 2, 3, \ldots . \tag{5.7}$$

We proceed by iteration. We first find a function f_1 such that $\|f_1\|_{-\alpha} \leq M_\alpha(\Gamma)$ and $(1 - |z_j|^2)^\alpha f_1(z_j) = w_j^{(0)}$ for all j. Set

$$w_j^{(1)} = w_j^{(0)} - (1 - |z'_j|^2)^\alpha f_1(z'_j), \qquad j = 1, 2, 3, \ldots .$$

By Lemma 5.1, we have $|w_j^{(1)}| \leq C_\alpha M_\alpha(\Gamma)\delta$. We can now find $f_2 \in \mathcal{A}^{-\alpha}$ such that $\|f_2\|_{-\alpha} \leq C_\alpha M_\alpha(\Gamma)^2\delta$ and $(1 - |z_j|^2)^\alpha f_2(z_j) = w_j^{(1)}$ for all j. Define

$$w_j^{(2)} = w_j^{(1)} - (1 - |z'_j|^2)^\alpha f_2(z'_j), \qquad j = 1, 2, 3, \ldots .$$

By Lemma 5.1 again, we have $|w_j^{(2)}| \leq C_\alpha^2 M_\alpha(\Gamma)^2\delta^2$ for $j = 1, 2, 3, \ldots$. Continuing this process, we get a sequence $\{f_n\}_n$ of functions in $\mathcal{A}^{-\alpha}$ with

$$\|f_n\|_{-\alpha} \leq C_\alpha^{n-1} M_\alpha(\Gamma)^n \delta^{n-1}$$

for all $n = 1, 2, 3, \ldots$, and a doubly indexed sequence $\{w_j^{(n)}\}_{j,n}$ with

$$|w_j^{(n)}| \leq C_\alpha^n M_\alpha(\Gamma)^{n+1}\delta^n$$

for all j and n, such that

$$(1 - |z'_j|^2)^\alpha f_n(z'_j) = w_j^{(n-1)} - w_j^{(n)}$$

for all j and n. If δ is chosen so small as to satisfy $\delta C_\alpha M_\alpha(\Gamma) < 1$, then the series $f = \sum_{n=1}^{+\infty} f_n$ will converge in the norm of $\mathcal{A}^{-\alpha}$, and the function f will solve the interpolation problem (5.7). Moreover,

$$M_\alpha(\Gamma') \leq \|f\|_{-\alpha} \leq \frac{M_\alpha(\Gamma)}{1 - C_\alpha \delta M_\alpha(\Gamma)}. \tag{5.8}$$

The proof is complete. ∎

We are going to combine the basic properties above with the properties of $\mathcal{A}^{-\alpha}$-zero sets studied in Chapter 4 to obtain necessary and sufficient conditions for

$\mathcal{A}^{-\alpha}$-interpolation sets. As a matter of fact, we will show that the density conditions that more or less described the zero sets for $\mathcal{A}^{-\alpha}$, if made Möbius invariant, will completely describe the $\mathcal{A}^{-\alpha}$-interpolation sets.

Given a sequence $\Gamma = \{z_j\}_j$ in $\mathbb{D} \setminus \{0\}$ and a finite subset F of $\mathbb{T}$, we recall from Chapter 4 the logarithmic summation function

$$\Lambda(\Gamma, \tau_F) = \sum_j \left\{ \log \frac{1}{|z_j|} : z_j \in \tau_F \right\},$$

where τ_F is the union of radii from 0 to points in F.

LEMMA 5.4 *Suppose* $\Gamma = \{z_j\}_j$ *is the zero set of a function* $f \in \mathcal{A}^{-\alpha}$ *with* $f(0) = 1$. *Then, for every* $\varepsilon > 0$,

$$\Lambda(\Gamma, \tau_F) \leq (\alpha + \varepsilon) \widehat{\kappa}(F) + 2 \log \|f\|_{-\alpha} + O(1),$$

where $O(1)$ *expresses a bounded quantity that only depends on* α *and* ε.

Proof. This is an immediate consequence of Theorem 4.23.   ∎

For a sequence $\Gamma = \{z_j\}_j$ of distinct points in $\mathbb{D}$ (not containing the origin), recall from Chapter 4, Section 4.2, that the upper asymptotic κ-density $D^+(\Gamma)$ is the infimum of all positive real numbers ϱ for which

$$\Lambda(\Gamma, \tau_F) \leq \varrho \widehat{\kappa}(F) + O(1),$$

uniformly in all the finite subset F of $\mathbb{T}$. We want to define a Möbius invariant uniform density. To this end, we introduce the Möbius maps

$$\Phi_n(z) = \frac{z_n - z}{1 - \bar{z}_n z}, \qquad z_n \in \Gamma,$$

and consider the shifted sequence $\Phi_n(\Gamma)$ with the origin removed:

$$\Gamma_n = \Phi_n(\Gamma) \setminus \{0\} = \Phi_n(\Gamma \setminus \{z_n\}), \qquad n = 1, 2, 3, \ldots.$$

DEFINITION 5.5 *The uniform separating upper asymptotic* κ-density $D_u^+(\Gamma)$ *is the infimum of all positive real numbers* ϱ *for which:*

$$\sup_n \Lambda(\Gamma_n, \tau_F) \leq \varrho \widehat{\kappa}(F) + O(1)$$

holds uniformly in all the finite subset F *of* $\mathbb{T}$.

Observe that the logarithmic singularity at the origin in the definition of $\Lambda(\Gamma, \tau_F)$ puts a severe penalty on points of Γ too close to 0, and as we insert this into the Möbius invariant form with the shifted sequences Γ_n, we see that any sequence Γ with $D_u^+(\Gamma) < +\infty$ is separated. Conversely, any separated sequence Γ has $D_u^+(\Gamma) < +\infty$. See Exercises 16 and 17 for details. Like the upper asymptotic κ-density $D^+(\Gamma)$ of Chapter 4 (see Section 4.2), the uniform density $D_u^+(\Gamma)$ can also be expressed as a lim sup:

$$D_u^+(\Gamma) = \limsup_{\widehat{\kappa}(F) \to +\infty} \frac{\sup_n \Lambda(\Gamma_n, \tau_F)}{\widehat{\kappa}(F)}.$$

We can now give a characterization of $\mathcal{A}^{-\alpha}$-interpolation sequences in terms of the uniform separating upper asymptotic κ-density.

THEOREM 5.6 *Let $\Gamma = \{z_j\}_j$ be a sequence in $\mathbb{D}$. Then the following conditions are equivalent:*

(i) *Γ is an $\mathcal{A}^{-\alpha}$-interpolation sequence.*

(ii) *There is some $\delta > 0$ such that Γ is an $\mathcal{A}^{-\beta}$-interpolation sequence for all $\beta > \alpha - \delta$.*

(iii) *$D_u^+(\Gamma) < \alpha$.*

Proof. Clearly, it is enough to prove the equivalence of (i) and (iii).

We first prove that (i) implies (iii). For each $n = 1, 2, 3, \ldots$, the identity $M_\alpha(\Gamma) = M_\alpha(\Phi_n(\Gamma))$ – which expresses the Möbius invariance of the interpolation problem – implies that there is a function $f_n \in \mathcal{A}^{-\alpha}$ such that

$$\|f_n\|_{-\alpha} \le M_\alpha(\Gamma), \qquad f_n(0) = 1, \quad \text{and} \quad f_n|_{\Gamma_n} = 0.$$

Using Lemma 5.4, we see that for each $\varepsilon > 0$ and finite subset $F \subset \mathbb{T}$,

$$\sup_n \Lambda(\Gamma_n, \mathfrak{r}_F) \le (\alpha + \varepsilon)\widehat{\kappa}(F) + O(1),$$

where the bound $O(1)$ only depends on α, ε, and the interpolation constant $M_\alpha(\Gamma)$. We proceed to improve this estimate.

For each $n = 1, 2, 3, \ldots$, write $\Gamma_n = \{z_{k,n}\}_{k=1}^{+\infty}$ and construct for positive parameter δ a perturbed sequence $\Gamma_n^\delta = \{z_{k,n}^\delta\}_{k=1}^{+\infty}$ by setting

$$z_{k,n}^\delta = |z_{k,n}|^\delta z_{k,n}$$

for all k. Note that on the left hand side, δ is only a superscript, whereas on the right, it is an exponent. The points of the sequence Γ_n^δ are pushed inward into the disk compared with Γ. We see that

$$\rho(z_{k,n}, z_{k,n}^\delta) = \frac{|z_{k,n}| - |z_{k,n}|^{1+\delta}}{1 - |z_{k,n}|^{2+\delta}} \le \sup_{0<t<1} \frac{t - t^{1+\delta}}{1 - t^{2+\delta}} = \frac{\delta}{2+\delta} < \frac{\delta}{2}.$$

Let C_α be the constant of (5.8), which comes from Lemma 5.1. By Möbius invariance,

$$M_\alpha(\Gamma_n \cup \{0\}) = M_\alpha(\Phi_n(\Gamma)) = M_\alpha(\Gamma),$$

so that if δ is so small that

$$\frac{\delta}{2} \le \frac{1}{2 C_\alpha M_\alpha(\Gamma)},$$

then the estimate (5.8) implies the following: for each n, there exists a function $f_n \in \mathcal{A}^{-\alpha}$ (not the previous f_n, of course) with

$$\|f_n\|_{-\alpha} \le 2M_\alpha(\Gamma), \qquad f_n(0) = 1, \quad \text{and} \quad f_n|_{\Gamma_n^\delta} = 0.$$

By the definition of the logarithmic sum and the perturbed sequence,

$$\Lambda(\Gamma_n^\delta, \tau_F) = (1 + \delta)\, \Lambda(\Gamma_n, \tau_F)$$

holds for all n. We apply Lemma 5.4 to these new functions f_n, and obtain, for every $\varepsilon > 0$,

$$(1 + \delta)\, \Lambda(\Gamma_n, \tau_F) \le (\alpha + \varepsilon)\,\widehat{\kappa}(F) + O(1),$$

where the bound $O(1)$ only depends on α, ε, and the interpolation constant $M_\alpha(\Gamma)$. This implies that

$$\sup_n \Lambda(\Gamma_n, \tau_F) \le (1 + \delta)^{-1}(\alpha + \varepsilon)\,\widehat{\kappa}(F) + O(1),$$

so that if we let δ be as big as allowed,

$$\delta = \frac{1}{C_\alpha M_\alpha(\Gamma)},$$

we get

$$\sup_n \Lambda(\Gamma_n, \tau_F) \le \frac{C_\alpha M_\alpha(\Gamma)}{1 + C_\alpha M_\alpha(\Gamma)}\,(\alpha + \varepsilon)\,\widehat{\kappa}(F) + O(1),$$

uniformly in the finite subsets F of $\mathbb{T}$. As ε is an arbitrarily small positive number, it follows that

$$D_u^+(\Gamma) \le \frac{C_\alpha M_\alpha(\Gamma)}{1 + C_\alpha M_\alpha(\Gamma)}\,\alpha < \alpha.$$

This proves that (i) implies (iii).

To prove the reverse implication, the following observation is essential. If we replace the radial stars τ_F with Stolz stars s_F, the crucial estimate of Lemma 5.4 remains unaffected. This follows from the methods we developed in Chapter 4. Thus, we may assume that the definition of $D_u^+(\Gamma)$ is based on $\Lambda(\Gamma, s_F)$ instead of $\Lambda(\Gamma, \tau_F)$.

We now prove that (iii) implies (i). Assume that $\varrho = D_u^+(\Gamma) < \alpha$. By the proof of Theorem 4.31 and Möbius invariance, there exist analytic functions $g_k \in \mathcal{A}^{-(\varrho+\alpha)/2}$, for $k = 1, 2, 3, \ldots$, such that

$$g_k(z_k) = (1 - |z_k|)^{-(\varrho+\alpha)/2}, \qquad g_k(z_j) = 0 \quad \text{for} \quad j \ne k,$$

and

$$|g_k(z)| \le C\,(1 - |z|^2)^{-(\varrho+\alpha)/2}, \qquad z \in \mathbb{D},$$

where C is a positive constant independent of k. The interpolation problem is then solved explicitly by the function

$$f(z) = \sum_k w_k\,(1 - |z_k|^2)^{(\varrho+\alpha)/2} g_k(z) \left(\frac{1 - |z_k|^2}{1 - \bar{z}_k z}\right)^\theta, \tag{5.9}$$

where θ is a real parameter with $\theta > 1 + (\alpha - \varrho)/2$. To see that $f \in A^{-\alpha}$, we observe that

$$|f(z)| \leq C \sup_k \left\{ (1 - |z_k|^2)^\alpha |w_k| \right\} (1 - |z|^2)^{-(\alpha+\varrho)/2}$$

$$\times \sum_j \frac{(1 - |z_j|^2)^{\theta-(\alpha-\varrho)/2}}{|1 - \overline{z}_j z|^\theta}, \qquad z \in \mathbb{D}.$$

Since $\{z_j\}_j$ is separated, an application of (5.6) yields a positive constant M such that

$$\sum_j \frac{(1 - |z_j|^2)^{\theta-(\alpha-\varrho)/2}}{|1 - \overline{z}_j z|^\theta} \leq M \int_\mathbb{D} \frac{(1 - |w|^2)^{\theta-(\alpha-\varrho)/2-2}}{|1 - \overline{w}z|^\theta} \, dA(w)$$

for all $z \in \mathbb{D}$. Combining this with Theorem 1.7, we can find a positive constant C such that

$$|f(z)| \leq C (1 - |z|^2)^{-\alpha}, \qquad z \in \mathbb{D}.$$

This completes the proof of the theorem. ∎

Note that the proof of Theorem 5.6 implies an additional property of $A^{-\alpha}$-interpolation sequences Γ, namely that $M_\beta(\Gamma)$ is bounded for β in some interval $[\alpha - \delta, \alpha]$.

THEOREM 5.7 *Every $A_0^{-\alpha}$-interpolation sequence is also a sequence of interpolation for $A^{-\alpha}$, and vice versa.*

Proof. First, let $\Gamma = \{z_j\}_j$ be an $A_0^{-\alpha}$-interpolation sequence and let $a = \{a_j\}_j$ be a sequence in $C^{-\alpha}(\Gamma)$. Consider the truncated sequences

$$a^{(N)} = \{a_1, \dots, a_N, 0, 0, \dots\}, \qquad N = 1, 2, 3, \dots.$$

For each N, there exists an $f_N \in A_0^{-\alpha}$ such that $f_N|_\Gamma = a^{(N)}$ and

$$\|f_N\|_{-\alpha} \leq M_{0.\alpha}(\Gamma) \|a\|_{-\alpha}.$$

By a normal family argument, we can extract from $\{f_N\}_N$ a subsequence converging to some $f \in A^{-\alpha}$ uniformly on compact sets. Clearly, $f|_\Gamma = a$ and $\|f\|_{-\alpha} \leq M_{0.\alpha}(\Gamma)$. This shows that Γ is an $A^{-\alpha}$-interpolation sequence with $M_\alpha(\Gamma) \leq M_{0.\alpha}(\Gamma)$.

Next, let $\Gamma = \{z_j\}_j$ be an $A^{-\alpha}$-interpolation sequence and let $a = \{a_j\}_j$ be a sequence in $C_0^{-\alpha}(\Gamma)$. We can assume $\|a\|_{-\alpha} \leq 1$ and pick a sequence of natural numbers

$$1 = N_1 < N_2 < N_3 < \cdots$$

such that the tails

$$a^{(k)} = \{0, \dots, 0, a_{N_k}, a_{N_k+1}, \dots\}$$

satisfy $\|a^{(k)}\|_{-\alpha,\Gamma} \leq 2^{-k+1}$ for all $k = 1, 2, 3, \ldots$. Consider now the finite sequences

$$b^{(k)} = \{0, \ldots, 0, a_{N_k}, a_{N_k+1} \ldots, a_{N_{k+1}-1}, 0, \ldots\}, \qquad k = 1, 2, 3, \ldots.$$

Each $b^{(k)}$ belongs to all $C_0^{-\beta}(\Gamma)$, $\beta > 0$. Also, we can choose $\beta'_k < \alpha$ so close to α that $\|b^{(k)}\|_{-\beta_k} < 2^{-k+1}$ and $M_{-\beta_k}(\Gamma) \leq C$ for all k, where C is some constant. For each k, there exists a function $f_k \in \mathcal{A}^{-\beta_k} \subset \mathcal{A}_0^{-\alpha}$ with $f_k|_\Gamma = b^{(k)}$ and

$$\|f_k\|_{-\alpha} \leq \|f_k\|_{-\beta_k} \leq 2^{-k+1}C.$$

It follows that the function $f = \sum_k f_k$ is in $\mathcal{A}^{-\alpha}$ with $\|f\|_{-\alpha} \leq 2C$. Clearly, the function f solves the interpolation problem $f|_\Gamma = a$.   ∎

The rest of the section will be devoted to proving that the uniform upper asymptotic κ-density $D_u^+(\Gamma)$ is equivalent to a simpler and more elegant notion of density. This will result in a more transparent characterization of $\mathcal{A}^{-\alpha}$-interpolation sets than that provided by Theorem 5.6.

It is sometimes necessary to distinguish between the notion of a countable set $\Gamma \subset \mathbb{D}$ and that of an associated sequence (or arrangement), although we may use the same notation for both. Two sequences associated with the same set will be called rearrangements of each other. A sequence $\Gamma = \{z_j\}_j$ is naturally ordered if $|z_1| \leq |z_2| \leq \cdots \leq |z_n| \leq \cdots$.

For a (countable) subset Γ of $\mathbb{D}$ that is separated in the pseudohyperbolic metric, we define

$$\rho(\Gamma) = \inf \{\rho(a, b) : a, b \in \Gamma, a \neq b\}.$$

Suppose $\{\Gamma^{(n)}\}_n$ is a sequence of subsets in $\mathbb{D}$ such that $\rho(\Gamma^{(n)}) \geq \delta > 0$ for all n. We say that $\{\Gamma^{(n)}\}$ converges weakly to a separated set Γ, and then write $\Gamma^{(n)} \to \Gamma$ as $n \to +\infty$, if there are naturally ordered arrangements $\Gamma^{(n)} = \{z_j^{(n)}\}_j$ and some $1 \leq N \leq +\infty$ such that

$$\lim_{n \to +\infty} z_j^{(n)} = z_j \in \mathbb{D}, \qquad 1 \leq j < N,$$

and

$$\lim_{n \to +\infty} |z_j^{(n)}| = 1, \qquad j \geq N.$$

The limit set Γ is then defined as $\{z_j\}_{j=1}^{N-1}$ for $N > 1$ and as the empty set for $N = 1$.

LEMMA 5.8 *Every sequence* $\{\Gamma^{(n)}\}_n$ *of sets satisfying* $\rho(\Gamma^{(n)}) \geq \delta > 0$ *for all* n *contains a subsequence that converges weakly to a separated set* Γ *(which may be empty).*

Proof. Arrange each $\Gamma^{(n)}$ into a naturally ordered sequence $\Gamma^{(n)} = \{z_j^{(n)}\}_j$. Pick a subsequence $\{\Gamma^{(n_k)}\}_k$ such that $z_1^{(n_k)}$ converges to some $z_1 \in \mathbb{D}$ or $|z_1^{(n_k)}| \to 1$ as $k \to +\infty$. In the latter case we stop, since the weak limit of $\{\Gamma^{(n_k)}\}_k$ is empty.

In the former case, we pick a subsequence $\{\Gamma^{(m_k)}\}_k$ of $\{\Gamma^{(n_k)}\}_k$ such that either $z_2^{(m_k)}$ converges to some z_2 in $\mathbb{D}$ or $|z_2^{(m_k)}| \to 1$ as $k \to +\infty$. If this process does not stop after a finite number of steps, then the resulting diagonal sequence of sets converges weakly to $\Gamma = \{z_1, z_2, z_3, \dots\}$. ∎

LEMMA 5.9 *If* $\{\Gamma^{(n)}\}$ *converges weakly to* Γ, *then*

$$M_\alpha(\Gamma) = \liminf_{n \to +\infty} M_\alpha(\Gamma^{(n)}).$$

Proof. We may assume that $\{M_\alpha(\Gamma^{(n)})\}_n$ converges; otherwise, we could replace $\{\Gamma^{(n)}\}_n$ by a suitable subsequence. There are arrangements $\Gamma^{(n)} = \{z_j^{(n)}\}_j$ and $\Gamma = \{z_j\}_j$ such that $z_j^{(n)} \to z_j$ as $n \to +\infty$ for each j. Let $\{b_j\}_j$ be a sequence in the unit ball of l^∞. Then for each n there is a solution $f_n \in \mathcal{A}^{-\alpha}$ to the interpolation problem

$$(1 - |z_j^{(n)}|^2)^\alpha f_n(z_j^{(n)}) = b_j, \qquad j = 1, 2, 3, \dots,$$

such that $\|f_n\|_{-\alpha} \le M_\alpha(\Gamma^{(n)})$. By a normal family argument, there is a subsequence $\{f_{n_k}\}_k$ that converges to some $f \in \mathcal{A}^{-\alpha}$ uniformly on compact subsets of $\mathbb{D}$. Clearly, $\|f\|_{-\alpha} \le \lim_n M_\alpha(\Gamma^{(n)})$, and

$$(1 - |z_j|^2)^\alpha f(z_j) = \lim_n (1 - |z_j^{(n_k)}|^2)^\alpha f_{n_k}(z_j^{(n_k)}) = b_j, \qquad j = 1, 2, 3, \dots.$$

This shows that Γ is an $\mathcal{A}^{-\alpha}$-interpolation set. ∎

LEMMA 5.10 *Given* δ_0, l_0, *and* α, *there exists a positive constant* c *such that if* Γ *is an* $\mathcal{A}^{-\alpha}$-*interpolation set with* $M_\alpha(\Gamma) \le l_0$, *and if* z_0 *is a point in* $\mathbb{D}$ *with* $\rho(z_0, \Gamma) \ge \delta_0$, *then there is an element* $f \in \mathcal{A}^{-\alpha}$ *for which* $f|_\Gamma = 0$, $\|f\|_{-\alpha} \le 1$, *and* $(1 - |z_0|^2)^\alpha |f(z_0)| \ge c$.

Proof. By Möbius invariance, we may assume that $z_0 = 0$. If the assertion is false, then there exists a sequence of sets $\Gamma^{(n)} = \{z_j^{(n)}\}_j$ ($n = 1, 2, \dots$) such that $\rho(0, \Gamma^{(n)}) \ge \delta_0$, $M_\alpha(\Gamma^{(n)}) \le l_0$, and

$$\sup\{|f(0)| : f \in \mathcal{A}^{-\alpha}, \|f\|_{-\alpha} \le 1, f|_{\Gamma^{(n)}} = 0\} \to 0 \qquad (5.10)$$

as $n \to +\infty$. By Lemma 5.8, there is a subsequence $\{\Gamma^{(n_k)}\}_k$ that converges weakly to a set $\Gamma' = \{z_j'\}_j$. For notational simplicity, assume that $\{\Gamma^{(n_k)}\}$ is the original sequence $\{\Gamma^{(n)}\}_n$. Since $\Gamma^{(n)}$ is an $\mathcal{A}^{-\alpha}$-interpolation sequence with $M_\alpha(\Gamma^{(n)}) \le l_0$ for each n, an obvious normal family argument shows that Γ' is also an $\mathcal{A}^{-\alpha}$-interpolation sequence with $M_\alpha(\Gamma') \le l_0$. By Theorem 5.7, Γ' is also an $\mathcal{A}_0^{-\alpha}$-interpolation set and therefore an $\mathcal{A}_0^{-\alpha}$-zero set. Choose an arbitrary $\varphi \in \mathcal{A}_0^{-\alpha}$ such that $\|\varphi\|_{-\alpha} = 1$, $\varphi|_{\Gamma'} = 0$, and $\varphi(0) = \gamma > 0$. We can solve the interpolation problem

$$f(z_k^{(n)}) = \varphi(z_k^{(n)}), \qquad k = 1, 2, 3, \dots,$$

for each n by a function ψ_n with

$$\|\psi_n\|_{-\alpha} \le l_0 \sup_k \left\{ (1 - |z_k^{(n)}|^2)^\alpha |\varphi(z_k^{(n)})| \right\}.$$

Since $\varphi(z_k^{(n)}) \to \varphi(z_k) = 0$ as $n \to +\infty$ for each k, and because $\varphi \in \mathcal{A}_0^{-\alpha}$, the above supremum tends to 0 and $\|\psi_n\|_{-\alpha} \to 0$ as $n \to +\infty$. Now, the function

$$f_n = \frac{\varphi - \psi_n}{\|\varphi - \psi_n\|_{-\alpha}}$$

vanishes on $\Gamma^{(n)}$, $\|f_n\|_{-\alpha} = 1$, and $f_n(0) \to \gamma$, which is a contradiction to (5.10). ∎

We will need a modification of the notion of uniform separated upper asymptotic κ-density $D_u^+(\Gamma)$. More specifically, we want to extend the family $\{\Phi_n\}_n$ of Möbius shifts involved in the definition of $D_u^+(\Gamma)$ to the entire Möbius group $\text{Aut}(\mathbb{D})$ of $\mathbb{D}$. We modify the Stolz stars $\mathfrak{s}_F$ slightly:

$$\mathfrak{s}_F^* = \mathfrak{s}_F \setminus \mathbb{D}(0, \tfrac{1}{2}),$$

where $\mathbb{D}(0, \tfrac{1}{2}) = \{z \in \mathbb{C} : |z| < \tfrac{1}{2}\}$ is the Euclidean disk of radius $\tfrac{1}{2}$ about the origin.

DEFINITION 5.11 *The uniform upper asymptotic κ-$*$-density $D_{*u}^+(\Gamma)$ of Γ is the infimum of all positive real numbers ϱ for which:*

$$\sup_\Phi \Lambda(\Phi(\Gamma), \mathfrak{s}_F^*) \le \varrho \widehat{\kappa}(F) + O(1)$$

holds uniformly in all the finite subset F of $\mathbb{T}$. Here, the supremum ranges over all $\Phi \in \text{Aut}(\mathbb{D})$.

We check the relation to the density $D_u^+(\Gamma)$.

LEMMA 5.12 *If Γ is separated in the pseudohyperbolic metric, then $D_{*u}^+(\Gamma) = D_u^+(\Gamma)$.*

Proof. First note that whereas $D_u^+(\Gamma) < +\infty$ implies that Γ is separated, it is not the case for $D_{*u}^+(\Gamma)$.

A comparison of the definitions in terms of radial and modified Stolz stars immediately reveals that $D_u^+(\Gamma) \le D_{*u}^+(\Gamma)$, since Γ is assumed separated.

Assume now that $D_u^+(\Gamma) < D_{*u}^+(\Gamma)$, and pick a real parameter α between $D_u^+(\Gamma)$ and $D_{*u}^+(\Gamma)$:

$$D_u^+(\Gamma) < \alpha < D_{*u}^+(\Gamma).$$

By Theorem 5.6, Γ is an $\mathcal{A}^{-\alpha}$-interpolation set. Then, by Lemma 5.10, there is a positive constant c such that

$$\sup \left\{ |f(0)| : \|f\|_{-\alpha} = 1, f|_{\Phi(\Gamma)^*} = 0 \right\} \ge c$$

for all $\Phi \in \text{Aut}(\mathbb{D})$, where

$$\Phi(\Gamma)^* = \Phi(\Gamma) \setminus \mathbb{D}(0, \tfrac{1}{2}).$$

Thus, for every $\Phi \in \mathrm{Aut}\,(\mathbb{D})$ there is a function $f = f_\Phi \in \mathcal{A}^{-\alpha}$ with

$$|f(0)| > \frac{c}{2}, \qquad \|f\|_{-\alpha} = 1, \quad \text{and} \quad f|_{\Phi(\Gamma)^*} = 0.$$

Applying Lemma 5.4 to the function $f/f(0)$, with $\Lambda(\Gamma, \mathfrak{r}_F)$ replaced by $\Lambda(\Gamma, \mathfrak{s}_F^*)$, we find that

$$\Lambda(\Phi(\Gamma), \mathfrak{s}_F^*) \le (\alpha + \varepsilon)\widehat{\kappa}(F) + O(1), \tag{5.11}$$

where the bound $O(1)$ only depends on α, ε, and c. The small positive number ε is arbitrary, and hence it follows from the definition of the uniform upper asymptotic κ-$*$-density that $D_{*u}^+(\Gamma) \le \alpha$, which contradict how we defined α. The proof is complete. ∎

We are now ready to introduce a more transparent notion of density and show that it is equivalent to the uniform separated upper asymptotic κ-density for separated sequences.

Suppose $\Gamma = \{z_j\}_j$ is separated and $r \in (\frac{1}{2}, 1)$. Let

$$D(\Gamma, r) = \frac{\sum_j \left\{ \log \frac{1}{|z_j|} : \frac{1}{2} < |z_j| < r \right\}}{\log \frac{1}{1-r}}. \tag{5.12}$$

For every $z \in \mathbb{D}$, we form a new sequence

$$\Gamma_z = \left\{ \frac{z_j - z}{1 - \bar{z}_j z} \right\}_j.$$

The *upper Seip density* of Γ is then defined as

$$D_s^+(\Gamma) = \limsup_{r \to 1^-} \sup_{z \in \mathbb{D}} D(\Gamma_z, r). \tag{5.13}$$

Note that because of rotational symmetry, $D_s^+(\Gamma)$ can also be defined as

$$D_s^+(\Gamma) = \limsup_{r \to 1^-} \sup_{\Phi \in \mathrm{Aut}\,(\mathbb{D})} D(\Phi(\Gamma), r). \tag{5.14}$$

THEOREM 5.13 *If $\Gamma = \{z_j\}_j$ is separated, then $D_u^+(\Gamma) = D_s^+(\Gamma)$.*

Proof. For every $r \in (\frac{1}{2}, 1)$, we can construct a standard Stolz star

$$\mathfrak{s}_F = \bigcup \{\mathfrak{s}_\zeta : \zeta \in F\}$$

containing the disk $|z| \le r$, so that $F = F_r$ consists of a minimum number N_r of points placed equidistantly on $\mathbb{T}$. A simple computation shows that N_r is approximately $C/(1-r)$ as $r \to 1^-$, for some positive constant C. Therefore,

$$\widehat{\kappa}(F_r) = \log N_r + 1 = \log \frac{1}{1-r} + O(1).$$

By Lemma 5.12, we have $D_{*u}^+(\Gamma) = D_u^+(\Gamma)$. We recall the estimate from the definition of $D_{*u}^+(\Gamma)$,

$$\sup_\Phi \Lambda(\Phi(\Gamma), \mathfrak{s}_F^*) \le \varrho\,\widehat{\kappa}(F) + O(1),$$

for all fixed ϱ with $D^+_{*u}(\Gamma) < \varrho < +\infty$. If we apply this to equidistant sets $F = F_r$, using the separation of the sequence Γ, the assertion $D^+_s(\Gamma) \leq D^+_{*u}(\Gamma) = D^+_u(\Gamma)$ is immediate.

To prove the reverse inequality $D^+_u(\Gamma) \leq D^+_s(\Gamma)$, we need the notion of the κ-area of a (Borel) measurable set $S \subset \mathbb{D}$:

$$\kappa A(S) = \int_S \frac{dA(z)}{1 - |z|^2}. \tag{5.15}$$

In the rest of the proof, we will use the notation s_F to denote only standard regions made up of the standard Stolz angles s_ζ with aperture $\pi/2$. If other Stolz angles are used with an aperture $\varphi \in (\frac{\pi}{2}, \pi)$, we will indicate this by writing

$$s'_F = s'_{F,\varphi} = \bigcup \left\{ s'_{\zeta,\varphi} : \zeta \in F \right\}.$$

Now, an elementary computation shows that

$$\left| \kappa A(s'_F) - \widehat{\kappa}(F) \right| \leq C, \tag{5.16}$$

where the constant C depends on the aperture φ but not on the finite set $F \subset \mathbb{T}$; see Exercise 25.

Let Γ be an arbitrary set in $\mathbb{D}$, with separation constant $\rho(\Gamma) = \delta > 0$, and Seip density $D^+_s(\Gamma) = \gamma < +\infty$. For any $\varepsilon > 0$, let $r_1 = r_1(\varepsilon)$ be so large that $D(\Phi(\Gamma), r) \leq \gamma + \varepsilon$ for all $\Phi \in \mathrm{Aut}(\mathbb{D})$ and all r with $r_1 < r < 1$. Choose the aperture φ so large that for each finite subset $F \subset \mathbb{T}$ and each $\zeta \in \partial s'_F$, $|\zeta| < 1$, we have $\rho(\zeta, s_F) \geq r_1$; it is easy to check that this is possible (φ will of course have to depend on $r_1 = r_1(\varepsilon)$).

For any finite subset $F \subset \mathbb{T}$, we consider the set $\Phi(\Gamma) \cap s_F$, where $\Phi \in \mathrm{Aut}(\mathbb{D})$ is a Möbius automorphism. For each $k = 0, 1, 2, \ldots$, consider the radii

$$r_k = \tanh\left(\frac{k}{2} \log \frac{1+r_1}{1-r_1}\right) = \frac{1 - \left(\frac{1-r_1}{1+r_1}\right)^k}{1 + \left(\frac{1-r_1}{1+r_1}\right)^k};$$

note that this is consistent for index $k = 1$. For $k = 0$, we have $r_0 = 0$, and with increasing k, r_k increases up to 1. The consecutive pseudohyperbolic distance is constant: $\rho(r_k, r_{k+1}) = r_1$. For each $k = 0, 1, 2, \ldots$, we introduce an annulus

$$\mathfrak{a}_k = \{ z \in \mathbb{C} : r_k \leq |z| < r_{k+2} \},$$

and consider the associated Blaschke product

$$B_k(w) = \prod_{z \in \Phi(\Gamma) \cap s_F \cap \mathfrak{a}_k} \frac{|z|}{z} \frac{z - w}{1 - \overline{z}w}, \qquad w \in \mathbb{D},$$

so that

$$\log \frac{1}{|B_k(w)|} = \sum_z \left\{ \log \frac{1}{\rho(z, w)} : z \in \Phi(\Gamma) \cap s_F \cap \mathfrak{a}_k \right\}.$$

We shall apply Jensen's formula to the function B_k in the disk

$$\mathbb{D}(0, r_{k+1}) = \{z \in \mathbb{C} : |z| < r_{k+1}\}.$$

First, recall that $ds(w) = |dw|/(2\pi)$ is the normalized length measure, and for $0 < r < 1$, introduce the notation $\mathbb{T}(0, r)$ for the circle $\{w \in \mathbb{C} : |w| = r\}$. Jensen's formula then results in

$$\sum_{z \in \Phi(\Gamma) \cap s_F \cap a_k} \log \frac{1}{|z|} = \sum_{z \in \Phi(\Gamma) \cap s_F \cap a_k \cap a_{k-1}} \log \frac{r_{k+1}}{|z|}$$

$$+ \frac{1}{r_{k+1}} \int_{\mathbb{T}(0, r_{k+1})} \log \frac{1}{|B_k(\xi)|} \, ds(\xi),$$

and considering that $a_k = (a_k \cap a_{k-1}) \cup (a_k \cap a_{k+1})$, we see that

$$\sum_{z \in \Phi(\Gamma) \cap s_F \cap a_k \cap a_{k+1}} \log \frac{1}{|z|} \leq \frac{1}{r_{k+1}} \int_{\mathbb{T}(0, r_{k+1})} \log \frac{1}{|B_k(\xi)|} \, ds(\xi). \tag{5.17}$$

We shall study the sets $\mathbb{T}(0, r_{k+1}) \cap s'_{F,\varphi}$, and their "complementary sets" $\mathbb{T}(0, r_{k+1}) \setminus s'_{F,\varphi}$. For $\xi \in \mathbb{T}(0, r_{k+1}) \setminus s'_{F,\varphi}$, we have

$$\log \frac{1}{|B_k(\xi)|} = \sum_z \left\{ \log \frac{1}{\rho(\xi, z)} : z \in \Phi(\Gamma) \cap s_F \cap a_k \right\}$$

$$\leq C (1 - r_{k+1}^2) \sum_{z \in \Phi(\Gamma) \cap s_F \cap a_k} \frac{1 - |z|^2}{|\xi - z|^2},$$

for some absolute constant C. Integrating the above inequality term by term and using the fact that

$$|\xi - z| > \sigma \frac{1 - r_{k+1}}{1 - r_1}, \qquad \text{for} \quad z \in a_k, \ \xi \in \mathbb{T}(0, r_{k+1}) \setminus s'_{F,\varphi},$$

where σ is a small but positive constant, we find

$$\frac{1}{r_{k+1}} \int_{\mathbb{T}(0, r_{k+1}) \setminus s'_{F,\varphi}} \log \frac{1}{|B_k(\xi)|} \, ds(\xi) \leq C (1 - r_1) \sum_{z \in \Phi(\Gamma) \cap s_F \cap a_k} (1 - |z|^2).$$

Applying the estimate following Lemma 2.14 in the same fashion as we did for (5.6), using the separation of Γ, we see that

$$\sum_{z \in \Phi(\Gamma) \cap s_F \cap a_k} (1 - |z|^2) \leq C(\delta) \int_{a_k \cap s'_F} \frac{dA(z)}{1 - |z|^2} = C(\delta) \kappa A(a_k \cap s'_F);$$

δ is the separation constant, and $C(\delta)$ a positive constant which depends on δ. Putting things together, we obtain

$$\frac{1}{r_{k+1}} \int_{\mathbb{T}(0, r_{k+1}) \setminus s'_{F,\varphi}} \log \frac{1}{|B_k(\xi)|} \, ds(\xi) \leq C(\delta) (1 - r_1) \kappa A(a_k \cap s'_F), \tag{5.18}$$

where the value of the constant has changed.

Now, consider the part of the integral on the right hand side of (5.17) over $\mathbb{T}(0, r_{k+1}) \cap s'_{F,\varphi}$. We need separate estimates for the quantities

$$Q_1 = \sum_z \left\{ 1 - \rho(z, \xi) : z \in \Phi(\Gamma) \cap s_F \cap a_k, \ \rho(z, \xi) < r_1 \right\}$$

and

$$Q_2 = \sum_z \left\{ 1 - \rho(z, \xi) : z \in \Phi(\Gamma) \cap s_F \cap a_k, \ r_1 \le \rho(z, \xi) < 1 \right\},$$

for $\xi \in \mathbb{T}(0, r_{k+1}) \cap s'_{F,\varphi}$. The first sum Q_1 differs from

$$\sum_z \left\{ \log \frac{1}{\rho(z, \xi)} : z \in \Phi(\Gamma) \cap s_F, \ \frac{1}{2} \le \rho(z, \xi) < r_1 \right\}$$

by a bounded quantity. Therefore, by the choice of the radius $r_1 = r_1(\varepsilon)$, we have

$$Q_1 \le (\gamma + \varepsilon) \log \frac{1}{1 - r_1} + O(1).$$

To estimate Q_2, first observe that $1 - t \le 1 - t^2$ for $0 \le t < 1$, so that

$$Q_2 \le \sum_z \left\{ \frac{(1 - |z|^2)(1 - |\xi|^2)}{|1 - z\bar{\xi}|^2} : z \in \Phi(\Gamma) \cap s_F \cap a_k, \ r_1 \le \rho(z, \xi) < 1 \right\}$$

and if we again use the estimate following Lemma 2.14, we find

$$Q_2 \le C(\delta) \int_{S(\xi, r_1)} \left(1 - \rho^2(w, \xi) \right) \frac{dA(w)}{(1 - |w|^2)^2},$$

for some positive constant $C(\delta)$, where

$$S(\xi, r_1) = \left\{ w \in \mathbb{D} : r_1 < \rho(w, \xi), \ |w| < \frac{|\xi| + r_1}{1 + |\xi| r_1} \right\}.$$

The above integral is bounded; to obtain the bound we apply a Möbius change of variables that takes ξ to 0:

$$\int_{S(\xi, r_1)} \left(1 - \rho^2(w, \xi) \right) \frac{dA(w)}{(1 - |w|^2)^2} \le \int_{S(r_1)} \frac{dA(w)}{1 - |w|^2} = \kappa A(S(r_1));$$

here

$$S(r_1) = \left\{ w \in \mathbb{D} : |w| > r_1, \ \left| w - \frac{1 - r_1}{2} \right| < \frac{1 + r_1}{2} \right\}.$$

A straightforward computation then shows that the κ-area of $S(r_1)$ is bounded by a constant C_0 that is independent of r_1.

We now combine the above estimates, and obtain

$$\frac{1}{r_{k+1}} \int_{\mathbb{T}(0, r_{k+1}) \cap s'_F} \log \frac{1}{|B_k(\xi)|} \, ds(\xi)$$

$$\le \frac{|\mathbb{T}(0, r_{k+1}) \cap s'_F|}{r_{k+1}} \left((\gamma + \varepsilon) \log \frac{1}{1 - r_1} + C_0 + C_1(\delta) \right), \quad (5.19)$$

where $|\cdot|_s$ is the normalized length (measure) of the set in question and $C_1(\delta)$ corresponds to the contribution from the points $z \in \Phi(\Gamma) \cap s_F$ for which $\rho(z, \xi) \leq \frac{1}{2}$; it depends on the separation constant δ.

For $k = 1, 2, 3, \ldots$ and any $0 < r_1 < 1$, we have

$$\log \frac{1}{1 - r_1} \leq \log \frac{1 - r_k}{1 - r_{k+1}} + O(1)$$

and

$$\frac{|\mathbb{T}(0, r_{k+1}) \cap s'_F|_s}{r_{k+1}} \log \frac{1 - r_k}{1 - r_{k+1}} \leq \kappa A(s'_F \cap a_k \cap a_{k-1}).$$

With these estimates in our pocket, we see from (5.19) that

$$\frac{1}{r_{k+1}} \int_{\mathbb{T}(0, r_{k+1}) \cap s'_F} \log \frac{1}{|B_k(\xi)|} \, ds(\xi)$$

$$\leq \left(\gamma + \varepsilon + \frac{C_2(\delta)}{\log \frac{1}{1-r_1}} \right) \kappa A(s'_F \cap a_k \cap a_{k-1}). \tag{5.20}$$

Adding (5.18) and (5.20), we then deduce from (5.17) that

$$\sum_{z \in \Phi(\Gamma) \cap s_F \cap a_k \cap a_{k+1}} \log \frac{1}{|z|} \leq C'(\delta) \, (1 - r_1) \kappa A(a_k \cap s'_F)$$

$$+ \left(\gamma + \varepsilon + \frac{C_2(\delta)}{\log \frac{1}{1-r_1}} \right) \kappa A(s'_F \cap a_k \cap a_{k-1}).$$

Summing over all indices $k = 1, 2, 3, \ldots$, we get

$$\sum_{z \in \Phi(\Gamma) \cap s_F \setminus \mathbb{D}(0, r_1)} \log \frac{1}{|z|} \leq 2C'(\delta) \, (1 - r_1) \kappa A(a_k \cap s'_F)$$

$$+ \left(\gamma + \varepsilon + \frac{C_2(\delta)}{\log \frac{1}{1-r_1}} \right) \kappa A(s'_F)$$

$$\leq \left(2C'(\delta) \, (1 - r_1) + \gamma + \varepsilon + \frac{C_2(\delta)}{\log \frac{1}{1-r_1}} \right) \widehat{\kappa}(F) + C_3(\delta, \varphi),$$

where $\mathbb{D}(0, r_1) = \{z \in \mathbb{C} : |z| < r_1\}$ and $C_3(\delta, \varphi)$ is a constant.

Since $\rho(\Phi(\Gamma)) = \rho(\Gamma) = \delta > 0$ for each $\Phi \in \text{Aut}(\mathbb{D})$, the number of points from $\Phi(\Gamma)$ in the annular region $\mathbb{D}(0, r_1) \setminus \mathbb{D}(0, \frac{1}{2})$ is bounded by a constant that depends only on r_1 and δ but not on the Möbius automorphism Φ; we can therefore include in the sum above all the points $z \in \Phi(\Gamma) \cap s_F \setminus \mathbb{D}(0, \frac{1}{2})$ and obtain for all finite $F \subset \mathbb{T}$ the following estimate of the modified logarithmic summation function:

$$\sup_{\Phi \in \text{Aut}(\mathbb{D})} \Lambda(\Phi(\Gamma), s_F^*) \leq \left(\gamma + \varepsilon + c(\varepsilon, \delta) \right) \widehat{\kappa}(F) + C(\varepsilon, \delta);$$

here, we recall the notation $\mathfrak{s}_F^* = \mathfrak{s}_F \setminus \mathbb{D}(0, \frac{1}{2})$. We replaced the κ-area of $\mathfrak{s}_F'$ by $\hat{\kappa}(F)$, because we know the two are comparable. The constants have the following properties: $c(\varepsilon, \delta) \to 0$ as $\varepsilon \to 0$, and $C(\varepsilon, \delta)$ is independent of F but may grow to $+\infty$ if we let $\varepsilon \to 0$. Comparing this with the definition of the uniform upper asymptotic κ-*-density $D_{*u}^+(\Gamma)$, we conclude that $D_{*u}^+(\Gamma) \leq \gamma = D_s^+(\Gamma)$. In view of Lemma 5.12, which states that for separated Γ, $D_u^+(\Gamma) = D_{*u}^+(\Gamma)$, this completes the proof of the identity $D_u^+(\Gamma) = D_s^+(\Gamma)$.   ∎

COROLLARY 5.14 *A sequence Γ in $\mathbb{D}$ is an $\mathcal{A}^{-\alpha}$-interpolation sequence if and only if Γ is separated and $D_s^+(\Gamma) < \alpha$.*

5.2   Sampling Sets for $\mathcal{A}^{-\alpha}$

A relatively closed subset Γ of $\mathbb{D}$ is called an $\mathcal{A}^{-\alpha}$-sampling set (or a sampling set for $\mathcal{A}^{-\alpha}$) if there exists a positive constant L such that

$$\|f\|_{-\alpha} \leq L \sup\{(1 - |z|^2)^\alpha |f(z)| : z \in \Gamma\} \tag{5.21}$$

for all $f \in \mathcal{A}^{-\alpha}$. The smallest such constant L will be denoted $L(\Gamma) = L_\alpha(\Gamma)$; we will write $L(\Gamma) = +\infty$ if Γ is not not a set of sampling for $\mathcal{A}^{-\alpha}$. It is easy to check that $L(\Gamma) = L(\Phi(\Gamma))$ for every Möbius map Φ of the disk.

For two relatively closed subsets Γ and Γ' of $\mathbb{D}$, we define

$$\rho(\Gamma, \Gamma') = \sup_{w \in \Gamma} \left\{\rho(w, \Gamma')\right\} = \sup_{w \in \Gamma} \inf_{w' \in \Gamma'} \rho(w, w').$$

The quantity

$$[\Gamma, \Gamma'] = \max \left\{\rho(\Gamma, \Gamma'), \rho(\Gamma', \Gamma)\right\}$$

is called the *Hausdorff pseudohyperbolic distance* between Γ and Γ'.

LEMMA 5.15 *If Γ and Γ' are two relatively closed subsets of $\mathbb{D}$ with*

$$\rho(\Gamma, \Gamma') < \min \left\{\frac{1}{2}, \frac{1}{C_\alpha L_\alpha(\Gamma)}\right\},$$

then

$$L_\alpha(\Gamma') \leq \frac{L_\alpha(\Gamma)}{1 - C_\alpha L_\alpha(\Gamma)\rho(\Gamma, \Gamma')},$$

where C_α is the constant in Lemma 5.1.

Proof. For any $f \in \mathcal{A}^{-\alpha}$ and any $w, w' \in \mathbb{D}$ with $\rho(w, w') < \frac{1}{2}$, Lemma 5.1 says that

$$\left|(1 - |w|^2)^\alpha |f(w)| - (1 - |w'|^2)^\alpha |f(w')|\right| \leq C_\alpha \|f\|_{-\alpha}\rho(w, w'),$$

which implies that

$$(1 - |w'|^2)^\alpha |f(w')| \geq (1 - |w|^2)^\alpha |f(w)| - C_\alpha \|f\|_{-\alpha}\rho(w, w').$$

For every $w \in \Gamma$, the definition of $\rho(\Gamma, \Gamma')$ implies that there is some $w' \in \Gamma'$ with $\rho(w, w') < \rho(\Gamma, \Gamma') + \varepsilon$, where $\varepsilon > 0$ is arbitrary.

Now, by assumption, there is some $w \in \Gamma$ such that

$$L_\alpha(\Gamma)(1 - |w|^2)^\alpha |f(w)| \geq \|f\|_\alpha - \varepsilon.$$

Thus, there exists some $w' \in \Gamma'$ with

$$(1 - |w'|^2)^\alpha |f(w')| \geq \frac{1}{L_\alpha(\Gamma)}\big(\|f\|_{-\alpha} - \varepsilon\big) - C_\alpha \|f\|_{-\alpha}\big[\rho(\Gamma, \Gamma') + \varepsilon\big].$$

Since ε can be arbitrarily small, we have

$$\sup\left\{(1 - |w'|^2)^\alpha |f(w')| : w' \in \Gamma'\right\} \geq \|f\|_{-\alpha}\left(\frac{1}{L_\alpha(\Gamma)} - C_\alpha \rho(\Gamma, \Gamma')\right),$$

whence the assertion follows. ∎

COROLLARY 5.16 *If Γ is an $\mathcal{A}^{-\alpha}$-sampling set, then Γ contains a separated sequence that is also sampling for $\mathcal{A}^{-\alpha}$.*

Since any superset of an $\mathcal{A}^{-\alpha}$-sampling set is also $\mathcal{A}^{-\alpha}$-sampling, the corollary above tells us that to characterize sampling sets for $\mathcal{A}^{-\alpha}$, it suffices to consider separated sequences.

COROLLARY 5.17 *If $\Gamma = \{z_j\}_j$ is a separated $\mathcal{A}^{-\alpha}$-sampling sequence, then there exists a constant $\delta > 0$ such that every sequence $\Gamma' = \{z_j'\}_j$ with $\rho(z_j, z_j') < \delta$ for all j is also an $\mathcal{A}^{-\alpha}$-sampling sequence. Moreover, $L_\alpha(\Gamma') \leq C$, where C depends only on δ and $L_\alpha(\Gamma)$.*

The above corollary states that separated $\mathcal{A}^{-\alpha}$-sampling sequences are stable under small perturbations in the pseudohyperbolic metric. Recall that sequences of interpolation for $\mathcal{A}^{-\alpha}$ also have this property.

The $\mathcal{A}^{-\alpha}$-sampling sets will be characterized in terms of a certain notion of lower density. More specifically, if $\Gamma = \{z_j\}_j$ is separated, then the *lower Seip density* of Γ is defined as

$$D_s^-(\Gamma) = \liminf_{r \to 1^-} \inf_{z \in \mathbb{D}} D(\Gamma_z, r), \tag{5.22}$$

where Γ_z and $D(\Gamma, r)$ are the same as in the definition of the upper Seip density; see equations (5.12) and (5.13) in the preceding section.

We can now characterize the sampling sets for $\mathcal{A}^{-\alpha}$.

THEOREM 5.18 *A set $\Gamma \subset \mathbb{D}$ is $\mathcal{A}^{-\alpha}$-sampling if and only if it contains a separated sequence Γ' with $\alpha < D_s^-(\Gamma')$. In particular, if Γ itself is separated, then Γ is sampling if and only if $\alpha < D_s^-(\Gamma)$.*

Proof. We first prove the necessity of the condition. By Corollary 5.16, we may assume that Γ is a separated sequence. Put $\beta = D_s^-(\Gamma)$ and assume $L(\Gamma) < +\infty$. Let $\{\varepsilon_j\}_j$ be a sequence of positive numbers approaching zero, and pick a sequence

$\{w_j\}_j$ of points in $\mathbb{D}$ and a sequence $\{r_j\}_j$ (with $1 - \varepsilon_j < r_j < 1$ for all j) such that

$$D(\Gamma_{w_j}, r_j) \le \beta + \varepsilon_j, \qquad j = 1, 2, 3, \ldots, \tag{5.23}$$

where Γ_{w_j} is the Möbius shifted sequence, as above. For each w_j in the sequence, put $\Gamma_{w_j} = \Gamma_j = \{z_{k.j}\}_k$ and construct a new sequence of points $\Gamma'_j = \{z'_{k.j}\}_k$ as follows. If $z_{k.j} \ne 0$, set $z'_{k.j} = z_{k.j}|z_{k.j}|^{-\delta_0}$; otherwise, set $z'_{k.j} = \delta_0$. Here, $\delta_0 \in (0, \frac{1}{2})$ is so small that $\sup_j L(\Gamma'_j) < +\infty$. By Corollary 5.17, this can be accomplished, because an easy computation shows that $\rho(z, |z|^{-\delta}z) \to 0$ as $\delta \to 0$ uniformly in $z \in \mathbb{D}$. We now have

$$D(\Gamma'_j, r_j) \le (1 - \delta_0)\, D(\Gamma_j, r_j) + \frac{C}{\log\frac{1}{1-r_j}} \tag{5.24}$$

for all j. On the other hand, the modified finite Blaschke products

$$f_j(z) = \prod_{k:|z'_{k.j}|<r_j} \frac{z'_{k.j} - z}{|z'_{k.j}|(1 - \bar{z}'_{k.j}z)}$$

satisfy $\|f_j\|_{-\alpha} \ge 1$ and

$$\sup_{z \in \Gamma'_j} (1 - |z|^2)^\alpha |f_j(z)| \le \exp\left(\sum_{k:|z'_{k.j}|<r_j} \log\frac{1}{|z'_{k.j}|} - \alpha\log\frac{1}{1 - r_j^2} \right),$$

which by (5.23) and (5.24) implies that

$$(1 - \delta_0)(\beta + \varepsilon_j) + \frac{C}{\log\frac{1}{\varepsilon_j}} \ge \alpha,$$

where C depends only on $L(\Gamma)$ and δ_0. Since $\varepsilon_j \to 0$ as $j \to +\infty$, we have proved that $\beta > \alpha$.

To prove the sufficiency part of the condition, we assume now that a separated sequence Γ with $D_s^-(\Gamma) > \alpha$ is not sampling for $\mathcal{A}^{-\alpha}$. Then there exists a sequence of functions $\{f_n\}_n$ in $\mathcal{A}^{-\alpha}$ and a corresponding sequence of points $\{a_n\}_n$ in $\mathbb{D}$, such that $\|f_n\|_{-\alpha} = n$ and

$$(1 - |a_n|^2)^\alpha |f_n(a_n)| \ge \frac{n}{2},$$

but

$$\sup_{z \in \Gamma}(1 - |z|^2)^\alpha |f_n(z)| \le C,$$

where C is a constant (independent of n). We now apply the unitary operator T_n to f_n, where

$$T_n f(z) = \Phi'_n(z)^\alpha \, f(\Phi_n(z)) \quad \text{and} \quad \Phi_n(z) = \frac{z + a_n}{1 + \bar{a}_n z}.$$

The resulting functions

$$g_n(z) = \frac{(1 - |a_n|^2)^\alpha}{(1 + \bar{a}_n z)^{2\alpha}} \, f_n \left(\frac{z + a_n}{1 + \bar{a}_n z} \right)$$

have the properties that $\|g_n\|_{-\alpha} = n$, $|g_n(0)| \geq \frac{n}{2}$, and

$$\sup \left\{ (1 - |z|^2)^\alpha |g_n(z)| : z \in \Gamma_{-a_n} \right\} \leq C,$$

where $\Gamma_{-a_n} = \Phi_n(\Gamma)$.

By Lemma 5.8 and a normal family argument, there exists a sequence of indices $n_1 < n_2 < \cdots$ such that $\Gamma_{-a_{n_k}} \to \Gamma'$ and $f_{n_k}/n_k \to h$ uniformly on compact subsets of $\mathbb{D}$, as $k \to +\infty$, where Γ' is a separated sequence with $D_s^-(\Gamma') > \alpha$. Also, $\|h\|_{-\alpha} \leq 1$, $|h(0)| \geq \frac{1}{2}$, and $h|_{\Gamma'} = 0$. Thus, Γ' is a zero sequence for $\mathcal{A}^{-\alpha}$, which, according to Corollary 4.27, implies that $D^+(\Gamma') \leq \alpha$. On the other hand, repeating the argument used in the first part of the proof of Theorem 5.13, we obtain $D^+(\Gamma') \geq D_s^-(\Gamma') > \alpha$ and arrive at a contradiction. This completes the proof of the theorem. ∎

Note that it is also easy to prove that Γ' is not a zero set for $\mathcal{A}^{-\alpha}$, when $D_s^-(\Gamma') > \alpha$, by using the classical Jensen's formula.

5.3 Interpolation and Sampling in A_α^p

In this section, we show how the techniques of the previous two sections can be adapted to yield characterizations of interpolating and sampling sequences for the spaces A_α^p. We begin with A_α^p-interpolation sequences.

A sequence $\Gamma = \{z_j\}_j$ of distinct points in $\mathbb{D}$ is called an A_α^p-*interpolation sequence* (or a sequence of interpolation for A_α^p) if for every sequence $\{w_j\}_j$ of complex numbers satisfying the condition

$$\sum_j (1 - |z_j|^2)^{2+\alpha} |w_j|^p < +\infty,$$

there exists a function $f \in A_\alpha^p$ such that $f(z_j) = w_j$ for all j. The compatibility condition above follows easily from estimate (5.6); that any A_α^p-interpolation sequence is separated will be proved shortly.

It will be convenient for us later if we introduce the *weighted restriction operator* $R_\Gamma = R_{\Gamma, p, \alpha}$, which is defined by assigning to every analytic function f in $\mathbb{D}$ the numerical sequence

$$R_\Gamma f = \{(1 - |z_j|^2)^{(2+\alpha)/p} f(z_j)\}_j.$$

In terms of this restriction operator, we see that Γ is an A_α^p-interpolation sequence if and only if $l^p \subset R_\Gamma(A_\alpha^p)$.

Recall from Exercise 19 of Chapter 2 that the restriction operator R_Γ maps A_α^p boundedly into l^p if and only if the sequence Γ is the union of finitely many separated sequences. Thus, if we can show that every A_α^p-interpolation sequence is separated, then for every A_α^p-interpolation sequence Γ we actually have $R_\Gamma(A_\alpha^p) = l^p$.

It is important to realize that the space A_α^p possesses a natural group of unitary operators. More specifically, if Φ is a Möbius map of the disk and U_Φ is the operator defined by

$$U_\Phi f(z) = f \circ \Phi(z) \left(\Phi'(z)\right)^{(2+\alpha)/p},$$

then U_Φ is an isometric isomorphism of A_α^p.

LEMMA 5.19 *If Γ is an A_α^p-interpolation sequence, then Γ is separated, and so R_Γ maps A_α^p boundedly onto l^p.*

Proof. Assume that $\Gamma = \{z_j\}_j$ is not separated. Then there is a sequence $\{(w_n, w_n')\}_n$ of pairs of distinct points from Γ such that $\rho(w_n, w_n') \le 2^{-n}$ for all n. From the interpolation assumption on Γ, we see that there must exist a function $f \in A_\alpha^p$ such that

$$(1 - |w_n|^2)^{(2+\alpha)/p} f(w_n) = 2^{-\sqrt{n}} \qquad \text{for all } n,$$

and $f(z) = 0$ for all $z \in \Gamma \setminus \{w_n\}_n$.

Consider now the functions $g_n = U_n f$, where U_n is the unitary operator on A_α^p introduced earlier that corresponds to the Möbius map

$$\varphi_n(z) = \frac{w_n - z}{1 - \overline{w}_n z}, \qquad z \in \mathbb{D}.$$

For each n, we have $\|g_n\| = \|f\|$, $g_n(0) = 2^{-\sqrt{n}}$, and $g_n(\zeta_n) = 0$, where

$$\zeta_n = \varphi_n(w_n') = \frac{w_n - w_n'}{1 - \overline{w}_n w_n'}.$$

In particular, the sequences $\{g_n\}_n$ and $\{g_n'\}_n$ are both uniformly bounded on compact subsets of $\mathbb{D}$.

For each n, we have

$$|\zeta_n| = \rho(w_n, w_n') \le 2^{-n}.$$

Now, write $g_n(\zeta_n) - g_n(0) = -2^{-\sqrt{n}}$ as an integral of the derivative g_n' along the line segment joining 0 and ζ_n, and apply the triangle inequality. The result is that for each n, there exists some $\theta_n \in [0, 1]$ such that $|g_n'(\theta_n \zeta_n)| \ge 2^{n-\sqrt{n}}$. This contradicts the earlier conclusion that $\{g_n'\}_n$ is uniformly bounded on compact subsets in $\mathbb{D}$. ∎

If $\Gamma = \{z_j\}_j$ is an A_α^p-interpolation sequence, then Γ is clearly an A_α^p-zero set, and the invariant subspace $I_\Gamma = \{f \in A_\alpha^p : f|_\Gamma = 0\}$ is the kernel of the weighted restriction operator $R_\Gamma : A_\alpha^p \to l^p$. Since R_Γ is bounded and onto, the quotient map $R_\Gamma : A_\alpha^p / I_\Gamma \to l^p$ has a bounded inverse with the norm

$$M(\Gamma) = \sup \left\{ \inf \left\{ \|f\|_{p,\alpha} : f(z_j) = w_j \right\} : \sum_j (1 - |z_j|^2)^{2+\alpha} |w_j|^p \le 1 \right\}.$$

A normal family argument shows that the infimum above is always achieved. The quantity $M(\Gamma)$ is Möbius invariant, that is, $M(\Gamma) = M(\Phi(\Gamma))$ for every Möbius map Φ. We will call $M(\Gamma)$ the A_α^p-interpolation constant of Γ. By convention, we are going to write $M(\Gamma) = +\infty$ if Γ is not an A_α^p-interpolation sequence.

We need an estimate for the l^p-distance between $R_\Gamma f$ and $R_{\Gamma'} f$, where $f \in A_\alpha^p$, and Γ and Γ' are two sequences close in the hyperbolic metric. This will enable us to prove that A_α^p-interpolation sequences are stable under small perturbations with respect to the hyperbolic metric.

LEMMA 5.20 *Let $\Gamma = \{z_j\}_j$ be a separated sequence with separation constant*

$$\rho_0 = \rho(\Gamma) = \inf\{\rho(a, b) : a, b \in \Gamma, a \neq b\} > 0.$$

Then there exists a positive constant $C = C(p, \alpha, \rho_0)$ with the property that if $\Gamma' = \{z_j'\}_j$ is another sequence in $\mathbb{D}$ with $\rho(z_j, z_j') \leq \delta \leq \rho_0/8$ for all j, then for each $f \in A_\alpha^p$ we have

$$\|R_\Gamma f - R_{\Gamma'} f\|_{l^p} \leq C\delta \|f\|_{p,\alpha}.$$

Proof. For each j, let D_j, D_j', and D_j'' be the closed pseudohyperbolic disks "centered" at z_j with "radius" $\rho_0/2$, $\rho_0/4$, and $\rho_0/8$, respectively. Also let

$$m_j = m_j(f) = \max\{|f(z)| : z \in D_j'\}$$

and

$$m_j' = m_j'(f) = \max\{|f(z)| : z \in D_j''\}.$$

Using the fact that $\rho(a, b)$ is comparable to $|a - b|/(1 - |a|^2)$ for $\rho(a, b) \leq \frac{1}{2}$, we get

$$\left| (1 - |z_j|^2)^{(2+\alpha)/p} f(z_j) - (1 - |z_j'|^2)^{(2+\alpha)/p} f(z_j') \right|$$
$$\leq (1 - |z_j|^2)^{(2+\alpha)/p} |f(z_j) - f(z_j')|$$
$$\times \left| (1 - |z_j|^2)^{(2+\alpha)/p} - (1 - |z_j'|^2)^{(2+\alpha)/p} \right| |f(z_j')|$$
$$\leq C \left[(1 - |z_j|^2)^{1+(2+\alpha)/p} m_j' + (1 - |z_j'|^2)^{(2+\alpha)/p} m_j \right] \rho(z_j, z_j').$$

By Cauchy's formula, $m_j' \leq C_1 m_j/(1 - |z_j|^2)$. Thus,

$$\left| (1 - |z_j|^2)^{(2+\alpha)/p} f(z_j) - (1 - |z_j'|^2)^{(2+\alpha)/p} f(z_j') \right|$$
$$\leq C_2 (1 - |z_j|^2)^{(2+\alpha)/p} m_j \delta.$$

Here, the constants C, C_1, and C_2 depend only on p, α, and ρ_0. It follows that

$$\|R_\Gamma f - R_{\Gamma'} f\|_{l^p}^p \leq C_2^p \delta^p \sum_j (1 - |z_j|^2)^{2+\alpha} m_j^p.$$

By Lemma 2.14 and the remark thereafter, there exists another positive constant C_3, depending only on p, α, and ρ_0, such that

$$m_j^p \leq \frac{C_3}{\rho_0^2 (1 - |z_j|^2)^{2+\alpha}} \int_{D_j} |f(z)|^p \, dA_\alpha(z)$$

for all j. The desired inequality now follows immediately.   ∎

LEMMA 5.21 *Let* $\Gamma = \{z_j\}_j$ *be an* A_α^p-*interpolation sequence with* $\rho_0 = \rho(\Gamma) > 0$. *Then there exists a positive constant* δ *such that any other sequence* $\Gamma' = \{z_j'\}_j$ *satisfying* $\rho(z_j, z_j') \leq \delta$ *for all* j *is also a sequence of interpolation for* A_α^p.

Proof. The proof is similar to that of Proposition 5.3. However, since for $0 < p < 1$ the "norms" $\|\cdot\|_{p,\alpha}$ and $\|\cdot\|_{l^p}$ do not satisfy the triangle inequality, we have to use the metric $d(f, g) = \|f - g\|_{p,\alpha}^p$ in A_α^p and $d(a, b) = \|a - b\|_{l^p}^p$ in l^p, rather than the norms, in proving the convergence of the iteration process. Because of this complication, we obtain two different estimates for $M(\Gamma')$, namely, if

$$\delta \leq \min\left(\frac{\rho_0}{8}, \frac{1}{CM(\Gamma)}\right),$$

then

$$M(\Gamma') \leq \frac{M(\Gamma)}{1 - C\delta M(\Gamma)}$$

for $1 \leq p < +\infty$, and

$$M(\Gamma') \leq \frac{M(\Gamma)}{\left[1 - (C\delta M(\Gamma))^p\right]^{1/p}}$$

for $0 < p < 1$. Here, C is the constant in Lemma 5.20.   ∎

We can now characterize the A_α^p-interpolating sequences in terms of the upper densities $D_u^+(\Gamma)$ and $D_s^+(\Gamma)$.

THEOREM 5.22 *Suppose* $0 < p < +\infty$, $-1 < \alpha < +\infty$, *and* Γ *is a sequence of distinct points in* $\mathbb{D}$. *Then the following conditions are equivalent:*

(i) Γ *is a sequence of interpolation for* A_α^p.

(ii) Γ *is separated and* $D_s^+(\Gamma) < (\alpha + 1)/p$.

(iii) $D_u^+(\Gamma) < (\alpha + 1)/p$.

Proof. According to Theorem 5.13, we only need to prove the equivalence of (i) and (iii). Essentially we are going to modify the proof of Theorem 5.6 to suit the present situation.

First assume that $\Gamma = \{z_j\}_j$ is a sequence of interpolation for A_α^p. For each n, the identity $M(\Gamma) = M(\Phi_n(\Gamma))$, where $\Phi_n(z) = (z_n - z)/(1 - \bar{z}_n z)$, implies that there is a function $f_n \in A_\alpha^p$ such that

$$\|f_n\|_{p,\alpha} \leq M(\Gamma), \qquad f_n(0) = 1, \quad \text{and} \quad f_n|_{\Gamma_n} = 0;$$

recall that $\Gamma_n = \Phi_n(\Gamma) \setminus \{0\} = \Phi_n(\Gamma \setminus \{z_n\})$. On the other hand, the extremal function G_n for the invariant subspace I_{Γ_n} maximizes $|f(0)|$ on the unit ball of I_{Γ_n}, so that $G_n(0) \geq M(\Gamma)^{-1}$ for all n. According to Theorem 3.3, the function G_n belongs to $\mathcal{A}^{-(1+\alpha)/p}$ with $\|G_n\|_{-(1+\alpha)/p} \leq 1$. Using the Jensen-type estimate of Lemma 5.4, we see that for every $\varepsilon > 0$,

$$\sup_n \Lambda(\Gamma_n, \tau_F) \leq \left(\frac{1+\alpha}{p} + \varepsilon \right) \widehat{\kappa}(F) + O(1),$$

uniformly in the finite subsets F of $\mathbb{T}$. In other words, $D_u^+(\Gamma) \leq (1+\alpha)/p$.

To prove that we actually have the strict inequality $D_u^+(\Gamma) < (\alpha+1)/p$, we write $\Gamma_n = \{z_{k,n}\}_k$ and construct a new sequence $\Gamma_n' = \{z_{k,n}'\}_k$, where $z_{k,n}' = z_{k,n}|z_{k,n}|^\delta$, and proceed exactly as we did in the proof of Theorem 5.6, except that we use Lemma 5.21 here instead of Proposition 5.3. This completes the proof that (i) implies (iii).

Next, we assume that (iii) holds. By Theorem 5.6 and Corollary 5.2, the sequence Γ is separated. We now fix any β such that

$$D_u^+(\Gamma) < \beta < (\alpha + 1)/p.$$

By Corollary 4.32 and Möbius invariance, there exists a sequence $\{g_k\}_k$ of functions in $\mathcal{A}^{-\beta}$ such that for each k,

$$g_k(z_k) = (1 - |z_k|^2)^{-\beta}$$

and

$$g_k(z_j) = 0, \qquad j = 1, 2, 3, \ldots, j \neq k,$$

which meet the growth restriction

$$|g_k(z)| \leq C (1 - |z|^2)^{-\beta}, \qquad z \in \mathbb{D},$$

for all k, where C is some constant independent of z and k.

If $\{w_k\}_k$ is a sequence of complex numbers satisfying the compatibility condition

$$\sum_k (1 - |z_k|^2)^{\alpha+2}|w_j|^p < +\infty,$$

we can construct a function $f \in A_\alpha^p$ such that $f(z_j) = w_j$ for all j. In fact, if we fix a sufficiently large number θ and define

$$f(z) = \sum_k w_k (1 - |z_k|^2)^\beta g_k(z) \left(\frac{1 - |z_k|^2}{1 - \overline{z}_k z} \right)^\theta, \qquad z \in \mathbb{D},$$

then it is clear that f is analytic in $\mathbb{D}$ with $f(z_j) = w_j$ for all j. It remains to show that $f \in A_\alpha^p$.

If $0 < p \le 1$, then

$$|f(z)|^p \le \sum_k |w_k|^p (1 - |z_k|^2)^{\beta p} |g_k(z)|^p \left| \frac{1 - |z_k|^2}{1 - \bar{z}_k z} \right|^{\theta p}, \qquad z \in \mathbb{D}.$$

Using the growth condition that each g_k satisfies, we can find a positive constant C such that

$$|f(z)|^p \le C \sum_k |w_k|^p (1 - |z_k|^2)^{p(\beta+\theta)} \frac{(1 - |z|^2)^{-\beta p}}{|1 - \bar{z}_k z|^{p\theta}}, \qquad z \in \mathbb{D}.$$

An application of Theorem 1.7 then yields that $f \in A_\alpha^p$.

If $1 < p < +\infty$, we let q be the conjugate exponent with $p^{-1} + q^{-1} = 1$. By the growth constraint of each g_k, we can find a positive constant C_1 such that

$$(1 - |z|^2)^\beta |f(z)| \le C_1 \sum_k |w_k| (1 - |z_k|^2)^a \frac{(1 - |z_k|^2)^b}{|1 - \bar{z}_k z|^\theta},$$

where a and b are constants satisfying $a + b = \beta + \theta$. We then apply Hölder's inequality to get

$$(1 - |z|^2)^\beta |f(z)| \le C_1 \left[\sum_k \frac{|w_k|^p (1 - |z_k|^2)^{pa}}{|1 - \bar{z}_k z|^2} \right]^{\frac{1}{p}} \left[\sum_k \frac{(1 - |z_k|^2)^{bq}}{|1 - \bar{z}_k z|^\theta} \right]^{\frac{1}{q}}.$$

The second sum above can be estimated as follows:

$$\sum_k \frac{(1 - |z_k|^2)^{bq}}{|1 - \bar{z}_k z|^\theta} \le C_2 \int_\mathbb{D} \frac{(1 - |w|^2)^{bq-2} \, dA(w)}{|1 - \bar{w} z|^\theta} \le C_3 (1 - |z|^2)^{bq-\theta};$$

here the first inequality follows from (5.6) and the second from Theorem 1.7, provided that θ is sufficiently large and $bq > 1$ (which are easy to achieve). We can now find a positive constant C_3 such that

$$|f(z)|^p \le C_3 (1 - |z|^2)^{p(b-\beta-\theta/q)} \sum_k \frac{|w_k|^p (1 - |z_k|^2)^{pa}}{|1 - \bar{z}_k z|^\theta}, \qquad z \in \mathbb{D}.$$

If θ is sufficiently large and

$$p\left(b - \beta - \frac{\theta}{q}\right) + \alpha > -1,$$

which are again easy to achieve, another application of Theorem 1.7 shows that

$$\int_\mathbb{D} |f(z)|^p \, dA_\alpha(z) \le C \sum_k (1 - |z_k|^2)^{2+\alpha} |w_k|^p,$$

where C is some other constant. This completes the proof of Theorem 5.22. ∎

We now turn to the study of A_α^p-sampling sequences.

A sequence $\Gamma = \{z_j\}_j$ of (not necessarily distinct) points in $\mathbb{D}$ is called an A_α^p-*sampling sequence* (or a sequence of sampling for A_α^p) if there exists a positive constant C such that

$$C^{-1} \int_{\mathbb{D}} |f(z)|^p \, dA_\alpha(z) \leq \sum_{j=1}^{+\infty} (1 - |z_j|^2)^{2+\alpha} |f(z_j)|^p \leq C \int_{\mathbb{D}} |f(z)|^p \, dA_\alpha(z)$$

(5.25)

for all $f \in A_\alpha^p$. Once again, the second inequality above implies that Γ is the union of finitely many separated sequences.

THEOREM 5.23 *Suppose $1 \leq p < +\infty$, $-1 < \alpha < +\infty$, and Γ is a sequence of points in $\mathbb{D}$. Then Γ is a sequence of sampling for A_α^p if and only if Γ is the union of finitely many separated sequences and contains a separated sequence Γ' such that $D_s^-(\Gamma') > (\alpha + 1)/p$. In particular, if Γ itself is separated, the criterion for sampling is $D_s^-(\Gamma) > (\alpha + 1)/p$.*

Proof. Assume that Γ is an A_α^p-sampling sequence. Then Γ is the union of finitely many, say n, separated sequences. We show that Γ contains an A_α^p-sampling sequence that is the union of $n - 1$ separated sequences. By induction, it will then follow that every A_α^p-sampling sequence contains a separated A_α^p-sampling sequence.

Let $\Gamma = \Gamma_1 \cup \Gamma_2$, where Γ_2 is separated with $\rho(\Gamma_2) = \rho_2 > 0$ and Γ_1 is the union of $n - 1$ separated sequences:

$$\Gamma_1 = \bigcup_{j=1}^{n-1} \Gamma_{1,j}.$$

If

$$\delta = \inf \left\{ \rho(z_1, z_2) : z_1 \in \Gamma_1, z_2 \in \Gamma_2 \right\} > 0,$$

there is nothing to prove because

$$\Gamma = (\Gamma_{1,1} \cup \Gamma_2) \cup \Gamma_{1,2} \cup \cdots \cup \Gamma_{1,n-1},$$

and so Γ is the union of $n - 1$ separated sequences. If $\delta = 0$, we split Γ_2 into two sequences, $\Gamma_2 = \Gamma_2' \cup \Gamma_2''$, where

$$\Gamma_2' = \{z \in \Gamma_2 : \rho(z, \Gamma_1) < \varepsilon\}$$

and $\Gamma_2'' = \Gamma_2 \setminus \Gamma_2'$; here ε is any fixed positive number less than or equal to $\rho_2/8$.

Write $\Gamma_2' = \{z_{2,j}'\}$. For each j, pick some $z_{1,j}' \in \Gamma_1$ with $\rho(z_{1,j}', z_{2,j}') < \varepsilon$. Let $\Gamma_1' = \{z_{1,j}'\}_j$ be the resulting sequence. It follows from Lemma 5.20 that for every $f \in A_\alpha^p$ we have

$$\left| \|R_{\Gamma_1'} f\|_p - \|R_{\Gamma_2'} f\|_p \right| \leq \|R_{\Gamma_1'} f - R_{\Gamma_2'} f\|_p \leq C(p, \alpha, \rho_2) \|f\|_{p,\alpha} \varepsilon.$$

For $\|f\|_{p,\alpha} = 1$, we obtain

$$\left| \|R_{\Gamma_1'} f\|_p^p - \|R_{\Gamma_2'} f\|_p^p \right| \leq C \|R_{\Gamma_1'} f - R_{\Gamma_2''} f\|_p \leq C'(p, \alpha, \rho_2) \varepsilon,$$

and so by homogeneity,

$$\left| \|R_{\Gamma_1'} f\|_p^p - \|R_{\Gamma_2'} f\|_p^p \right| \leq C'(p, \alpha, \rho_2)\varepsilon \|f\|_{p,\alpha}^p$$

for all $f \in A_\alpha^p$. This implies that

$$\sum_z \left\{ (1 - |z|^2)^{2+\alpha} |f(z)|^p : z \in \Gamma \right\} = \|R_\Gamma f\|_p^p$$

$$= \|R_{\Gamma_2''} f\|_p^p + \|R_{\Gamma_2'} f\|_p^p + \|R_{\Gamma_1'} f\|_p^p + \|R_{\Gamma_1 - \Gamma_1'} f\|_p^p$$

$$\leq 2\|R_{\Gamma_1 \cup \Gamma_2''} f\|_p^p + C' \|f\|_{p,\alpha}^p \varepsilon.$$

Since Γ is A_α^p-sampling, the above inequality implies that $\Gamma_1 \cup \Gamma_2''$ is also A_α^p-sampling, provided that ε is small enough. On the other hand,

$$\inf\{\rho(z_1, z_2) : z_1 \in \Gamma_1, z_2 \in \Gamma_2''\} > 0.$$

Therefore, $\Gamma_1 \cup \Gamma_2''$ is the union of $n - 1$ separated sequences. This completes the proof that every A_α^p-sampling sequence contains a separated sampling subsequence.

Next, we show that if Γ is a separated A_α^p-sampling sequence, then $D_s^-(\Gamma) > (\alpha + 1)/p$. The proof is very similar to that of the necessity part of Theorem 5.18; it involves the following steps.

Step 1. *Stability and Möbius invariance.* For a separated sequence $\Gamma = \{z_j\}_j$, let $L(\Gamma) = L(\Gamma; p, \alpha)$ denote the smallest constant L such that

$$\int_{\mathbb{D}} |f(z)|^p \, dA_\alpha(z) \leq L \sum_{j=1}^{+\infty} (1 - |z_j|^2)^{\alpha+2} |f(z_j)|^p \tag{5.26}$$

for all $f \in A_\alpha^p$; we put $L(\Gamma) = +\infty$ if Γ is not A_α^p-sampling. Stability means that if $L(\Gamma) < +\infty$ then there exist positive constants δ and C such that every sequence $\Gamma' = \{z_j'\}_j$ with $\rho(z_j, z_j') < \delta$ for all j satisfies $L(\Gamma') \leq C$; moreover, δ and C depend only on $\rho(\Gamma)$ and $L(\Gamma)$. Möbius invariance means that $L(\Gamma) = L(\Phi(\Gamma))$, where Φ is any Möbius map of the disk. The proof of Möbius invariance is based on the unitary transformations of A_α^p; see the paragraph preceding Lemma 5.19. The proof of stability is essentially a replica of the proof of Corollary 5.17.

Step 2. By moving radially every z_j away from the origin and replacing it with $\zeta_j = z_j |z_j|^{-\delta}$, where δ is sufficiently small, we obtain a new sequence $\Gamma' = \{\zeta_j\}_j$ that is also A_α^p-sampling (by stability) but has smaller local densities:

$$D(\Gamma', r) \leq (1 - \delta)D(\Gamma, r) + \frac{C}{\log \frac{1}{1-r}}.$$

Step 3. The desired result then follows by constructing a Blaschke product B with zeros from Γ' lying in a disk of radius r and substituting B for f in (5.26).

The above is an outline of the proof of the necessity part of Theorem 5.23. Details are left out to avoid repetition.

To prove the sufficiency part of Theorem 5.23, let us assume that $\Gamma = \{z_j\}_j$ is a separated sequence with $D_s^-(\Gamma) > (\alpha + 1)/p$. Write $\beta = (\alpha + 1)/p$ and pick some $\varepsilon > 0$ such that $\beta + \varepsilon < D_s^-(\Gamma)$. By Theorem 5.18, Γ is a sequence of sampling for $A^{-(\beta+\varepsilon)}$. This implies that the linear transformation

$$f \mapsto Tf = \left\{(1 - |z_j|^2)^{\beta+\varepsilon} f(z_j)\right\}_j,$$

is a bounded invertible operator from $A_0^{-(\beta+\varepsilon)}$ onto a closed subspace of the sequence space c_0 (consisting of sequences that converge to 0). Denote this subspace by a_0. Then any bounded linear functional φ on $A_0^{-(\beta+\varepsilon)}$ induces a bounded linear functional $\widetilde{\varphi}$ on a_0 via T, with $\|\widetilde{\varphi}\| \leq K\|\varphi\|$. For each $\zeta \in \mathbb{D}$, let e_ζ denote the normalized functional on $A_0^{-(\beta+\varepsilon)}$ of point evaluation at ζ, that is, $e_\zeta(f) = (1 - |\zeta|^2)^{\beta+\varepsilon} f(\zeta)$. We have $\|e_\zeta\| = 1$. Since the dual space of c_0 is l^1, an application of the Hahn-Banach extension theorem shows that for each $\zeta \in \mathbb{D}$, there exists a sequence $\{g_j(\zeta)\}_j$ in l^1 such that

$$(1 - |\zeta|^2)^{\beta+\varepsilon} f(\zeta) = \sum_{z_j \in \Gamma} (1 - |z_j|^2)^{\beta+\varepsilon} f(z_j) g_j(\zeta), \tag{5.27}$$

with

$$\sum_j |g_j(\zeta)| \leq K. \tag{5.28}$$

The factors $g_j(\zeta)$ in (5.27) are not uniquely determined, and we shall see that we can use this uncertainty to make them behave like $O[(1 - |\zeta|^2)]$ as $|\zeta| \to 1^-$, and in the process improve the convergence of (5.27). Fix an arbitrary number c, and for each $\zeta \in \mathbb{D}$, define

$$\Lambda(j, \zeta, c) = \left\{z \in \Gamma : \rho(z, \zeta) > \frac{1}{2}; \ |g_j(\zeta)| > c \frac{(1 - |\zeta|^2)(1 - |z|^2)}{|1 - \overline{\zeta}z|^2}\right\}.$$

By (5.28), each $\Lambda(j, \zeta, c)$ is a Blaschke sequence, so that we may apply (5.27) to $B_\Lambda f$, where $f \in A_0^{-(\beta+\varepsilon)}$ and B_Λ is the Blaschke product associated with the set $\Lambda = \Lambda(j, \zeta, c)$. Thus,

$$(1 - |\zeta|^2)^{\beta+\varepsilon} B_\Lambda(\zeta) f(\zeta) = \sum_j (1 - |z_j|^2)^{\beta+\varepsilon} f(z_j) \widetilde{g}_j(\zeta), \tag{5.29}$$

where $\widetilde{g}_j(\zeta) = B_\Lambda(z_j) g_j(\zeta)$. It is easily seen that for all j, we have

$$|\widetilde{g}_j(\zeta)| \leq C \frac{(1 - |\zeta|^2)(1 - |z_j|^2)}{|1 - \overline{z}_j\zeta|^2}, \qquad \zeta \in \mathbb{D},$$

where C is independent of ζ. In fact, if $z_j \in \Lambda(j, \zeta, c)$, then $\widetilde{g}_j(\zeta) = 0$; if $z_j \notin \Lambda(j, \zeta, c)$ with $\rho(z_j, \zeta) > \frac{1}{2}$, then by the definition of $\Lambda(j, \zeta, c)$,

$$|g_j(\zeta)| \leq c \frac{(1 - |\zeta|^2)(1 - |z_j|^2)}{|1 - \overline{\zeta}z_j|^2}:$$

and if $z_j \notin \Lambda(j, \zeta, c)$ with $\rho(z_j, \zeta) \le \frac{1}{2}$, then

$$|\widetilde{g}_j(\zeta)| \le K \le \frac{3}{4}K\left[1 - \rho(z_j, \zeta)^2\right] = \frac{3}{4}K\frac{(1 - |\zeta|^2)(1 - |z_j|^2)}{|1 - \bar{\zeta}z_j|^2}.$$

Observe that

$$|B_\Lambda(\zeta)| = \prod_{z_j \in \Lambda(j,\zeta,c)} \rho(z_j, \zeta) \ge C,$$

where C is independent of ζ. This follows from the facts that all factors in the above product are greater than or equal to $\frac{1}{2}$ and that

$$\sum_j \left\{1 - \rho(z_j, \zeta)^2 : z_j \in \Lambda(j, \zeta, c)\right\} \le \frac{K}{c}.$$

We thus obtain the following sharper form of (5.27):

$$(1 - |\zeta|^2)^{\beta+\varepsilon} f(\zeta) = \sum_{z_j \in \Gamma} (1 - |z_j|^2)^{\beta+\varepsilon} f(z_j) h_j(\zeta), \qquad (5.30)$$

where $h_j(\zeta) = \widetilde{g}_j(\zeta)/B_\Lambda(\zeta)$ satisfies

$$\sum_j |h_j(\zeta)| \le C, \qquad |h_j(\zeta)| \le C\frac{(1 - |\zeta|^2)(1 - |z_j|^2)}{|1 - \bar{\zeta}z_j|^2}. \qquad (5.31)$$

To complete the proof of Theorem 5.23, we only need to verify (5.26) for $f \in A_0^{-(\beta+\varepsilon)} \cap A_\alpha^p$, because this space is dense in A_α^p.

For $1 < p < +\infty$ with $\frac{1}{p} + \frac{1}{q} = 1$, we use (5.30), (5.31), and Hölder's inequality to get

$$(1 - |\zeta|^2)^\alpha |f(\zeta)|^p \le (1 - |\zeta|^2)^{\alpha-(\beta+\varepsilon)p}\left[\sum_j (1 - |z_j|^2)^{\beta+\varepsilon}|f(z_j)h_j(\zeta)|\right]^p$$

$$\le (1 - |\zeta|^2)^{-1-\varepsilon p}\sum_j (1 - |z_j|^2)^{\alpha+1+p\varepsilon}|f(z_j)|^p|h_j(\zeta)|$$

$$\times\left[\sum_j |h_j(\zeta)|\right]^{p/q}$$

$$\le C\sum_j \frac{(1 - |\zeta|^2)^{-\varepsilon p}(1 - |z_j|^2)^{\alpha+2+p\varepsilon}}{|1 - \bar{\zeta}z_j|^2}|f(z_j)|^p.$$

We now integrate over $\mathbb{D}$ and use Theorem 1.7, to obtain

$$\int_{\mathbb{D}} |f(\zeta)|^p\, dA_\alpha(\zeta) \le C\sum_j (1 - |z_j|^2)^{2+\alpha}|f(z_j)|^p.$$

The case $p = 1$ follows from (5.30), (5.31), and Theorem 1.7 as well; we simply proceed as in the previous paragraph but omit the use of Hölder's inequality. ∎

5.4 Hyperbolic Lattices

In this section, we present a class of sequences in the disk for which the upper and lower Seip densities are computable. Basically, these sequences are lattices in the hyperbolic metric. Because the hyperbolic lattices are easier to describe and easier to visualize in the upper half plane, we will need to switch between the unit disk and the upper half plane via the Cayley transform.

For a positive Borel measure μ on $\mathbb{D}$ and $r \in (0, 1)$, we use $n_\mu(r)$ to denote the μ-measure of the disk $|z| < r$. Define

$$N_\mu(r) = \int_0^r n_\mu(t)\, dt, \qquad 0 < r < 1.$$

It is easy to see that

$$N_\mu(r) = \int_{|z|<r} (r - |z|)\, d\mu(z).$$

If Γ is a sequence in $\mathbb{D}$ and

$$d\mu = \sum_{\gamma \in \Gamma} d\delta_\gamma$$

is corresponding atomic measure, where each $d\delta_\gamma$ is a unit point mass at γ, then n_μ and N_μ are the classical counting functions associated with Γ (except that N_μ is slightly different here); see Section 4.1.

We introduce two more counting type functions. Thus, for $0 < r < 1$ we define

$$B_\mu(r) = \int_{\frac{1}{2}<|z|<r} \log \frac{1}{|z|}\, d\mu(z)$$

and

$$C_\mu(r) = \int_{\frac{1}{2}<|z|<r} \log \frac{r}{|z|}\, d\mu(z).$$

Observe again that if μ is the atomic measure associated with a sequence Γ in $\mathbb{D}$, then

$$B_\mu(r) = \sum_z \left\{ \log \frac{1}{|z|} : z \in \Gamma, \frac{1}{2} < |z| < r \right\}$$

and

$$C_\mu(r) = \sum_z \left\{ \log \frac{r}{|z|} : z \in \Gamma, \frac{1}{2} < |z| < r \right\}.$$

The counting function B_μ appears as the numerator in the definition of the Seip densities; see the previous two sections. The denominator in the definition of the Seip densities, $\log \frac{1}{1-r}$, has the same magnitude as $N_\mu(r)$, $B_\mu(r)$, and $C_\mu(r)$, if we take

$$d\mu(z) = \frac{dA(z)}{(1 - |z|^2)^2},$$

the Möbius invariant area measure on $\mathbb{D}$.

LEMMA 5.24 *Suppose μ is a positive Borel measure on $\mathbb{D}$ such that*

$$n_\mu(r) = O\left(\frac{1}{1-r}\right), \qquad r \to 1^-.$$

Then

$$N_\mu(r) = O\left(\log\frac{1}{1-r}\right), \qquad r \to 1^-.$$

Furthermore, we have

$$B_\mu(r) = N_\mu(r) + O(1), \qquad r \to 1^-,$$

and

$$C_\mu(r) = N_\mu(r) + O(1), \qquad r \to 1^-.$$

Proof. The estimate on $N_\mu(r)$ follows immediately from integrating the growth bound on $n_\mu(r)$. Also, after a second integration, we have

$$\int_{|z|<r} (r - |z|)^2 \, d\mu(z) = O(1), \qquad r \to 1^-. \tag{5.32}$$

Since

$$B_\mu(r) - C_\mu(r) = \log\frac{1}{r} \int_{\frac{1}{2}<|z|<r} d\mu(z) = \left[n_\mu(r) - n_\mu\left(\tfrac{1}{2}^+\right)\right]\log\frac{1}{r},$$

the assumption on the growth of n_μ gives

$$B_\mu(r) = C_\mu(r) + O(1), \qquad r \to 1^-.$$

Finally, we can write

$$
\begin{aligned}
C_\mu(r) - N_\mu(r) &= \int_{\frac{1}{2}<|z|<r} \left(\log\frac{r}{|z|} - 1 + \frac{|z|}{r}\right) d\mu(z) \\
&\quad + (1-r)\int_{\frac{1}{2}<|z|<r} \left(1 - \frac{|z|}{r}\right) d\mu(z) \\
&\quad - \int_{|z|\leq\frac{1}{2}} (r - |z|) \, d\mu(z).
\end{aligned}
$$

By the growth conditions on n_μ and N_μ, the last two terms above are $O(1)$ as $r \to 1^-$; also, combining the estimate (5.32) with

$$\log\frac{1}{t} = 1 - t + O\left((1-t)^2\right), \qquad t \to 1^-,$$

we see that the third term is $O(1)$ as $r \to 1^-$. ∎

Note that the estimates in the lemma above are uniform in $\varphi \in$ Aut $(\mathbb{D})$ if μ is replaced by the atomic measures associated with $\varphi(\Gamma)$, where Γ is a separated sequence in $\mathbb{D}$ and Aut $(\mathbb{D})$ is the full Möbius group. This is because the counting function n_μ associated with $\varphi(\Gamma)$ has the estimate

$$n_\mu(r) = O\left(\frac{1}{1-r}\right), \qquad r \to 1^-,$$

uniformly in φ. As a consequence, we see that when computing the Seip densities $D_s^+(\Gamma)$ and $D_s^-(\Gamma)$, we may use $N_\mu(r)$ or $C_\mu(r)$ in place of $B_\mu(r)$, with μ being the atomic measure associated with Γ. This helps us develop better geometric intuition for the Seip densities.

We now begin the construction of hyperbolic lattices in $\mathbb{D}$.

A sequence of points $\{x_j\}_j$ on the real line $\mathbb{R}$ has *pure density* ϱ provided that the points are separated (in the usual Euclidean metric) and

$$\#\big(\{x_j\}_j \cap [A, B]\big) = (B - A)\varrho + O(1),$$

where $O(1)$ stands for a quantity bounded by a constant that is independent of A, B; and # counts the number of points in a set.

If we have a sequence of sequences $\{x_j^{(1)}\}_j, \{x_j^{(2)}\}_j, \{x_j^{(3)}\}_j, \ldots$, each of which has pure density ϱ, we say that they have pure density ϱ *uniformly* provided that

$$\inf_{j,k,l} \left\{ \left| x_j^{(k)} - x_l^{(k)} \right| : j \neq l \right\} > 0$$

and

$$\sup_k \left| \#\big(\{x_j^{(k)}\}_j \cap [A, B]\big) + (A - B)\varrho \right| = O(1),$$

where again $O(1)$ stands for a quantity that is bounded independently of A and B.

Let $\mathbb{U}$ be the open upper half plane. Recall that the hyperbolic metric on $\mathbb{U}$ is given by

$$\beta'(z, w) = \frac{1}{2} \log \frac{1 + \rho'(z, w)}{1 - \rho'(z, w)},$$

where ρ' is the pseudo-hyperbolic metric on $\mathbb{U}$:

$$\rho'(z, w) = \left| \frac{z - w}{z - \overline{w}} \right|.$$

We shall first construct a sequence $\Gamma' = \{\zeta_n\}_n$ in $\mathbb{U}$, and then map it to a sequence Γ in $\mathbb{D}$ by the *Cayley transform*

$$\phi(\zeta) = \frac{\zeta - i}{\zeta + i}, \qquad \zeta \in \mathbb{U}.$$

THEOREM 5.25 *Fix a real parameter $\beta \in (1, +\infty)$. For each integer k, let $\{x_j^{(k)}\}_j$ be a sequence in $\mathbb{R}$ with pure density ϱ. Also, assume that these sequences have pure density ϱ uniformly in k. Let $\Gamma' = \{\zeta_{j,k}\}_{j,k}$ be the doubly indexed* *sequence*

$$\zeta_{j,k} = \beta^k \left(x_j^{(k)} + i \right)$$

in $\mathbb{U}$, and let $\Gamma = \{z_{j,k}\}_{j,k}$ be the image sequence in $\mathbb{D}$ under the Cayley transform. Then Γ is separated, and

$$D_s^+(\Gamma) = D_s^-(\Gamma) = \frac{2\pi\varrho}{\log\beta}.$$

Proof. Clearly, the sequence Γ is separated, as the hyperbolic distance between two points of Γ' is bounded from below by a positive number and the Cayley transform preserves the hyperbolic metric.

For $\xi \in \mathbb{U}$, let

$$\phi_\xi(\zeta) = \frac{\zeta - \xi}{\zeta - \bar{\xi}}, \qquad \zeta \in \mathbb{U},$$

be the associated conformal mapping $\mathbb{U} \to \mathbb{D}$. We need to estimate $N_{\phi_\xi(\Gamma')}(r)$ for $r \in (0,1)$ close to 1, because this is equivalent to $N_{\varphi(\Gamma)}(r)$ for Möbius maps φ of $\mathbb{D}$. We do this by first estimating $n_{\phi_\xi(\Gamma')}(r)$, which equals the number of points of Γ' in the pseudohyperbolic disk

$$D_{\mathbb{U}}(\xi, r) = \{z \in \mathbb{U} : \rho'(z, w) < r\}.$$

The Euclidean center of this disk is located at the point

$$\operatorname{Re}\xi + i\,\frac{1 + r^2}{1 - r^2}\,\operatorname{Im}\xi,$$

and the Euclidean radius is $2r(1 - r^2)^{-1}\operatorname{Im}\xi$.

The points of Γ' are located along the horizontal lines $\operatorname{Im}\zeta = \beta^k$, with k an integer. The number of different such lines intersecting the disk $D_{\mathbb{U}}(\xi, r)$ is approximately given by

$$\frac{2\log\frac{1+r}{1-r}}{\log\beta}.$$

If we, instead of counting the number of points on each such line $\operatorname{Im}\zeta = \beta^k$ in $\Gamma_{\mathbb{U}} \cap D_{\mathbb{U}}(\xi, r)$, just calculate $\varrho\,\beta^{-k}$ times the Euclidean length of the line segment, the error we make each time will be $O(1)$ uniformly in k, by the assumptions we made on the sequence Γ'. Adding up the errors, we obtain a total error of the order of magnitude

$$O\left(\log\frac{1}{1-r}\right),$$

which is negligible compared with the total number of points in the disk $D_{\mathbb{U}}(\xi, r)$, the latter being of the order of magnitude $(1 - r)^{-1}$, for r close to 1. Let us write this down more carefully.

Let μ be the positive Borel measure on $\mathbb{U}$ defined by

$$\int_{\mathbb{U}} f(\zeta) \, d\mu(\zeta) = \sum_{n=-\infty}^{+\infty} \frac{1}{\beta^n} \int_{-\infty}^{+\infty} f(x + i\beta^n) \, dx,$$

for compactly supported continuous functions f. Then

$$n_{\varphi(\Gamma)}(r) = n_{\phi_\xi(\Gamma_{\mathbb{U}})}(r) = \varrho \, n_{\phi_\xi^* \mu}(r) + O\left(\log \frac{1}{1-r} \right) \qquad \text{as} \quad r \to 1,$$

where $\phi_\xi^* \mu$ is the positive Borel measure on $\mathbb{D}$ defined by

$$d(\phi_\xi^* \mu)(z) = d\mu(\phi_\xi^{-1}(z)),$$

provided that $\xi \in \mathbb{U}$ is chosen such that $\phi_\xi(\Gamma')$ equals $\varphi(\Gamma)$ modulo a rotation. After an integration, we obtain

$$N_{\varphi(\Gamma)}(r) = N_{\phi_\xi(\Gamma')}(r) = \varrho \, N_{\phi_\xi^* \mu}(r) + O(1) \qquad \text{as} \quad r \to 1. \tag{5.33}$$

We turn to the function $C_{\phi_\xi^* \mu}(r)$, which is equivalent to $N_{\phi_\xi^* \mu}(r)$ by Lemma 5.24. It is easy to see that

$$C_{\phi_\xi^* \mu}(r) = \int_{D_{\mathbb{U}}(\xi,r) \setminus \overline{D}_{\mathbb{U}}(\xi,\frac{1}{2})} \log \left| \frac{r(\zeta - \bar{\xi})}{\zeta - \xi} \right| \, d\mu(\zeta).$$

We show that

$$\int_{\overline{D}_{\mathbb{U}}(\xi,\frac{1}{2})} \log \left| \frac{\zeta - \bar{\xi}}{\zeta - \xi} \right| \, d\mu(\zeta) = O(1), \tag{5.34}$$

uniformly in $\xi \in \mathbb{U}$. In fact, it follows from the construction that the μ-mass of the pseudohyperbolic disk $\overline{D}_{\mathbb{U}}(\xi, \frac{1}{2})$ is uniformly bounded in ξ. The kernel

$$\log \left| \frac{\zeta - \bar{\xi}}{\zeta - \xi} \right|$$

is bounded in the variable ζ except for a logarithmic singularity at $\zeta = \xi$; the logarithmic singularity is leveled out in the integral by the fact that $d\mu$ is so smooth in the $\operatorname{Re} \zeta$ direction. To simplify the calculations, we switch from integrating over the disk $\overline{D}_{\mathbb{U}}(\xi, \frac{1}{2})$ to the larger square (after all, the integrand is positive)

$$\left\{ z = x + iy \in \mathbb{U} : |x| \leq \frac{4}{3} \operatorname{Im} \xi, \; \frac{1}{3} \operatorname{Im} \xi \leq y \leq 3 \operatorname{Im} \xi \right\}.$$

After some simplifications, it remains to show that

$$\frac{1}{\theta} \sum_{n: \frac{1}{3} \leq \theta \beta^n \leq 3} \beta^{-n} \int_{-\frac{4}{3}}^{\frac{4}{3}} \log \frac{t^2 + (1 + \theta \beta^n)^2}{t^2 + (1 - \theta \beta^n)^2} \, dt$$

is uniformly bounded in θ, $0 < \theta < +\infty$. This is easily verified, and the claim (5.34) follows.

We now introduce the function

$$F(\xi, \eta; r) = \int_{D_{\mathbb{U}}(\xi, r)} \log \left| \frac{r(\zeta - \bar{\eta})}{\zeta - \eta} \right| d\mu(\zeta),$$

which, according to (5.34) and the above observation that the μ-mass of $\overline{D}_{\mathbb{U}}(\xi, \frac{1}{2})$ is uniformly bounded in ξ, satisfies

$$C_{\phi_{\xi}^* \mu}(r) = F(\xi, \xi; r) + O(1), \qquad r \to 1^-, \tag{5.35}$$

uniformly in ξ. The advantage with the function $F(\xi, \eta; r)$ is that in η, it solves a boundary value problem on $D_{\mathbb{U}}(\xi, r)$: it vanishes on the boundary $\partial D_{\mathbb{U}}(\xi, r)$, and inside, its Laplacian is, in the sense of distribution theory,

$$\Delta_\eta F(\xi, \eta; r) = -\frac{1}{2} d\mu(\eta).$$

We will return to this function shortly.

Let H be the *Heaviside function*, so that $H(x) = 1$ for $0 < x < +\infty$ and $H(x) = 0$ for $-\infty < x \leq 0$, and consider the function

$$U(x) = \sum_{n=-\infty}^{+\infty} \beta^{-n} \int_1^x H(\beta^n - t) \, dt, \qquad 0 < x < +\infty.$$

The series converges, because only finitely many terms with negative index n actually occur in the sum. This function has the functional property

$$U(\beta x) = U(x) + 1, \qquad 0 < x < +\infty;$$

the verification involves two manoeuvres, the first one being to check that the two sides have the same derivative, and the second to obtain that $U(\beta) = 1$ whereas $U(1) = 0$. Using the above functional equation, we easily establish that

$$0 \leq \frac{\log x}{\log \beta} - U(x) \leq 3, \qquad 0 < x < +\infty. \tag{5.36}$$

The second derivative of U is

$$U''(x) = -\sum_{n=-\infty}^{+\infty} \beta^{-n} \delta_{\beta^n}(x), \qquad 0 < x < +\infty,$$

where $\delta_{\beta^n}(x)$ represents the unit point mass at $x = \beta^n$. In view of the definition of the Borel measure μ, it follows that

$$\Delta_\eta \left(U(\operatorname{Im} \eta) \right) = -\frac{1}{4\pi} d\mu(\eta), \qquad \eta \in \mathbb{U}.$$

The factor π^{-1} comes from our choice of interpreting locally integrable functions u on a domain Ω as distributions via the duality

$$\langle f, u \rangle = \int_\Omega f(z) \, u(z) \, dA(z).$$

We return to the function $F(\xi, \eta; r)$, and conclude that

$$F(\xi, \eta; r) = 2\pi \left(U(\operatorname{Im} \eta) - \widehat{U}_\xi(\eta) \right), \qquad \eta \in D_{\mathbb{U}}(\xi, r),$$

where $\widehat{U}_\xi(\eta)$ stands for the harmonic function on $D_{\mathbb{U}}(\xi, r)$ which equals $U(\operatorname{Im} \eta)$ along the boundary. The harmonic extension operation respects inequalities, so that in view of (5.36), we can get $F(\xi, \eta; r)$ trapped:

$$V(\operatorname{Im} \eta) - \widehat{V}_\xi(\eta) - 6 \le \frac{F(\xi, \eta; r)}{2\pi} \le V(\operatorname{Im} \eta) - \widehat{V}_\xi(\eta) + 6, \qquad \eta \in D_{\mathbb{U}}(\xi, r),$$

where $V(x) = (\log x)/(\log \beta)$, and $\widehat{V}_\xi(\eta)$ stands for the harmonic function on $D_{\mathbb{U}}(\xi, r)$ which equals $V(\operatorname{Im} \eta)$ along the boundary. Since the Laplacian of $V(\operatorname{Im} \eta)$ is readily calculated, an application of Green's formula yields the representation

$$V(\operatorname{Im} \eta) - \widehat{V}_\xi(\eta) = \frac{1}{2 \log \beta} \int_{D_{\mathbb{U}}(\xi, r)} \log \left| \frac{r(\zeta - \bar{\eta})}{\zeta - \eta} \right| \frac{dA(\zeta)}{(\operatorname{Im} \zeta)^2}.$$

Plugging in $\eta = \xi$, we have

$$\frac{F(\xi, \xi; r)}{2\pi} = \frac{1}{2 \log \beta} \int_{D_{\mathbb{U}}(\xi, r)} \log \left| \frac{r(\zeta - \bar{\xi})}{\zeta - \xi} \right| \frac{dA(\zeta)}{(\operatorname{Im} \zeta)^2} + O(1).$$

Performing the integration on the unit disk instead, using the change of variables $z = \phi_\xi(\zeta)$, we obtain

$$\frac{F(\xi, \xi; r)}{2\pi} = \frac{2}{\log \beta} \int_{|z| < r} \log \frac{r}{|z|} \frac{dA(z)}{(1 - |z|^2)^2} + O(1).$$

An easy calculation then shows that

$$F(\xi, \xi; r) = \frac{2\pi}{\log \beta} \log \frac{1}{1 - r} + O(1).$$

Combining this with (5.33), (5.35), and Lemma 5.24, we conclude that

$$B_{\varphi(\Gamma)}(r) = \frac{2\pi \varrho}{\log \beta} \log \frac{1}{1 - r} + O(1),$$

uniformly in $\varphi \in \operatorname{Aut}(\mathbb{D})$. It is now immediate from the definitions of the Seip densities that

$$D_s^+(\Gamma) = D_s^-(\Gamma) = \frac{2\pi \varrho}{\log \beta},$$

and we are done.   ∎

5.5   Notes

The characterization of interpolation and sampling sequences for the spaces $\mathcal{A}^{-\alpha}$ is due to Seip [112], where the spaces A_α^2 are also discussed. The cases A_α^p, as presented here in Section 3, essentially follow from Seip's proof in [112] as well. Seip's work was strongly influenced by Beurling's results [22] on interpolation and sampling for the Banach space of functions of exponential type $\leq \alpha$, bounded on the real line.

The proofs of Theorems 5.18 and 5.23 supply no quantitative information concerning the size of the sampling constants; such information would be desirable.

We believe that Theorem 5.23 holds for $0 < p < 1$ as well. However, this cannot be proved using the line of ideas pursued in Section 3. It appears that a proof for $0 < p < 1$ can be built by suitably modifying the methods in [20].

In [130], Thomson applies Seip's sampling theorem to show that the closure of the polynomials in $L^p(\mathbb{D}, d\mu)$ can change quite dramatically with the parameter p: for a certain Borel probability measure μ this closure is a space of holomorphic functions on $\mathbb{D}$ if p is large, but for small p it becomes all of $L^p(\mathbb{D}, d\mu)$; see also Thomson's fundamental paper [129].

The section on hyperbolic lattices is new: it was left unexplored by Seip because he first characterized extremely regular lattices in terms of sampling and interpolation properties using very explicit methods [111], and only later in terms of counting functions for the sets [112]. Counting points in hyperbolic space is a bit tricky because the space expands faster than in Euclidean geometry: the hyperbolic area of the (hyperbolic) annulus between the disks of radii R and $R + 1$ is comparable to that of the whole disk of radius R.

5.6   Exercises and Further Results

1. If A is a classical interpolating sequence, then A is a sequence of interpolation for A^p.

2. There exist two sequences of interpolation for A^p such that their union is a sampling sequence for A^p. See [61].

3. Let $A = \{a_n\}_n$ be a sequence of distinct points in $\mathbb{D}$. Then A is sampling for A^2 if and only if atomic decomposition for A^2 holds on A, that is, an analytic function f in $\mathbb{D}$ belongs to A^2 if and only if

$$f(z) = \sum_{n=1}^{+\infty} c_n \frac{1 - |a_n|^2}{(1 - \bar{a}_n z)^2}$$

for some $\{c_n\}_n \in l^2$ and the series converges in norm. See [139].

4. For a sequence $A = \{a_n\}_n$ of distinct points in $\mathbb{D}$, let R_A denote the operator that sends a function $f \in A^2$ to the sequence $\{(1 - |a_n|^2)f(a_n)\}_n$. Recall from Exercise 19 of Chapter 2 that R_A maps A^2 into l^2 if and only if A is the union of finitely many separated sequences. For such a sequence A, show that R_A has closed range in l^2 if and only if A is a sampling sequence for A^2 or A is a sequence of interpolation for A^2. See [139].

5. Suppose A is sequence of distinct points in $\mathbb{D}$ such that the operator R_A defined above maps A^2 onto a closed subspace of l^2. Then A is sampling for A^2 if and only if A is not a zero sequence for A^2; and A is a sequence of interpolation for A^2 if and only if A is a zero sequence for A^2. See [139].

6. Let $A = \{a_n\}_n$ be a sequence of interpolation for A^2. Define a sequence $\{\varphi_n\}_n$ of functions in A^2 by

$$\varphi_n(z) = K_{A_n}(z, a_n)/K_{A_n}(a_n, a_n), \qquad z \in \mathbb{D}, \ n = 1, 2, 3, \ldots,$$

where $A_n = A \setminus \{a_n\}$ for all $n = 1, 2, 3, \ldots$, and K_{A_n} is the reproducing kernel function for the zero-based invariant subspace I_{A_n}. Show that $\varphi_n(a_n) = 1$ and $\varphi_n(a_m) = 0$ if $n \neq m$.

7. With the same assumption and notation as before, show that there exists a positive constant C such that

$$1 - |a_n|^2 \leq \|\varphi_n\| \leq C\,(1 - |a_n|^2)$$

for all $n = 1, 2, 3, \ldots$. For Problems 7–12, see [140].

8. With the same assumption and notation as before, show that the sequence $\{(1 - |a_n|^2)^{-3/2}\varphi_n(z)\}_n$ is uniformly bounded on every compact subset of $\mathbb{D}$.

9. With the same assumption and notation as before, show that for every sequence $\{w_n\}_n$ of complex numbers satisfying

$$\sum_n (1 - |a_n|^2)|w_n|^2 < +\infty,$$

the series $\sum_n w_n\varphi_n$ converges (in norm) to a function in A^2 that uniquely solves the minimal interpolation problem:

$$\inf\{\|f\| : \ f(a_n) = w_n, \ n \geq 1\}.$$

10. With the same assumption and notation as before, show that the reproducing kernel K_A of I_A admits the following partial fraction expansion:

$$K_A(z, w) = \frac{1}{(1 - z\overline{w})^2} - \sum_n \frac{\overline{\varphi_n(w)}}{(1 - \overline{a}_n z)^2}.$$

11. With the same assumption and notation as before, show that every function f in $I_A^{\perp}$ admits the following expansion:

$$f(z) = \sum_n \frac{\langle f, \varphi_n \rangle}{(1 - \overline{a}_n z)^2},$$

with the series converging in norm.

12. Let $A = \{a_n\}_n$ be a sequence of interpolation for A^2. Show that an analytic function f in $\mathbb{D}$ belongs to $I_A^\perp$ if and only if

$$f(z) = \sum_n c_n \frac{1 - |a_n|^2}{(1 - \bar{a}_n z)^2},$$

where $\{c_n\}_n \in l^2$ and the series converges in norm. In fact, the mapping that sends $\{c_n\}_n$ to f via the above series is an invertible operator from l^2 onto $I_A^\perp$.

13. Every A_α^p-sampling sequence is the union of finitely many A_α^p-interpolating sequences. In fact, every separated sequence is a finite union of A_α^p-interpolating sequences.

14. A sequence Γ of distinct points in $\mathbb{D}$ is called a weak $\mathcal{A}^{-\alpha}$-interpolating sequence if there exists a positive constant C such that for each k there is a function $g_k \in \mathcal{A}^{-\alpha}$ with $\|g_k\|_{-\alpha} \leq C$, $(1 - |z_k|^2)^\alpha g_k(z_k) = 1$, and $g_k(z_j) = 0$ for $j \neq k$. Show that Γ is a weak $\mathcal{A}^{-\alpha}$-interpolating sequence if and only if it is an $\mathcal{A}^{-\alpha}$-interpolating sequence. See [110].

15. Formulate and prove a result in the context of A_α^p that is similar to the above.

16. Show that if the sequence Γ is separated in $\mathbb{D}$, then $D_u^+(\Gamma) < +\infty$ and $D_s^+(\Gamma) < +\infty$.

17. Show that the condition $D_u^+(\Gamma) < +\infty$ implies that Γ is separated. Does this hold for $D_s^+(\Gamma)$?

18. Define and characterize interpolating sequences for the space $\mathcal{A}^{-\infty}$. See [32] and [96].

19. Let us call the sequence $\Gamma = \{z_j\}_j$ of distinct points in $\mathbb{D}$ a *type sampling sequence* if for each $f \in \mathcal{A}^{-\infty}$ we have

$$\limsup_{j \to +\infty} \frac{\log |f(z_j)|}{\log \frac{1}{1-|z_j|}} = \limsup_{|z| \to 1^-} \frac{\log |f(z)|}{\log \frac{1}{1-|z|}} = t(f).$$

Characterize the type sampling sequences. This is an open problem; for partial results, see [78].

20. Suppose $A = \{a_n\}_n$ is a zero set of A^p consisting of distinct points. For $n = 1, 2, 3, \ldots$, define $A_n = A \setminus \{a_n\}$ and let G_n be the contractive zero divisor for A_n in A^p. Show that A is an interpolating sequence for A^p if and only if there exists a positive constant δ such that $\delta \leq G_n(a_n)$ for all n. See [109].

21. Give another proof, based on Lemma 5.1, Corollary 5.2, and the contractive imbedding $A_\alpha^p \subset \mathcal{A}^{-(2+\alpha)/p}$, that every A_α^p-interpolation sequence is separated.

22. Show that if Γ is a separated sequence, then there exists a positive constant C such that the number of points of Γ lying in $|z| < r$ is less than or equal to $C/(1-r)$, where $r \in (0, 1)$.

23. Suppose A and B are disjoint sequences in $\mathbb{D}$. Show that

$$D_s^+(A \cup B) \leq D_s^+(A) + D_s^+(B)$$

and

$$D_s^-(A \cup B) \geq D_s^-(A) + D_s^-(B).$$

24. Do the analysis of Section 5.4 for concentric circles about the origin in $\mathbb{D}$ instead of horizontal lines in the upper half plane $\mathbb{U}$.

25. Prove the estimate (5.16) relating the Beurling-Carleson characteristic to the κ-area of the corresponding Stolz star.

6
Invariant Subspaces

In this chapter we study several problems related to invariant subspaces of Bergman spaces. First, we show by explicit examples that there exist invariant subspaces of index n for all $0 \leq n \leq +\infty$. Then we prove a theorem that can be considered an analogue to the classical Beurling's theorem on invariant subspaces of the Hardy space. It states that in the spaces A_α^2, with $-1 < \alpha \leq 0$, each invariant subspace I is generated by $I \ominus zI$. In the classical Hardy space case, $I \ominus zI$ is one-dimensional, and spanned by a classical inner function (unless $I = \{0\}$). In A_α^2, the dimension may be bigger, but all elements of $I \ominus zI$ of unit norm are A_α^2-inner functions.

6.1 Invariant Subspaces of Higher Index

Let I be an invariant subspace of A_α^p. We say that I has index n (or the codimension n property) if $n = \dim(I/zI)$. In this section we show that for any $0 \leq n \leq +\infty$ there exists an invariant subspace I of A_α^p of index n. Of course, $n = 0$ occurs only for the trivial subspace $I = \{0\}$.

First observe that if I is a singly generated invariant subspace, or a zero-based invariant subspace of A^p, then I has index 1; see Exercise 13.

LEMMA 6.1 *If I is an invariant subspace of A_α^p with index 1, then there exists a nonzero continuous linear functional $\varphi : I \to \mathbb{C}$ such that φ vanishes exactly on zI.*

Proof. Let n be the smallest nonnegative integer such that there exists a function $f \in I$ with $f^{(n)}(0) \neq 0$. Define $\varphi : I \to \mathbb{C}$ by

$$\varphi(f) = f^{(n)}(0), \qquad f \in I.$$

Then φ has the desired properties. ∎

LEMMA 6.2 *Fix a positive integer n and let $I_1, I_2, \ldots, I_n$ be invariant subspaces of A_α^p, all having index 1. If there exists a positive number ε such that*

$$\|f_1 + \cdots + f_n\| \geq \varepsilon \left(\|f_1\| + \cdots + \|f_n\|\right)$$

for all $f_1 \in I_1, \ldots, f_n \in I_n$, then $I = I_1 + \cdots + I_n$ is an invariant subspace of A_α^p having index n.

Proof. It is easy to see that the reverse triangle inequality implies that $I = I_1 + \cdots + I_n$ is a direct sum and that I is closed in A_α^p. In particular, I is an invariant subspace of A_α^p.

For each $1 \leq k \leq n$, let $\varphi_k : I_k \to \mathbb{C}$ be a nontrivial continuous linear functional that vanishes on $z I_k$. Define $\varphi : I \to \mathbb{C}^n$ by

$$\varphi(f_1 + \cdots + f_n) = (\varphi_1(f_1), \ldots, \varphi_n(f_n)),$$

where $f_k \in I_k$ for $1 \leq k \leq n$. The reverse triangle inequality guarantees that φ is continuous. Since each φ_k is surjective, φ is also surjective. And since the kernel of each φ_k is $z I_k$, the kernel of φ must be $zI = z I_1 + \cdots + z I_n$. It follows that φ induces an isomorphism between $I/(zI)$ and $\mathbb{C}^n$, so that the quotient space $I/(zI)$ is n-dimensional. ∎

We proceed to construct zero-based invariant subspaces of A_α^p that satisfy the reverse triangle inequality Lemma 6.2. This will then produce invariant subspaces of arbitrary index.

LEMMA 6.3 *Suppose $0 < p < +\infty$, $-1 < \alpha < +\infty$, and n is a positive integer greater than or equal to 2. Then there exists a sampling sequence A for A_α^p, and a decomposition of it as a finite disjoint union $A = \cup_{j=1}^n A_j$, with the property that each $A \setminus A_j$ is an interpolating sequence for A_α^p.*

Proof. First recall from Chapter 5 that a sequence A is sampling for A_α^p if it is a finite union of separated sequences and contains a separated subsequence A_0 with

$$D_s^-(A_0) > \frac{\alpha + 1}{p}.$$

Similarly, a sequence A is interpolating for A_α^p if it is separated with

$$D_s^+(A) < \frac{\alpha + 1}{p}.$$

We say that a sequence A is *regular* if its upper and lower Seip densities coincide, in which case we write $D_s(A)$ in place of $D_s^+(A) = D_s^-(A)$.

Let A be a regular sequence as constructed in Section 5.4. More specifically, let β and ϱ be real parameters with $1 < \beta < +\infty$ and $0 < \varrho < +\infty$, let $A' = \{a'_{j,k}\}_{j,k}$ be the doubly indexed sequence in $\mathbb{U}$ defined by

$$a'_{j,k} = \beta^k \left(\varrho j + i \right),$$

where j and k run over all the integers, and let $A = \{a_{j,k}\}_{j,k}$ be the image sequence in $\mathbb{D}$ under the Cayley transform

$$\phi(\zeta) = \frac{\zeta - i}{\zeta + i}, \qquad \zeta \in \mathbb{U}.$$

By Theorem 5.25, the sequence A is regular with

$$D_s(A) = \frac{2\pi\varrho}{\log \beta}.$$

For $m = 1, 2, \dots, n$, let A_m be the image under the Cayley transform of A'_m, where A'_m consists of those points $a'_{j,k}$ in A' such that $j \equiv m \pmod{n}$. Clearly, $A = \cup_{m=1}^n A_m$, and the sets A_m, $m = 1, 2, \dots, n$, are disjoint. Each A_m contains every n-th element of A, and by the analysis of Chapter 5 (Theorem 5.25), each sequence A_m is regular with its Seip density given by

$$D_s(A_m) = \frac{D_s(A)}{n} = \frac{2\pi\varrho}{n \log \beta}.$$

Analogously, each sequence $A \setminus A_m$ is regular with its Seip density given by

$$D_s(A \setminus A_m) = \frac{n-1}{n} D_s(A) = \frac{2(n-1)\pi\varrho}{n \log \beta}.$$

It follows that

$$D_s\left(\bigcup_{m=1}^n A_m \right) = \sum_{m=1}^n D_s(A_m);$$

when this happens, we say that the decomposition $A = \cup_{m=1}^n A_m$ is *homogeneous*.

We need to fulfill the requirements

$$\frac{\alpha + 1}{p} < D_s(A) = \frac{2\pi\varrho}{\log \beta}$$

and

$$D_s(A \setminus A_m) = \frac{2(n-1)\pi\varrho}{n \log \beta} < \frac{\alpha + 1}{p}, \qquad m = 1, 2, \dots, n.$$

These amount to the condition

$$\frac{\alpha + 1}{p} < D_s(A) = \frac{2\pi\varrho}{\log \beta} < \frac{n}{n-1} \frac{\alpha + 1}{p}.$$

All that remains is for us to pick the density in this nonempty interval. This is easily done by suitably adjusting the parameters β and ϱ. ∎

THEOREM 6.4 *For $j = 1, \ldots, n$, let $I_j = I_{A \setminus A_j}$ be the invariant subspace of A_α^p consisting of all functions that vanish on $A \setminus A_j$, where A and A_j are as in the previous lemma. Then there exists some ε, $0 < \varepsilon < 1$, such that*

$$\| f_1 + \cdots + f_n \| \geq \varepsilon \left(\| f_1 \| + \cdots + \| f_n \| \right)$$

for all $f_k \in I_{A \setminus A_k}$, $1 \leq k \leq n$.

Proof. For each point $a \in A$, and for $f_1 \in I_1, \ldots, f_n \in I_n$, we have

$$\left| f_1(a) + \ldots + f_n(a) \right|^p = |f_1(a)|^p + \ldots + |f_n(a)|^p.$$

Since A is sampling, there are positive constants K_1 and K_2 such that

$$K_1 \int_{\mathbb{D}} |f|^p \, dA_\alpha \leq \sum_{a \in A} (1 - |a|^2)^{2+\alpha} |f(a)|^p \leq K_2 \int_{\mathbb{D}} |f|^p \, dA_\alpha, \qquad f \in A_\alpha^p.$$

Apply these inequalities to each f_k and then to $f = f_1 + \cdots + f_n$. We conclude that

$$
\begin{aligned}
K_1 \sum_{j=1}^{n} \| f_j \|^p &\leq \sum_{j=1}^{n} \sum_{a \in A} (1 - |a|^2)^{2+\alpha} |f_j(a)|^p \\
&= \sum_{a \in A} (1 - |a|^2)^{2+\alpha} |f(a)|^p \leq K_2 \| f \|^p.
\end{aligned}
$$

This is the sought-after reverse triangle inequality, except that it is with p-th powers of norms. But in finite-dimensional spaces, l^p and l^1 norms are equivalent, so the result follows.   ∎

COROLLARY 6.5 *For any positive integer n there exists an invariant subspace of A_α^p with index n.*

The technique used in this section does not allow us to find invariant subspaces of infinite index. However, the main scheme, to form the closure of the sum $I_1 + I_2 + I_3 + \cdots$, denoted $I_1 \vee I_2 \vee I_3 \vee \cdots$, where $I_1, I_2, I_3, \ldots$ are index-one invariant subspaces, still applies. Essentially, by requiring the subspaces I_j to be "far apart" from each other, we obtain an invariant subspace of infinite index in A_α^p; the details are worked out in [71]. An alternative construction, which is perhaps more flexible and applies to a large collection of (quasi) Banach spaces of holomorphic functions on the unit disk, can be found in [26]; see also [1]. It is known that the elements of invariant subspaces of index bigger than or equal to 2 must exhibit rather bad boundary behavior; for instance, if we restrict our attention to $p = 2$ and $\alpha = 0$, then at each point of $\mathbb{T}$, every such function must have the whole plane $\mathbb{C}$ as cluster set (part of this assertion can be found in [62]). A considerably more precise statement can be found in [8].

6.2   Inner Spaces in A_α^2

In the classical theory of Hardy spaces it was shown by Beurling that every invariant subspace I of H^2 is of the form $I = \varphi H^2$, where φ is inner. Furthermore, the invariant subspace I determines the inner function uniquely up to a unimodular constant. Therefore, Beurling's theorem can be restated as follows: The invariant subspaces of H^2 are in a one-to-one correspondence with the (one-dimensional) spaces $\mathbb{C}\varphi$, where φ is inner. In the next section we will show that an analogue holds in the Bergman space setting, except that here one-dimensional spaces will not be enough.

DEFINITION 6.6 *A closed subspace X of A_α^p is called an inner space if every unit vector in X is A_α^p-inner.*

The next result characterizes inner spaces in A_α^2, whose proof works in much more general settings (an injective bounded operator with closed range on separable Hilbert space).

THEOREM 6.7 *A closed subspace X of A_α^2 is inner if and only if there exists an invariant subspace I of A_α^2 such that $X = I \ominus zI$.*

Proof. If I is an invariant subspace of A_α^2, then it is obvious that every unit vector in $I \ominus zI$ is an A_α^2-inner function, so that $I \ominus zI$ is an inner space.

Next assume that X is an inner space in A_α^2. Let I be the invariant subspace generated by X. We proceed to show that $X = I \ominus zI$.

Let $\{e_n\}_n$ be an orthonormal basis for X. In particular, each vector e_n is A_α^2-inner. To prove $X \subset I \ominus zI$, or $X \perp zI$, it suffices to show that

$$\langle e_n, z^k e_m \rangle = 0, \qquad k = 1, 2, 3, \ldots,$$

for all indices m and n. If $n = m$, then the desired equalities follow from the fact that e_n is A_α^2-inner. If $n \neq m$, we use a polarization trick. The function

$$f(z) = \frac{a\, e_n(z) + b\, e_m(z)}{\sqrt{|a|^2 + |b|^2}}$$

is A^2-inner for all complex numbers a and b with $|a|^2 + |b|^2 > 0$, because the space X is inner. It follows that

$$\langle a\, e_n + b\, e_m, z^k (a\, e_n + b\, e_m) \rangle = 0$$

for all complex numbers a and b. Since

$$\langle e_n, z^k e_n \rangle = \langle e_m, z^k e_m \rangle = 0,$$

we obtain

$$a\overline{b}\, \langle e_n, z^k e_m \rangle + \overline{a}b\, \langle e_m, z^k e_n \rangle = 0$$

for all complex a and b. By first setting $a = b \neq 0$, and then setting $a = ib \neq 0$, we easily obtain

$$\langle e_n, z^k e_m \rangle = \langle e_m, z^k e_n \rangle = 0.$$

This proves $X \subset I \ominus zI$.

If X is not all of $I \ominus zI$, then there exists a unit vector $f \in I \ominus zI$ such that $f \perp X$. It is of the form

$$f = \lim_{m \to +\infty} \sum_{k=1}^{N_m} p_k^{(m)} e_k,$$

where each $p_k^{(m)}$ is a polynomial and N_m is a positive integer. Write

$$\sum_{k=1}^{N_m} p_k^{(m)} e_k = \sum_{k=1}^{N_m} (p_k^{(m)} - p_k^{(m)}(0)) e_k + \sum_{k=1}^{N_m} p_k^{(m)}(0) e_k,$$

and denote the two sums on the right-hand side above by $f_{1,m}$ and $f_{2,m}$, respectively. Then $f_{1,m} \perp f_{2,m}$, by what was proved in the previous paragraph. Also, $f \perp X$ implies that $f \perp f_{2,m}$, and $f \perp zI$ implies that $f \perp f_{1,m}$. Thus, by the Pythagorean theorem,

$$\left\| f - \sum_{k=1}^{N_m} p_k^{(m)} e_k \right\|^2 = \| f - f_{1,m} - f_{2,m} \|^2$$

$$= \| f \|^2 + \| f_{1,m} \|^2 + \| f_{2,m} \|^2 \geq \| f \|^2 = 1.$$

This contradicts the assumption that

$$f = \lim_{m \to +\infty} \sum_{k=1}^{N_m} p_k^{(m)} e_k,$$

and hence completes the proof of the theorem.   ■

Combining the theorem above with results from the previous section, we conclude that the dimension of an inner space in A_α^2 can assume any value in the set $\{0, 1, 2, 3, \ldots, +\infty\}$.

6.3   A Beurling-Type Theorem

The purpose of this section is to show that invariant subspaces of A_α^2 are in a one-to-one correspondence with inner spaces in A_α^2, provided that $-1 < \alpha \leq 0$. More specifically, we show that every invariant subspace I in A_α^2 is generated by $I \ominus zI$.

Actually, we will prove a stronger result in the context of general Hilbert spaces. Throughout this section we let $\mathcal{H}$ be a separable (infinite-dimensional) Hilbert space and let $\mathbf{T} : \mathcal{H} \to \mathcal{H}$ be a bounded linear operator satisfying

(a) $\|\mathbf{T}x + y\|^2 \leq 2\left(\|x\|^2 + \|\mathbf{T}y\|^2\right)$, $x, y \in \mathcal{H}$,

(b) $\bigcap \{\mathbf{T}^n \mathcal{H} : n \geq 0\} = \{0\}$.

First note that setting $x = 0$ in condition (a) above shows that $\mathbf{T}$ is bounded below. In particular, the operator $\mathbf{T}$ is one-to-one and has closed range, so that the operator $\mathbf{T}^*\mathbf{T}$ is invertible. The operator

$$\mathbf{T}_1 = \mathbf{T}(\mathbf{T}^*\mathbf{T})^{-1}$$

will play a vital role in our analysis. In the following, operator inequalities are given the standard interpretation in terms of positive definiteness.

LEMMA 6.8 *For all $x \in \mathcal{H}$, we have*

$$\|\mathbf{T}_1^2 x\|^2 + \|x\|^2 \leq 2\|\mathbf{T}_1 x\|^2.$$

Proof. Let $y = (\mathbf{T}^*\mathbf{T})^{-1/2}z$ in condition (a). Then

$$\|\mathbf{T}x + (\mathbf{T}^*\mathbf{T})^{-1/2}z\|^2 \leq 2\left(\|x\|^2 + \|\mathbf{T}(\mathbf{T}^*\mathbf{T})^{-1/2}z\|^2\right)$$

for all x and z in $\mathcal{H}$. An easy calculation shows that

$$\|\mathbf{T}(\mathbf{T}^*\mathbf{T})^{-1/2}z\|^2 = \|z\|^2$$

for any $z \in \mathcal{H}$. Thus,

$$\|\mathbf{T}x + (\mathbf{T}^*\mathbf{T})^{-1/2}z\|^2 \leq 2(\|x\|^2 + \|z\|^2)$$

for all $x, z \in \mathcal{H}$.

Consider the operator $\mathbf{S} : \mathcal{H} \oplus \mathcal{H} \to \mathcal{H}$ defined by

$$\mathbf{S}(x, z) = \mathbf{T}x + (\mathbf{T}^*\mathbf{T})^{-1/2}z,   (x, z) \in \mathcal{H} \oplus \mathcal{H}.$$

We have $\|\mathbf{S}\| \leq \sqrt{2}$, so that $\mathbf{S}\mathbf{S}^* \leq 2\mathbf{I}$, where $\mathbf{I}$ is the identity operator on $\mathcal{H}$. Since

$$\mathbf{S}^*(y) = (\mathbf{T}^*y, (\mathbf{T}^*\mathbf{T})^{-1/2}y),   y \in \mathcal{H},$$

it follows that

$$\mathbf{S}\mathbf{S}^* = \mathbf{T}\mathbf{T}^* + (\mathbf{T}^*\mathbf{T})^{-1},$$

and so

$$\mathbf{T}\mathbf{T}^* + (\mathbf{T}^*\mathbf{T})^{-1} \leq 2\mathbf{I}.$$

Multiplying both sides by $\mathbf{T}_1^*$ from the left and by $\mathbf{T}_1$ from the right, we arrive at

$$\mathbf{I} + \mathbf{T}_1^{*2}\mathbf{T}_1^2 \leq 2\mathbf{T}_1^*\mathbf{T}_1,$$

which is clearly what we wanted.   ∎

We will say that the operator $\mathbf{T}_1$ is *concave down*. This concavity of $\mathbf{T}_1$ is the key to the success of our analysis.

LEMMA 6.9 $\|\mathbf{T}_1 x\| \geq \|x\|$ *for all $x \in \mathcal{H}$.*

Proof. For any $x \in \mathcal{H}$, the sequence $\{\|\mathbf{T}_1^n x\|^2\}_n$ is positive and concave down, by Lemma 6.8. It follows that $\|\mathbf{T}_1^n x\|^2$ is increasing in n; in particular, $\|\mathbf{T}_1 x\| \geq \|x\|$. ∎

In what follows, we let

$$\mathcal{H}_1 = \bigcap \{\mathbf{T}_1^n \mathcal{H} : n \geq 0\}.$$

It is obvious that $\mathcal{H}_1$ is invariant under $\mathbf{T}_1$. Also, the operator $\mathbf{T}_1$ maps $\mathcal{H}_1$ onto $\mathcal{H}_1$. In fact, if $x \in \mathcal{H}_1$, then for any $n = 0, 1, 2, \ldots$, we can find $y_n \in \mathcal{H}$ such that $x = \mathbf{T}_1^n y_n$. In particular,

$$x = \mathbf{T}_1 y_1 = \mathbf{T}_1(\mathbf{T}_1^{n-1} y_n)$$

for all $n = 1, 2, 3, \ldots$. Since $\mathbf{T}$ is one-to-one on $\mathcal{H}$, so is $\mathbf{T}_1$. Therefore, $y_1 = \mathbf{T}_1^{n-1} y_n$ for all $n = 1, 2, 3, \ldots$, which implies that $y_1 \in \mathcal{H}_1$. Combining these with Lemma 6.9, we conclude that the restriction of $\mathbf{T}_1$ to $\mathcal{H}_1$ is an invertible operator.

LEMMA 6.10 *The restriction of $\mathbf{T}_1$ to $\mathcal{H}_1$ is unitary: $\mathbf{T}_1 \mathcal{H}_1 = \mathcal{H}_1$ and $\|\mathbf{T}_1 x\| = \|x\|$ for all $x \in \mathcal{H}_1$.*

Proof. Fix a point $x \in \mathcal{H}_1$. For any $n = 0, 1, 2, \ldots$, pick $y_n \in \mathcal{H}_1$ such that $x = \mathbf{T}_1^n y_n$. Since $\mathbf{T}_1$ is one-to-one, we easily obtain $y_{n-1} = \mathbf{T}_1 y_n$ for all $n = 1, 2, 3, \ldots$. An application of Lemma 6.9 then shows that the sequence $\{\|y_n\|\}_n$ is decreasing.

On the other hand, the concavity of $\mathbf{T}_1$ (see Lemma 6.8) gives

$$
\begin{aligned}
\|y_{n-1}\|^2 + \|y_{n+1}\|^2 &= \|\mathbf{T}_1^2 y_{n+1}\|^2 + \|y_{n+1}\|^2 \\
&\leq 2\|\mathbf{T}_1 y_{n+1}\|^2 = 2\|y_n\|^2,
\end{aligned}
$$

so that the sequence $\{\|y_n\|^2\}_n$ is positive and concave down, which implies that $\{\|y_n\|\}_n$ is increasing. We conclude that the sequence $\{\|y_n\|\}_n$ is constant. In particular,

$$\|\mathbf{T}_1 y_1\| = \|y_0\| = \|y_1\|.$$

Since x is arbitrary and $\mathbf{T}_1$ is invertible on $\mathcal{H}_1$, it follows that $y_1 = \mathbf{T}_1^{-1} x$ is arbitrary in $\mathcal{H}_1$, that is, we have shown that $\|\mathbf{T}_1 y\| = \|y\|$ for all $y \in \mathcal{H}_1$. ∎

DEFINITION 6.11 *Let $\mathbf{R}$ be a bounded linear operator on a general Hilbert space $\mathcal{H}$. A closed subspace $\mathcal{G}$ of $\mathcal{H}$ is said to be reducing for $\mathbf{R}$ if both $\mathcal{G}$ and its orthogonal complement $\mathcal{H} \ominus \mathcal{G}$ are invariant under $\mathbf{R}$.*

LEMMA 6.12 *The subspace $\mathcal{H}_1$ is reducing for the operator $\mathbf{T}_1$.*

Proof. We have already shown that the restriction of $\mathbf{T}_1$ to $\mathcal{H}_1$ is a unitary operator. Given any $x \in \mathcal{H}_1$, an easy calculation shows that $\mathbf{P}_1 \mathbf{T}_1^* \mathbf{T}_1 x = x$, where $\mathbf{P}_1$ is the orthogonal projection from $\mathcal{H}$ onto $\mathcal{H}_1$. It follows that

$$\langle (\mathbf{T}_1^* \mathbf{T}_1 - \mathbf{I})x, x \rangle = 0, \qquad x \in \mathcal{H}_1.$$

By Lemma 6.9, the operator $\mathbf{T}_1$ is norm expansive, which means that $\mathbf{T}_1^* \mathbf{T}_1 - \mathbf{I} \geq 0$. In view of the above, we conclude that $\mathbf{T}_1^* \mathbf{T}_1 x = x$ for all $x \in \mathcal{H}_1$; see Exercise 16. Combining this with the fact that $\mathbf{T}_1 \mathcal{H}_1 = \mathcal{H}_1$, we arrive at $\mathbf{T}_1^* \mathcal{H}_1 = \mathcal{H}_1$. In particular, $\mathcal{H}_1$ is invariant under $\mathbf{T}_1^*$, and hence $\mathcal{H} \ominus \mathcal{H}_1$ is invariant under $\mathbf{T}$. $\blacksquare$

In the remainder of this section, we let

$$\mathcal{E} = \ker(\mathbf{T}^*) = \mathcal{H} \ominus \mathbf{T}\mathcal{H}$$

and

$$\mathbf{L} = \mathbf{T}_1^* = (\mathbf{T}^* \mathbf{T})^{-1} \mathbf{T}^*.$$

It is easy to check that $\mathbf{LT} = \mathbf{I}$, the identity operator on $\mathcal{H}$, and that $\mathbf{I} - \mathbf{TL} = \mathbf{P}_{\mathcal{E}}$, where $\mathbf{P}_{\mathcal{E}}$ is the orthogonal projection from $\mathcal{H}$ onto $\mathcal{E}$.

LEMMA 6.13 *For* $n = 1, 2, 3, \ldots$, *we have*

$$\ker(\mathbf{L}^n) = \mathcal{E} + \mathbf{T}\mathcal{E} + \cdots + \mathbf{T}^{n-1}\mathcal{E}.$$

Proof. By the definitions of $\mathcal{E}$ and $\mathbf{L}$ we have $\mathbf{L}(\mathcal{E}) = \{0\}$, and so $\mathbf{L}^n(\mathcal{E}) = \{0\}$ for all $n = 1, 2, 3, \ldots$. Since $\mathbf{LT} = \mathbf{I}$, we have

$$\mathbf{L}^{k+1}\mathbf{T}^k(\mathcal{E}) = \mathbf{L}(\mathcal{E}) = \{0\}$$

for $k = 0, 1, 2, \ldots$, and so

$$\mathbf{L}^n(\mathcal{E} + \mathbf{T}\mathcal{E} + \cdots + \mathbf{T}^{n-1}\mathcal{E}) = \{0\}.$$

On the other hand,

$$\mathbf{I} - \mathbf{T}^n \mathbf{L}^n = \sum_{k=0}^{n-1} \mathbf{T}^k (\mathbf{I} - \mathbf{TL}) \mathbf{L}^k = \sum_{k=0}^{n-1} \mathbf{T}^k \mathbf{P}_{\mathcal{E}} \mathbf{L}^k.$$

Thus, $x \in \ker(\mathbf{L}^n)$ implies that

$$x = \sum_{k=0}^{n-1} \mathbf{T}^k \mathbf{P}_{\mathcal{E}} \mathbf{L}^k x \in \mathcal{E} + \mathbf{T}\mathcal{E} + \cdots + \mathbf{T}^{n-1}\mathcal{E}.$$

This proves the desired formula for $\ker(\mathbf{L}^n)$. $\blacksquare$

We are now ready to prove the main theorem of this section.

THEOREM 6.14 *If* $\mathbf{T} : \mathcal{H} \to \mathcal{H}$ *satisfies conditions (a) and (b), then*

$$\mathcal{H} = [\mathcal{E}] = \bigvee \left\{ \mathbf{T}^n x : x \in \mathcal{E}, n \geq 0 \right\},$$

where $\mathcal{E} = \ker(\mathbf{T}^*) = \mathcal{H} \ominus \mathbf{T}\mathcal{H}$.

Proof. It is easy to check that $\mathbf{T}_1^* \mathbf{T}_1$ is invertible on $\mathcal{H}$, and that $\mathbf{T} = \mathbf{T}_1 (\mathbf{T}_1^* \mathbf{T}_1)^{-1}$. Since $\mathcal{H}_1$ reduces $\mathbf{T}_1$, it also reduces $\mathbf{T}$.

If $\mathbf{R} = \mathbf{T}|_{\mathcal{H}_1}$ and $\mathbf{R}_1 = \mathbf{T}_1|_{\mathcal{H}_1}$, then we have

$$\mathbf{R} = \mathbf{R}_1 (\mathbf{R}_1^* \mathbf{R}_1)^{-1} = (\mathbf{R}_1^*)^{-1},$$

because $\mathcal{H}_1$ is reducing for both $\mathbf{T}$ and $\mathbf{T}_1$, and as $\mathbf{R}_1$ is invertible, we obtain that $\mathbf{R}$ is invertible.

Write $\mathcal{H} = \mathcal{H}_1 \oplus \mathcal{H}_1^{\perp}$ and use the fact that $\mathbf{T}(\mathcal{H}_1) = \mathcal{H}_1$. We obtain $\mathcal{H}_1 \subset \mathbf{T}^n \mathcal{H}$ for all $n = 1, 2, 3, \ldots$. It follows that

$$\mathcal{H}_1 \subset \bigcap_{n=1}^{+\infty} \mathbf{T}^n \mathcal{H} = \bigcap_{n=0}^{+\infty} \mathbf{T}^n \mathcal{H} = \{0\},$$

so that $\mathcal{H}_1 = \{0\}$. By the definition of $\mathcal{H}_1$, we have

$$\{0\} = \bigcap_{n=0}^{+\infty} \mathbf{T}_1^n \mathcal{H} = \bigcap_{n=1}^{+\infty} \mathbf{T}_1^n \mathcal{H}.$$

Taking the orthogonal complement, we get

$$\mathcal{H} = \left[\bigcap_{n=1}^{+\infty} \mathbf{T}_1^n \mathcal{H} \right]^{\perp} = \bigvee_{n=1}^{+\infty} (\mathbf{T}_1^n \mathcal{H})^{\perp} = \bigvee_{n=1}^{+\infty} \ker(\mathbf{L}^n).$$

The last identity holds because the kernel of the adjoint equals the orthocomplement of the range of the operator. The desired result now follows from Lemma 6.13. ∎

As an application of the above operator-theoretic result we obtain the following Beurling-type theorem for invariant subspaces of the Bergman space.

THEOREM 6.15 *Suppose* $-1 < \alpha \leq 0$ *and* I *is an invariant subspace of* A_{α}^2. *Then* I *is generated by* $I \ominus zI$.

Proof. Let S be the operator of multiplication by z on A_{α}^2. We shall see that condition (a) holds for S:

$$\|Sf + g\|^2 \leq 2 \left(\|f\|^2 + \|Sg\|^2 \right), \qquad f, g \in A_{\alpha}^2.$$

Scrutinizing the proof of Lemma 6.8, we see that this is in fact equivalent to the operator inequality

$$SS^* + (S^*S)^{-1} \leq 2\mathbf{I}.$$

Using Taylor expansions, we realize that the latter is equivalent to the concavity of the sequence $\{1/\omega_n\}_n$, where

$$\omega_n = \frac{n!\, \Gamma(2+\alpha)}{\Gamma(n+2+\alpha)},$$

plus the condition $\omega_0 \leq 2\omega_1$, which is satisfied for all $-1 < \alpha \leq 0$. The concavity condition reads

$$\frac{1}{\omega_{n-1}} + \frac{1}{\omega_{n+1}} \leq \frac{2}{\omega_n}, \qquad n = 1, 2, 3, \ldots,$$

which is equivalent to

$$n(n+1)\Gamma(n+1+\alpha) + \Gamma(n+3+\alpha) \leq 2(n+1)\Gamma(n+2+\alpha).$$

Using the well-known functional identity $\Gamma(x+1) = x\Gamma(x)$, we reduce the above to

$$n(n+1) + (n+2+\alpha)(n+1+\alpha) \leq 2(n+1)(n+1+\alpha), \qquad n = 1, 2, 3, \ldots,$$

which is easily seen to be the same as

$$\alpha + \alpha^2 \leq 0.$$

We conclude that the sequence $\{1/\omega_n\}_n$ is concave for $-1 < \alpha \leq 0$.

To treat the general case, let $\mathbf{T}$ be the restriction of S to the invariant subspace I. The property (a) holds for $\mathbf{T}$ since it does for S. Likewise, it is obvious that $\mathbf{T}$ satisfies condition (b). The result is now immediate from Theorem 6.14.   ∎

COROLLARY 6.16 *Fix α, $-1 < \alpha \leq 0$. If I is an invariant subspace of A_α^2 of index 1, then I is generated by its extremal function G_I, and $\|f/G_I\| \leq \|f\|$ for all $f \in I$.*

Proof. If I has index 1 and G_I is the extremal function of I, then it is easy to see that

$$I \ominus zI = \mathbb{C}G_I.$$

The assertion that I is generated by G_I then follows from Theorem 6.15. We omit the proof that G_I is a contractive divisor on I; this is a special case of Exercise 8 in Chapter 9.   ∎

COROLLARY 6.17 *If $-1 < \alpha \leq 0$, then every function $f \in A_\alpha^2$ admits a factorization $f = GF$, where G is A_α^2-inner and F is cyclic in A_α^2.*

Proof. Let G be the extremal function of I_f, where I_f is the invariant subspace generated by f. Then I_f is generated by G, since I_f has the index 1. It follows that $f = GF$ with $\|F\| \leq \|f\|$. Furthermore, if p is a polynomial, then

$$\|pF\| = \left\| \frac{pf}{G} \right\| \leq \|pf\|.$$

Now, if $\{p_n\}_n$ is a sequence of polynomials such that $p_n f \to G$ in norm, then $p_n(z)F(z) \to 1$ pointwise. But the inequality

$$\|p_n F - p_m F\| \leq \|p_n f - p_m f\|$$

shows that $\{p_n F\}_n$ is a Cauchy sequence. We must have $p_n F \to 1$ in norm, making F cyclic in A_α^2.   ∎

6.4   Notes

That an invariant subspace I of a Bergman space may have index greater than 1 was first proved by Apostol, Bercovici, Foiaş, and Pearcy in [10]; their proof, however, is only an existence result and gives no clue of what such invariant subspaces look like. Explicit construction of such spaces was first carried out by Hedenmalm in [61], based on Seip's work on interpolation and sampling sequences, in the case $n = \dim(I/zI) < +\infty$; this construction was later improved to cover the case $\dim(I/zI) = +\infty$ in the paper [71]. It should be mentioned that the reverse triangle inequality of Lemma 6.2 is by no means necessary for the index one invariant subspaces $I_1, \ldots, I_n$ to generate an invariant subspace of index n.

The notion of inner spaces was introduced by Zhu in [143], where another notion called maximal inner spaces was also introduced. Several characterizations and examples of maximal inner spaces can be found in [143] as well.

Theorem 6.15, which we call a Beurling-type theorem, was proved by Aleman, Richter, and Sundberg in [7] in the special case $\alpha = 0$. The proof given here, which covers the cases $-1 < \alpha \le 0$, is due to Shimorin; see [126]. The case $0 < \alpha \le 1$ remains open. We point out that, unlike the classical Beurling theorem for H^2, Theorem 6.15 does not imply a function-theoretic description of invariant subspaces of A_α^2. Such a description in the context of Bergman spaces is known only for $\mathcal{A}^{-\infty}$; see [84].

One of the reasons that Bergman spaces have attracted so much attention in recent years is that they are closely related to an old open problem in Operator Theory. More specifically, the *invariant subspace problem* (of whether every bounded linear operator on a separable Hilbert space of dimension greater than one has a nontrivial invariant subspace) is equivalent to the following question about z-invariant subspaces of the Bergman space A^2: Given two invariant subspaces I and J of A^2 with $I \subset J$ and $\dim(J \ominus I) = +\infty$, does there exist another invariant subspace L of A^2 lying strictly between I and J? See [71] for an explanation and references.

6.5   Exercises and Further Results

1. A maximal invariant subspace is an invariant subspace contained in no other invariant subspace than the whole space. Show that if I is a maximal invariant subspace of A_α^p, then $I = I_a$ for some point $a \in \mathbb{D}$, where I_a stands for the subspace of all functions that vanish at a. See [67].

2. If an invariant subspace I contains a Nevanlinna function, then I is generated by a Nevanlinna function. See [143].

3. If $\{I_n\}_n$ is an increasing sequence of invariant subspaces in A_α^p and each I_n has index 1, show that the closure of $\cup_n\{I_n\}$ also has index 1.

4. Let I be an invariant subspace of A_α^p. Then $J = \{f \in A_\alpha^p : zf \in I\}$ is also an invariant subspace of A_α^p. Furthermore, I and J have the same index. See [80].

5. For any function $f \in A^2$, the space $J_f = \{g \in A^2 : \mathbf{P}(f\bar{g}) = 0\}$ is an invariant subspace of A^2.

For Problems 5–8, see [143].

6. Let G be an A^2-inner function. With the notation from the previous problem, show that J_G either has index 1 or 2.

7. A maximal inner space in A^2 is an inner space contained in no larger inner space. Show that every inner space is contained in a maximal inner space. Hint: apply Zorn's lemma.

8. If G is an A^2-inner function, then the one-dimensional space generated by G is a maximal inner space if and only if J_G has the index 1, where J_G is as defined in Problem 5.

9. For an invariant subspace I in A^2, let $M_z[I]$ denote the multiplication operator on I induced by the coordinate function z. Show that $M_z[I]$ and $M_z[J]$ are unitarily equivalent if and only if $I = J$.

10. Let $I = A^2$ and J be an invariant subspace of A^2. Show that $M_z[I]$ and $M_z[J]$ are similar if and only if J is generated by a Blaschke product whose zero set is the union of finitely many interpolating sequences. See [29].

11. Let $I = A^2$, and let J be an invariant subspace of A^2. Show that $M_z[I]$ and $M_z[J]$ are quasi-similar if and only if J is generated by a bounded analytic function. See [70].

12. For any positive real number σ, let I_σ be the invariant subspace of A^2 generated by the singular inner function S_σ (with a single point mass σ at $z = 1$). Show that $M_z[I_\sigma]$ and $M_z[I_\tau]$ are similar for all positive σ and τ. See [141].

13. Show that if the invariant subspace I of A^2 is singly generated or if I is zero-based, then I has the index 1.

14. If A and B are disjoint regular sequences, then $A \cup B$ is regular, and the decomposition $A \cup B$ is homogeneous.

15. If I and J are invariant subspaces in A_α^p of index 1, with the properties that $I \subset J$ and $n = \dim(J/I) < +\infty$, then there exists a Blaschke product b with n zeros such that $I = bJ$. What if I, J have higher index, say 2?

16. Let $\mathbf{A}$ be a positive bounded operator on the (separable) Hilbert space $\mathcal{H}$ (over the scalar field $\mathbb{C}$, as usual), which means that $\langle \mathbf{A}x, x \rangle \geq 0$ for all $x \in \mathcal{H}$. Suppose $\mathcal{H}_1$ is a closed subspace of $\mathcal{H}$, and that $\langle \mathbf{A}x, x \rangle = 0$ for all $x \in \mathcal{H}_1$. Show that $\mathbf{A}x = 0$ for all $x \in \mathcal{H}_1$.

17. Fix $0 < p < +\infty$ and $-1 < \alpha < +\infty$. Recall that the index of an invariant subspace I in A_α^p is defined as the dimension of the quotient space I/zI. Show that for $\lambda \in \mathbb{D}$, $(z - \lambda)I$ is a closed subspace of I, and that the dimension of the quotient space $I/(z - \lambda)I$ does not depend on λ. Hint: prove that the dimension is a locally constant function of λ.

18. Fix $0 < p < +\infty$ and $-1 < \alpha < +\infty$. Suppose I is an invariant subspace of A_α^p other than the trivial one $\{0\}$. Let $Z(I)$ denote the common zero set of the functions in I; we think of it as a discrete subset of $\mathbb{D}$. Then I has index 1 if and only if it satisfies the following *division property*: for each $\lambda \in \mathbb{D} \setminus Z(I)$, the function

$$f_\lambda(z) = \frac{f(z)}{z - \lambda}, \qquad z \in \mathbb{D} \setminus \{\lambda\},$$

extended analytically across the point λ, is in I whenever $f \in I$ and $f(\lambda) = 0$.

19. Suppose I is an invariant subspace of A^2 having index 2. This means that $I \ominus zI$ is an inner space of dimension 2. Let φ_1, φ_2 be two orthogonal inner functions in $I \ominus zI$. Then each function $f \in I$ can be written $f = f_1\varphi_1 + f_2\varphi_2$, where $f_1, f_2 \in A^2$ and $\|f_1\|^2 + \|f_2\|^2 \le \|f\|^2$. What should be added to insure uniqueness? Can we get $f_1\varphi_1, f_2\varphi_2 \in A^2$? See [7].

7
Cyclicity

In this chapter, we study the cyclic functions in the Bergman spaces A^p. First, we identify them with the A^p-outer functions, which are defined in terms of a notion of domination, in a fashion analogous to what is done in the classical Hardy space setting. Second, we show that a function that belongs to a smaller space A^q, $p < q$, is cyclic in A^p if and only if it is cyclic in the growth space $\mathcal{A}^{-\infty}$. Then we characterize the cyclic vectors for $\mathcal{A}^{-\infty}$ in terms of boundary premeasures; this constitutes the bulk of the material in the chapter.

7.1   Cyclic Vectors as Outer functions

A function in A^p is said to be *cyclic* if it generates A^p as an invariant subpace, that is, the smallest invariant subspace of A^p containing the function is the whole space. Thus, f is cyclic in A^p if and only if the closed linear span of the vectors $f, zf, z^2 f, \ldots$ is all of A^p.

In the classical Hardy space theory, a function $f \in H^p$ ($0 < p < +\infty$) is said to be outer if it is zero-free in the disk $\mathbb{D}$, and if the harmonic function $\log |f|$ equals the Poisson integral of its boundary values; the latter condition may be formulated as

$$\log |f(0)| = \frac{1}{2\pi} \int_{-\pi}^{\pi} \log |f(e^{i\theta})| \, d\theta,$$

where the boundary values of f are obtained from nontangential approach regions. It is well known that a function in H^p is cyclic if and only if it is outer; see [37], [49], or [82].

An alternative characterization of classical outer functions is as follows: $f \in H^p$ is outer if and only if whenever the conditions $g \in H^p$, $|g| \leq |f|$ on $\mathbb{T}$ hold, we also have that $|g(0)| \leq |f(0)|$. This latter form is amenable to generalization. In fact, it is clear that the condition that $|g| \leq |f|$ on $\mathbb{T}$ (almost everywhere) is equivalent to requiring $\|gq\|_{H^p} \leq \|fq\|_{H^p}$ for all polynomials q. This motivates the following definitions.

DEFINITION 7.1 *For $f, g \in A^p$, we say that f dominates g in A^p, denoted $g \prec f$, if $\|gq\|_{A^p} \leq \|fq\|_{A^p}$ holds for all polynomials q. And we say that a function $f \in A^p$ is A^p-outer if $|g(0)| \leq |f(0)|$ whenever $g \prec f$ in A^p.*

THEOREM 7.2 *Let $f \in A^p$ with $0 < p < +\infty$. Then f is A^p-outer if and only if it is cyclic in A^p.*

Proof. We first assume that f is cyclic, with the intention to prove that it is A^p-outer. Let $g \in A^p$ with $g \prec f$. Since f is cyclic, there exists a sequence of polynomials $\{q_n\}_n$ such that fq_n converges to the constant function 1 in the norm of A^p. Since $g \prec f$,

$$|g(0)q_n(0)| \leq \|gq_n\|_{A^p} \leq \|fq_n\|_{A^p} \to 1 \quad \text{as } n \to +\infty,$$

and $q_n(0) \to 1/f(0)$ as $n \to +\infty$, it follows that $|g(0)| \leq |f(0)|$, and so f is A^p-outer.

For the reverse implication, we assume instead that f is A^p-outer. Let φ be the extremal function for the invariant subspace I_f generated by f in A^p. By Theorem 3.33, $I_\varphi = I_f$, and $\|qf/\varphi\|_{A^p} \leq \|qf\|_{A^p}$ for all polynomials q. Thus, $f/\varphi \prec f$, so by the definition of outer functions,

$$\left| \frac{f}{\varphi}(0) \right| \leq |f(0)|.$$

Since $I_\varphi = I_f$, the left-hand side cannot vanish, and hence we must have $1 \leq |\varphi(0)|$. Since φ has norm 1, we must then have $\varphi(z) \equiv 1$, so that f is cyclic. ∎

7.2 Cyclicity in A^p Versus in $\mathcal{A}^{-\infty}$

Just as in the case of the Bergman spaces A^p, a function f in $\mathcal{A}^{-\infty}$ is said to be *cyclic in $\mathcal{A}^{-\infty}$* if the functions $f, zf, z^2 f, z^3 f, \ldots$ span a dense subspace of $\mathcal{A}^{-\infty}$. Since the topology in $\mathcal{A}^{-\infty}$ is softer than that of any A^p, $0 < p < +\infty$, it is immediate that a cyclic function in A^p is cyclic in $\mathcal{A}^{-\infty}$, too. It is a meaningful question whether these two concepts are in fact equivalent, that is, whether a function $f \in A^p$ that is cyclic in $\mathcal{A}^{-\infty}$ is automatically cyclic in A^p. In the next chapter we will show that this is not true in general. However, if the function f belongs to a slightly smaller space than A^p, then cyclicity in $\mathcal{A}^{-\infty}$ implies cyclicity in A^p.

THEOREM 7.3 *Let $f \in A^q$ with $0 < p < q < +\infty$. Then f is cyclic in A^p if and only if it is cyclic in $\mathcal{A}^{-\infty}$.*

Proof. By the definition of the topology in $\mathcal{A}^{-\infty}$, the fact that f is cyclic translates to the requirement that for some sequence $\{Q_n\}_n$ of polynomials and some fixed r, $0 < r < +\infty$, $f Q_n$ tends to the constant function 1 in norm in A^r as $n \to +\infty$. The assertion holds automatically if $p \le r$, and hence we may restrict our attention to the case $r < p$.

If these polynomials Q_n were all zero-free in $\mathbb{D}$, we could run the following argument. For real ε, $0 < \varepsilon < 1$, we form the powers f^ε and q_n^ε to obtain well-defined holomorphic functions in $\mathbb{D}$, and in particular, f^ε belongs to $A^{q/\varepsilon}$. Moreover, if the power is chosen in such a way that $f^\varepsilon Q_n^\varepsilon$ tends to 1 uniformly on compact subsets of $\mathbb{D}$ as $n \to +\infty$, then one shows that the convergence is also in norm in $A^{r/\varepsilon}$. This is a consequence of the fact that in $A^{r/\varepsilon}$, the norm of $f^\varepsilon Q_n^\varepsilon$ tends to the norm of 1; see Lemma 3.17. For sufficiently small ε, the functions $f Q_n^\varepsilon = f^{1-\varepsilon} f^\varepsilon Q_n^\varepsilon$, which are elements of A^q, tend to $f^{1-\varepsilon}$ as $n \to +\infty$ in the norm of A^p. Scrutinizing the requirement for ε, we find that for the above to work, ε should satisfy

$$0 < \varepsilon \le \frac{r(q-p)}{p(q-r)}.$$

We conclude that $f^{1-\varepsilon}$ belongs to the invariant subspace I_f generated by f in A^p. Proceeding analogously, we also get $f^{1-2\varepsilon} \in I_f$. By choosing ε in such a way that $1/\varepsilon$ is an integer, we get eventually that 1 is in I_f, which clearly implies that f is cyclic.

Now we return to the real world where the polynomials Q_n need not be zero-free at all. Let Z_n denote the finitely many zeros the polynomial Q_n may have in $\mathbb{D}$ (with multiplicities), and let φ_n be the corresponding canonical zero divisor (extremal function) in A^r. It is the function of norm 1 in A^r that vanishes on Z_n and has biggest value in modulus at 0. Since $f Q_n$ for large n has norm close to 1, and the value at 0 close to 1, we conclude that $|\varphi_n(0)| \to 1$ as $n \to +\infty$. We now form $f \widetilde{Q}_n$, where $\widetilde{Q}_n = Q_n/\varphi_n$, and observe that it has smaller norm than $f Q_n$ in A^r, by the contractive division property of canonical divisors. Therefore,

$$\limsup_n \left\| f \widetilde{Q}_n \right\|_{A^r} \le 1,$$

whereas the limit of the values at the origin is

$$\lim_n \left| f(0) \widetilde{Q}_n(0) \right| = 1,$$

so that in a first step, $f \widetilde{Q}_n \to 1$ uniformly on compact subsets of $\mathbb{D}$, and in a second step, the convergence is in the norm of A^r; see Lemma 3.17. We are now able to work with the zero-free functions $\widetilde{Q}_n$ in place of the functions Q_n in the above argument; they are not polynomials, but certainly approximable by polynomials, since the extremal function φ_n extends holomorphically across the unit circle and is bounded away from 0 in a neighborhood of $\mathbb{T}$. ∎

Although a complete "geometric" characterization for cyclic vectors in A^p is still lacking, the corresponding problem for $\mathcal{A}^{-\infty}$ was solved over twenty years ago. The key idea of the solution is the notion of premeasures for functions in $\mathcal{A}^{-\infty}$.

7.3 Premeasures for Functions in $\mathcal{A}^{-\infty}$

Let $\mathcal{B}(\mathbb{T})$ be the set of all open, closed, and half-open arcs of $\mathbb{T}$, including $\mathbb{T}$, $\emptyset$, and all one-point sets. A real-valued function μ defined on $\mathcal{B}(\mathbb{T})$ is called a *premeasure* if

(i) $\mu(\mathbb{T}) = 0$.

(ii) $\mu(I_1 \cup I_2) = \mu(I_1) + \mu(I_2)$ for all $I_1, I_2 \in \mathcal{B}(\mathbb{T})$ with $I_1 \cap I_2 = \emptyset$ and $I_1 \cup I_2 \in \mathcal{B}(\mathbb{T})$.

(iii) $\mu(I_n) \to 0$ as $n \to +\infty$ whenever $\{I_n\}_n$ is a decreasing sequence in $\mathcal{B}(\mathbb{T})$ with empty intersection.

Every premeasure is immediately extended by finite additivity to the class of sets of the form

$$S = \bigcup_{k=1}^{n} I_k,$$

where each I_k is in $\mathcal{B}(\mathbb{T})$. In particular, if $\{I_k\}_k$ is a finite collection of mutually disjoint arcs in $\mathbb{T}$, then

$$\mu\left(\bigcup_k I_k\right) = \sum_k \mu(I_k).$$

For every premeasure μ, we define a real-valued function $\widehat{\mu}$ on $(0, 2\pi]$ as follows:

$$\widehat{\mu}(\theta) = \mu(I_\theta), \qquad \theta \in (0, 2\pi],$$

where

$$I_\theta = \{e^{it} : 0 \le t < \theta\}.$$

Thus, a one-to-one correspondence is established between the set of premeasures and the set of real-valued functions f on $(0, 2\pi]$ such that

(a) $f(\theta-)$ exists for all $\theta \in (0, 2\pi]$ and $f(\theta+)$ exists for all $\theta \in [0, 2\pi)$.

(b) $f(\theta) = f(\theta-)$ for all $\theta \in (0, 2\pi]$.

(c) $f(2\pi) = 0$.

It is clear that any function f satisfying the above conditions has only a countable number of discontinuities, all of which are jumps.

For an arc I on the unit circle $\mathbb{T}$, let $|I|_s = |I|/(2\pi)$ be its normalized length, as in Chapter 4. The *logarithmic entropy* of I is the quantity

$$\kappa(I) = |I|_s \log \frac{e}{|I|_s}.$$

A premeasure μ is said to be κ-*bounded above* if

$$\mu(I) \leq C\kappa(I)$$

for all arcs I, with some positive constant C independent of I. The least such constant C will be called the κ-bound of μ and will be denoted by $\|\mu\|^+$. The set of all premeasures μ with $\|\mu\|^+ < +\infty$ will be denoted by κB^+. For comparison, consider the case when $\mu(I) \leq C|I|_s$ for all arcs I. Then the premeasure μ is an ordinary real-valued Borel measure, with the property that $d\mu - C\,ds$ is a negative measure, where $ds(z) = |dz|/(2\pi)$ is the normalized arc length measure on $\mathbb{T}$, as usual.

It is clear that in general, $\|\mu\|^+ \geq 0$, and $\|\mu\|^+ = 0$ holds if and only if $\mu = 0$. The space κB^+ is not linear; it is only a cone. In fact, for all $\mu_1, \mu_2 \in \kappa B^+$ and reals $0 \leq t_1, t_2 < +\infty$, we have $\|t_1\mu_1\|^+ = t_1\|\mu_1\|^+$ and

$$\|t_1\mu_1 + t_2\mu_2\|^+ \leq t_1\|\mu_1\|^+ + t_2\|\mu_2\|^+.$$

To obtain a vector space we should instead consider premeasures of the form $\kappa B^+ - \kappa B^+$; these are the so-called premeasures of *bounded κ-variation* (see Exercise 4). This concept is important for the study of meromorphic functions of the class of quotients $\mathcal{A}^{-\infty}/\mathcal{A}^{-\infty}$ as well as for the description of all invariant subspaces of $\mathcal{A}^{-\infty}$. However, it will not be needed for the study of cyclicity in $\mathcal{A}^{-\infty}$.

We now prove that a large class of real-valued harmonic functions in $\mathbb{D}$ can be represented as the Poisson integral of premeasures that are κ-bounded above.

THEOREM 7.4 *Let U be a real-valued harmonic function in $\mathbb{D}$ with $U(0) = 0$, such that for some positive constants A and B,*

$$U(z) \leq A \log \frac{1}{1 - |z|^2} + B, \qquad z \in \mathbb{D}. \tag{7.1}$$

Then for every open arc I of $\mathbb{T}$ the following limit exists:

$$\mu(I) = \mu_U(I) = \lim_{r \to 1} \int_I U(rz)\,ds(z), \tag{7.2}$$

where ds is the normalized arc-length measure. Moreover, there exists an absolute constant C such that

$$\mu(I) \leq 2A\kappa(I) + (A + B)C\,|I|_s. \tag{7.3}$$

We can represent U in terms of the Poisson integral of μ:

$$U(z) = \int_{\mathbb{T}} \frac{1 - |z|^2}{|\zeta - z|^2}\,d\mu(\zeta), \qquad z \in \mathbb{D}. \tag{7.4}$$

Proof. The proof consists of three steps. First we establish a crude estimate for

$$\sup\left\{\left|\int_I U(rz)\,ds(z)\right| : 0 < r < 1,\ I \in \mathcal{B}(\mathbb{T})\right\},$$

where $\mathcal{B}(\mathbb{T})$ is the collection of all arcs in $\mathbb{T}$. In particular, this implies that the supremum is finite. Based on that crude estimate, we then refine it to obtain that for all arcs I on $\mathbb{T}$ and r with $0 < r < 1$,

$$\int_I U(r\zeta)\,ds(\zeta) \leq 2A\kappa(I) + C(A+B)\,|I|_s.$$

To derive from this the existence of the limit (7.2) we need a compactness property of premeasures that are uniformly κ-bounded above; this is analogous to the classical Helly selection theorem for functions of bounded variation. The representation (7.4), the uniqueness thereof, and the existence of the limit (7.2) then follow. Note that the integral in (7.4), although it involves a premeasure μ, can be understood as a classical Stieltjes integral involving the integrated function $\hat{\mu}$.

Step 1. Fix r, $0 < r < 1$, and let $I(\zeta_1, \zeta_2)$ (the arc running counterclockwise from ζ_1 to ζ_2) be a solution to the extremal problem

$$\min\left\{\int_I U(r\zeta)\,ds(\zeta) : I \in \mathcal{B}(\mathbb{T})\right\}.$$

Since U is harmonic and $U(0) = 0$, this minimum will be negative or 0; and if we denote it by $-M$, where $0 \leq M < +\infty$, then we see that

$$M = \max\left\{\int_J U(r\zeta)\,ds(\zeta) : J \in \mathcal{B}(\mathbb{T})\right\},$$

the extremum being attained for $J = \mathbb{T} \setminus I(\zeta_1, \zeta_2)$. It follows that

$$
\begin{aligned}
-M &= \int_{I(\zeta_1,\zeta_2)} U(r\zeta)\,ds(\zeta) \leq \int_I U(r\zeta)\,ds(\zeta) \\
&\leq \int_{\mathbb{T}\setminus I(\zeta_1,\zeta_2)} U(r\zeta)\,ds(\zeta) = M,
\end{aligned}
$$

for all $I \in \mathcal{B}(\mathbb{T})$.

Consider the harmonic function

$$
\begin{aligned}
V_0(z) &= \int_{\mathbb{T}\setminus I(\zeta_1,\zeta_2)} P(z,\zeta)\,U(r\zeta)\,ds(\zeta) \\
&= U(rz) - \int_{I(\zeta_1,\zeta_2)} P(z,\zeta)\,U(r\zeta)\,ds(\zeta),
\end{aligned}
$$

where

$$P(z,\zeta) = \frac{1 - |z\zeta|^2}{|1 - z\bar{\zeta}|^2}$$

is the Poisson kernel. Let $S(\zeta_1, \zeta_2)$ be the sector of $\mathbb{D}$ bounded by two radii and the arc $I(\zeta_1, \zeta_2)$. Using the symmetry and unimodality of the Poisson kernel, we obtain by applying integration by parts the estimate

$$V_0(z) \leq U(rz) + M \frac{1 - |z|^2}{d_{\mathbb{C}}(z, \{\zeta_1, \zeta_2\})^2}, \qquad z \in \mathbb{D} \setminus S(\zeta_1, \zeta_2), \qquad (7.5)$$

where as usual $d_{\mathbb{C}}$ is the Euclidean metric in $\mathbb{C}$. After all, in the remaining sector $\mathbb{D} \setminus S(\zeta_1, \zeta_2)$, the distance to the arc $I(\zeta_1, \zeta_2)$ coincides with the distance to the two end points $\{\zeta_1, \zeta_2\}$. Consider now the curve $\gamma(\varepsilon, \zeta_1, \zeta_2)$ joining ζ_1 and ζ_2:

$$\gamma(\varepsilon, \zeta_1, \zeta_2) = \left\{ z \in \mathbb{D} \setminus S(\zeta_1, \zeta_2) : \frac{1 - |z|^2}{d_{\mathbb{C}}(z, \{\zeta_1, \zeta_2\})^2} = \varepsilon \right\},$$

where $0 < \varepsilon < \frac{1}{4}$. The curve $\gamma(\varepsilon, \zeta_1, \zeta_2)$ consists of two arcs of congruent circles tangent to $\mathbb{T}$ at ζ_1 and ζ_2, respectively, which are joined at a point outside the sector $S(\zeta_1, \zeta_2)$. To get the general idea, take a look at Figure 4.1. An easy way to see this is to identify the circles to which the arcs belong as level sets for the Poisson kernels $P(\cdot, \zeta_1)$ and $P(\cdot, \zeta_2)$. We have from (7.5) that

$$V_0(z) \leq A \log \frac{1}{\varepsilon \, d_{\mathbb{C}}(z, \{\zeta_1, \zeta_2\})^2} + B + \varepsilon M, \qquad z \in \gamma(\varepsilon, \zeta_1, \zeta_2),$$

and $V_0(z) = 0$ for $z \in I(\zeta_1, \zeta_2)$. We can rewrite this in the form

$$
\begin{aligned}
V_0(z) \quad &\leq \quad 2A \max \left\{ \log \frac{2}{|z - \zeta_1|}, \log \frac{2}{|z - \zeta_2|} \right\} \\
&\quad + A \log \frac{1}{4\varepsilon} + B + \varepsilon M \\
&\leq \quad 2A \left(\log \frac{2}{|z - \zeta_1|} + \log \frac{2}{|z - \zeta_2|} \right) \\
&\quad + A \log \frac{1}{4\varepsilon} + B + \varepsilon M, \qquad (7.6)
\end{aligned}
$$

for $z \in \gamma(\varepsilon, \zeta_1, \zeta_2)$.

The function V_0 vanishes on $I(\zeta_1, \zeta_2)$, so that (7.6) holds on $I(\zeta_1, \zeta_2)$ as well. Since the right-hand side of (7.6) is a harmonic function, we can apply the maximum principle to the domain bounded by the closed contour $\gamma(\varepsilon, \zeta_1, \zeta_2) \cup I(\zeta_1, \zeta_2)$ to get that (7.6) holds there, and in particular at the origin, so that

$$V_0(0) = M \leq 4A \log 2 + A \log \frac{1}{4\varepsilon} + B + \varepsilon M = A \log \frac{4}{\varepsilon} + B + \varepsilon M,$$

which yields, for instance by fixing $\varepsilon = \frac{1}{8}$, that

$$M \leq 4(A + B). \qquad (7.7)$$

Step 2. Let I be an arbitrary arc of $\mathbb{T}$ and let $S(I)$ be the associated sector bounded by two radii and I. We want to establish a sharper upper estimate for

$$\int_I U(r\zeta)\,ds(\zeta),$$

valid for all r, $0 < r < 1$. We define, for $0 < \varepsilon < \frac{1}{4}$,

$$\gamma'(\varepsilon, I) = \left\{ z \in S(I) : \frac{1 - |z|^2}{d_{\mathbb{C}}(z, \{\xi_1, \xi_2\})^2} = \varepsilon \right\}, \tag{7.8}$$

where ξ_1 and ξ_2 are the endpoints of the arc I (ordered so that we go counterclockwise as we go from ξ_1 to ξ_2 along I), and consider the harmonic function

$$V_1(z) = \int_I P(z, \zeta)\,U(r\zeta)\,ds(\zeta), \qquad z \in \mathbb{D}.$$

Repeating the argument based on unimodality of the Poisson kernel, we obtain

$$V_1(z) \le U(rz) + M\,\frac{1 - |z|^2}{d_{\mathbb{C}}(z, \{\xi_1, \xi_2\})^2}, \qquad z \in S(I).$$

It follows that

$$
\begin{aligned}
V_1(z) \ &\le\ A \log \frac{1}{1 - |z|^2} + B + M\varepsilon \\
&=\ A \log \frac{1}{(d_{\mathbb{C}}(z, \{\xi_1, \xi_2\}))^2} + B + M\varepsilon \\
&\le\ 2A \log \frac{1}{d_{\mathbb{C}}(z, \{\xi_1, \xi_2\})} + A \log \frac{1}{\varepsilon} + B + M\varepsilon \\
&\le\ A \left(2 \log \frac{1}{d_{\mathbb{C}}(z, \{\xi_1, \xi_2\})} + \log \frac{1}{\varepsilon} + 4\varepsilon \right) + B\,(1 + 4\varepsilon),
\end{aligned}
$$

where have used (7.7).

Consider now a harmonic function V_2 on $\mathbb{D}$ with boundary values

$$V_2(\zeta) = \begin{cases} \log \frac{1}{d_{\mathbb{C}}(\zeta, \{\xi_1, \xi_2\})}, & \zeta \in I, \\ 0, & \zeta \in \mathbb{T} \setminus I. \end{cases}$$

In terms of an integral formula, it is expressed by

$$V_2(z) = \int_I P(z, \zeta) \log \frac{1}{d_{\mathbb{C}}(\zeta, \{\xi_1, \xi_2\})}\,ds(\zeta), \tag{7.9}$$

so that

$$
\begin{aligned}
V_2(0) \ &=\ \frac{1}{\pi} \int_0^{\pi |I|_s} \log \frac{1}{t}\,dt + o(|I|_s) \\
&=\ |I|_s \log \frac{e}{|I|_s} + O(|I|_s) = \kappa(I) + O(|I|_s) \qquad \text{as} \quad |I|_s \to 0.
\end{aligned}
$$

One shows that

$$V_2(z) \geq \log \frac{1}{d_{\mathbb{C}}(z, \{\xi_1, \xi_2\})} - C_1, \qquad z \in \gamma'(\varepsilon, I), \qquad (7.10)$$

with some constant C_1 independent of I and ε. This can be seen either directly from (7.9), which is somewhat messy, or by comparing $V_2(z)$ with the harmonic function (recall that the real part times the imaginary part of a holomorphic function is always harmonic)

$$Q(z) = \log \frac{1}{|I|_s} + \log \left| \frac{(\xi_0 - z)^2}{(\xi_1 - z)(\xi_2 - z)} \right| \cdot \frac{1}{\pi} \arg \left(\frac{\xi_0 - z)^2}{(\xi_1 - z)(\xi_2 - z)} \right), \quad (7.11)$$

where ξ_0 is the midpoint of the arc $I = I(\xi_1, \xi_2)$, and the branch of the argument in the third term is chosen such that its value at 0 equals 0. The boundary values of the third term are 1 on $I(\xi_1, \xi_0)$, -1 on $I(\xi_0, \xi_2)$, and 0 on the remaining $\mathbb{T} \setminus I$. A close inspection of the second term (the first term is a constant) then leads to the conclusion that $Q(\zeta) \leq V_2(\zeta)$ on $\mathbb{T}$. Hence $Q(z) \leq V_2(z)$ throughout $\mathbb{D}$. Moreover, from geometric considerations, it is evident that $Q(z)$ satisfies

$$Q(z) \geq \log \frac{1}{d_{\mathbb{C}}(z, \{\xi_1, \xi_2\})} - C_2$$

on the part of $\gamma'(\varepsilon, I)$ corresponding to the first quarter of I starting from ξ_1, where C_2 is a constant. Thus, (7.10) holds on that part of $\gamma'(\varepsilon, I)$. A similar argument establishes (7.10) for the quarter of I ending at ξ_2, and finally for the middle half of I, too. Of course, we need different harmonic functions for comparison in each case.

We also need a harmonic function $V_3(z)$ with boundary values 1 on I and 0 on $\mathbb{T} \setminus I$. It is given by the explicit formula

$$V_3(z) = \frac{1}{\pi} \arg \frac{\xi_2 - z}{\xi_1 - z} - |I|_s,$$

and a simple geometric argument shows that

$$\min \left\{ V_3(z) : z \in \gamma'(\varepsilon, I) \right\} \geq 1 - C(\varepsilon) |I|_s,$$

with $C(\varepsilon) \to 0$ as $\varepsilon \to 0$. We also have $V_3(0) = |I|_s$.

Combining all the above facts, we see that the harmonic function

$$
\begin{aligned}
V_4(z) \; &= \; V_1(z) - 2A\, V_2(z) \\
&\quad - \left(AC_1 + A \log \frac{1}{\varepsilon} + A\, C(\varepsilon) + B + B\, C(\varepsilon) \right) \left(V_3(z) + C(\varepsilon)|I|_s \right)
\end{aligned}
$$

is negative in the domain bounded by $\mathbb{T} \setminus I$ and $\gamma'(\varepsilon, I)$, which contains the origin. Therefore, fixing some value of ε, we get

$$V_4(0) = \int_{\mathbb{T}} U(r\zeta)\, ds(\zeta) - 2A\, |I|_s \log \frac{e}{|I|_s} - C_4 A |I|_s - C_5 B |I|_s < 0,$$

which is equivalent to

$$\int_I U(r\zeta)\,ds(\zeta) \leq 2A\,\kappa(I) + C(A+B)|I|_s, \tag{7.12}$$

as asserted.

Step 3. For r with $0 < r < 1$, we can represent the function $U_r(z) = U(rz)$ as a Poisson integral

$$U_r(z) = \int_{\mathbb{T}} P(z,\zeta)\,d\mu_r(\zeta), \tag{7.13}$$

where $d\mu_r(\zeta) = U_r(\zeta)\,ds(\zeta)$ is the corresponding boundary measure. This measure is also a premeasure, and by (7.12),

$$\mu_r(I) = \int_I U(r\zeta)\,ds(\zeta) \leq 2A\,\kappa(I) + C\,(A+B)\,|I|_s, \tag{7.14}$$

which makes the family of premeasures μ_r, $0 < r < 1$, uniformly κ-bounded above. By the above-mentioned compactness principle, there is a sequence of radii $r_1 < r_2 < \cdots \to 1$ and a premeasure μ that is κ-bounded above such that $\mu_n = \mu_{r_n} \to \mu$ weakly (as premeasures) as $n \to +\infty$. Weak convergence (as premeasures) of μ_n to μ means that $\mu_n(I) \to \mu(I)$ for all arcs I, with the possible exception of a countable set of arcs with the property that μ carries nonzero mass at at least one of the endpoints of the arc. This weak convergence permits us to go to the limit in the Poisson integral (7.13):

$$U(z) = \int_{\mathbb{T}} P(z,\zeta)\,d\mu(\zeta),$$

which is (7.4). Clearly, μ satisfies (7.3). The existence of the limit (7.2) then follows from (7.4) by integration by parts, computation of the integral in (7.2), and subsequent transition to the limit as $r \to 1$. ∎

Note that the term $C(A+B)|I|_s$ in the theorem above can be discarded at the cost of increasing the coefficient of the principal quantity $\kappa(I)$. In particular, the μ in the theorem is a premeasure that is κ-bounded above.

A converse statement to Theorem 7.4 is supplied by Exercise 1 of Chapter 4: if $\mu \in \kappa B^+$ with $\|\mu\|^+ = A$, then its Poisson extension has the bound

$$P[\mu](z) \leq A\log\frac{1}{1-|z|^2} + B, \qquad z \in \mathbb{D},$$

for some constant B.

Given a premeasure $\mu \in \kappa B^+$, it is natural for us to ask whether μ can be reasonably defined for sets $F \subset \mathbb{T}$ that are more general than finite unions of arcs. If F is closed, then it is natural to define

$$\mu(F) = -\sum_k \mu(I_k),$$

where $\{I_k\}_k$ is the collection of complementary arcs of F in $\mathbb{T}$. For this definition to be useful (or just for it to be meaningful), the series above should be absolutely convergent. If, in addition, we want the corresponding series for any closed subset of F also to be absolutely convergent for all $\mu \in \kappa B^+$, then F must have finite entropy: $\widehat{\kappa}(F) < +\infty$. In particular, this is so for Beurling-Carleson sets F, which in addition have zero length.

Now suppose $\mu \in \kappa B^+$ and F is a Beurling-Carleson set in $\mathbb{T}$. Then μ generates a negative Borel measure $\sigma_{\mu,F}$ on the σ-algebra of all Borel measurable subsets of F; see Exercise 7. Furthermore, if F_1 and F_2 are two Beurling-Carleson sets, then the measures σ_{μ,F_1} and σ_{μ,F_2} coincide on $F_1 \cap F_2$. The totality of all $\sigma_{\mu,F}$, where F is Beurling-Carleson in $\mathbb{T}$, form the κ-*singular part* of μ, and is denoted by σ_μ; it is a negative Borel measure defined on the σ-algebra generated by Beurling-Carleson sets.

Recall that the logarithmic entropy of an arc I on the circle $\mathbb{T}$ is the quantity

$$\kappa(I) = |I|_s \, \log \frac{e}{|I|_s}.$$

The definition extends to unions of disjoint *open* arcs:

$$\kappa\left(\cup_j I_j\right) = \sum_j \kappa(I_j),$$

so that for a closed subset F of $\mathbb{T}$, we have the identity

$$\widehat{\kappa}(F) = \kappa(\mathbb{T} \setminus F).$$

The κ-singular part σ_μ of a premeasure $\mu \in \kappa B^+$ satisfies

$$-\|\mu\|^+ \widehat{\kappa}(F) \leq \sigma_\mu(F) \leq 0,$$

for all Beurling-Carleson sets F. Recall from Section 4.2 that the entropy of F can also be written as

$$\widehat{\kappa}(F) = \int_{\mathbb{T}} \log \frac{\pi}{d_{\mathbb{T}}(\zeta, F)} \, ds(\zeta),$$

provided that F is a Beurling-Carleson set. Here, for z and w in $\mathbb{T}$, the distance on the circle is

$$d_{\mathbb{T}}(z, w) = \left| \arg \frac{z}{w} \right|,$$

with arg taking values in $(-\pi, \pi]$, and

$$d_{\mathbb{T}}(\zeta, F) = \inf\{d_{\mathbb{T}}(\zeta, w) : w \in F\}.$$

The κ-singular part σ_μ of μ is supported on a "small" set. In fact, there exists an increasing sequence $\{F_n\}_n$ of Beurling-Carleson sets such that σ_μ is supported on $E = \cup_n F_n$. Note that the total κ-singular measure of $\mathbb{T}$ might be infinite:

$$\sigma_\mu(\mathbb{T}) = \sigma_\mu(E) = -\infty.$$

For a closed set F, let

$$F^\delta = \{\zeta \in \mathbb{T} : d_{\mathbb{T}}(\zeta, F) < \delta\}$$

be the δ-*neighborhood* of F, which is an open set. It can be shown that

$$\kappa(F^\delta) = \int_{F^\delta} \log \frac{\pi}{d_{\mathbb{T}}(\zeta, F)} \, ds(\zeta) \to 0 \qquad \text{as } \delta \to 0^+,$$

for all Beurling-Carleson sets F; see Exercise 8. From this and the definition of $\mu(F)$, we conclude that

$$\sigma_\mu(F) = \lim_{\delta \to 0^+} \mu(F^\delta) = \lim_{\delta \to 0^+} \mu(\overline{F^\delta}),$$

which then implies that

$$\sigma_{\mu_1 + \mu_2} = \sigma_{\mu_1} + \sigma_{\mu_2}$$

for all μ_1 and μ_2 in κB^+.

A premeasure μ in κB^+ will be called κ-*smooth*, or κ-*absolutely continuous*, if there exists a sequence $\{\mu_n\}_n$ of premeasures in κB^+ such that $\mu + \mu_n$ is in κB^+ for all n,

$$\sup_n \|\mu + \mu_n\|^+ < +\infty,$$

and

$$\sup_I |(\mu + \mu_n)(I)| \to 0$$

as $n \to +\infty$, where the second supremum is taken over all arcs in $\mathbb{T}$.

The rest of the section is devoted to the following approximation theorem for premeasures.

THEOREM 7.5 *A premeasure μ in κB^+ is κ-absolutely continuous if and only if its κ-singular part, σ_μ, is zero.*

This theorem is critical for describing the cyclic vectors of $\mathcal{A}^{-\infty}$. To better understand it, let us first consider the classical setting of H^∞ and explain the underlying ideas.

A function $f \in H^\infty$ is called *weakly cyclic* if there exists a sequence $\{f_n\}_n$ of functions in H^∞ such that $\sup_n \|f_n f\|_\infty < +\infty$ and $f_n(z)f(z) \to 1$, as $n \to +\infty$, uniformly on compact sets of $\mathbb{D}$. One obvious condition for f to be weakly cyclic is that f be zero-free. The other condition is that the boundary Herglotz measure $\mu = \mu_f$ of $\log|f(z)|$, determined by

$$\log|f(z)| = \int_{\mathbb{T}} \frac{1 - |z|^2}{|\zeta - z|^2} \, d\mu_f(\zeta),$$

be absolutely continuous with respect to Lebesgue measure, that is, the (negative) singular part σ_μ of μ must be zero.

The classical singular part σ_μ is "indestructible", in the sense that $\sigma_{\mu_{fg}} \leq \sigma_{\mu_f}$ for all $g \in H^\infty$. Therefore, we cannot "whittle down" the singular part of μ_f. On the other hand, we can do that with an absolutely continuous measure

$$d\mu_f(\zeta) = \log|f(\zeta)| \, ds(\zeta)$$

by truncating the L^1-function $\log|f(\zeta)|$ and gradually reducing it to zero; namely, we can achieve this by multiplying f by suitable functions $g_n \in H^\infty$.

In the $\mathcal{A}^{-\infty}$ setting, the truncation process is not available for premeasures. Therefore, to show that the condition $\sigma_\mu = 0$, where $\mu = \mu_f$ is the premeasure of the harmonic function $\log|f|$, is sufficient for f to be cyclic in $\mathcal{A}^{-\infty}$, we must find an alternative method.

We now begin the proof of Theorem 7.5.

Proof. The necessity is relatively simple. Thus, we assume that μ is κ-absolutely continuous, so that there exists a sequence $\{\mu_n\}_n$ in κB^+ such that $\|\mu + \mu_n\|^+ \leq C$ for all n and $(\mu + \mu_n)(I) \to 0$, as $n \to +\infty$, uniformly in I, where C is a positive constant and I is any arc in $\mathbb{T}$.

Take an arbitrary Beurling-Carleson set F and let $\{I_n\}_n$ be its complementary arcs in $\mathbb{T}$. We have

$$
\begin{aligned}
-\sigma_{\mu+\mu_n} &= \sum_{n \leq N}(\mu + \mu_n)(I_n) + \sum_{n > N}(\mu + \mu_n)(I_n) \\
&\leq \sum_{n \leq N}(\mu + \mu_n)(I_n) + C\sum_{n > N}\kappa(I_n),
\end{aligned}
$$

and so

$$
-\liminf_n \sigma_{\mu+\mu_n}(F) \leq C\sum_{n > N}\kappa(I_n).
$$

Letting $N \to +\infty$ and using $\widehat{\kappa}(F) < +\infty$, we obtain

$$
\liminf_n \sigma_{\mu+\mu_n}(F) \geq 0.
$$

Since $\mu_n \in \kappa B^+$, its κ-singular part is negative, and so $\sigma_{\mu+\mu_n}(F) \geq \sigma_\mu(F)$. It follows that $\sigma_\mu(F) \geq 0$ for all Beurling-Carleson sets F. On the other hand, $\mu \in \kappa B^+$ implies that $\sigma_\mu(F) \leq 0$. Thus, $\sigma_\mu(F) = 0$ for all Beurling-Carleson sets F, and we have proved the necessity of the condition $\sigma_\mu = 0$ in order for μ to be κ-absolutely continuous.

We will need several auxiliary results before we can prove the sufficiency part of Theorem 7.5. The following is a kind of "normal families" result for Beurling-Carleson sets.

LEMMA 7.6 *Let $\{F_n\}_n$ be a sequence of sets, each of which is the union of a finite number of closed arcs. Suppose that $|F_n|_s \to 0$ and $\kappa(\mathbb{T} \setminus F_n) = O(1)$ as $n \to +\infty$. Then there exists a subsequence $\{F_{n_k}\}_k$ and a Beurling-Carleson set F_∞ with the following property: For every $\delta > 0$, there is some $N = N_\delta$ with $F_{n_k} \subset F_\infty^\delta$ and $F_\infty \subset F_{n_k}^\delta$ for all $k > N$, where we recall the notation F^δ for the δ-neighborhood of F.*

Proof. For each n, let $\{I_{k,n} : k = 1, 2, \ldots\}$ be the complementary arcs of F_n arranged so that $|I_{k,n}|_s$ is decreasing in k. First, we show that the sequence

$\{|I_{1,n}|_s\}_n$ is bounded away from zero. In fact,

$$\kappa(\mathbb{T} \setminus F_n) = \sum_k |I_{k,n}|_s \log \frac{e}{|I_{k,n}|_s} \geq |\mathbb{T} \setminus F_n|_s \log \frac{e}{|I_{1,n}|_s},$$

and therefore,

$$\log \frac{e}{|I_{1,n}|_s} \leq \frac{\kappa(\mathbb{T} \setminus F_n)}{|\mathbb{T} \setminus F_n|_s}.$$

Since $|\mathbb{T} \setminus F_n|_s \to 1$ and $\kappa(\mathbb{T} \setminus F_n) = O(1)$ as $n \to +\infty$, the inequality above shows that the sequence of normalized lengths $\{|I_{1,n}|_s\}_n$ is bounded away from zero.

Next, choose a subsequence $\{F'_k\}_k = \{F_{n_k}\}_k$ such that $I'_{1,n} \to J_1$ as $n \to +\infty$, where $I'_{1,n}$ are the complementary arcs of F'_n and J_1 is some open arc of positive length. If $|J_1|_s = 1$, then $\{F'_n\}_n$ is the desired subsequence and

$$F_\infty = \mathbb{T} \setminus J_1$$

is the desired Beurling-Carleson set.

If $|J_1|_s < 1$, then the same argument shows that

$$\log \frac{e}{|I'_{2,n}|_s} \leq \frac{\kappa(\mathbb{T} \setminus F'_n)}{|\mathbb{T} \setminus F'_n|_s - |I'_{1,n}|_s}.$$

Since $|\mathbb{T} \setminus F'_n|_s - |I'_{1,n}|_s \to 1 - |J_1|_s > 0$ as $n \to +\infty$, the sequence of normalized lengths $\{|I_{2,n}|_s\}_n$ must be bounded away from zero. We can then choose a subsequence $\{F''_k\}_k = \{F'_{n_k}\}_k$ such that $I''_{2,n} \to J_2$ as $n \to +\infty$; here $I''_{k,n}$ are the complementary arcs of F''_n.

Continuing this process, we then arrive at two scenarios. Either

(1) after a finite number of steps, we obtain a subsequence $\{F_n^{(l)}\}_n$, such that $I_{k,n}^{(l)} \to J_k$ as $n \to +\infty$ for all $1 \leq k \leq l$ and $|J_1|_s + \cdots + |J_l|_s = 1$, in which case $\{F_n^{(l)}\}_n$ is the desired sequence and

$$F_\infty = \mathbb{T} \setminus \cup_{k=1}^l J_k$$

is the desired Beurling-Carleson set, or

(2) the number of steps is infinite, in which case we must have

$$\sum_{k=1}^{+\infty} |J_k|_s = 1,$$

which follows from

$$\log \frac{e}{|J_l|_s} \leq \frac{A}{1 - \sum_{k=1}^{l-1} |J_k|_s}$$

and the fact that $|J_l|_s \to 0$ as $l \to +\infty$. Here, A is any upper bound for the sequence $\{\kappa(\mathbb{T} \setminus F_n)\}_n$. Taking the diagonal subsequence $\{F_n^{(n)}\}_n$, we then get the required result, with $F_\infty = \mathbb{T} \setminus \cup_{k=1}^{+\infty} J_k$.

The proof of the lemma is complete.   ∎

To state the next lemma, we need to introduce a certain notion of covering for intervals of integers.

For any integers k and l with $k \leq l$, we let $I = [k, l]$ denote the set of integers n with $k \leq n \leq l$ and call I an interval of integers. In particular, $[n, n] = \{n\}$.

A simple covering of $[p, q]$ is a system of intervals $\{I_n\}_n$ of integers such that

$$\sum_n \chi_{I_n} = \chi_{[p,q]},$$

where χ_I denotes the characteristic function of the interval I.

LEMMA 7.7 *Consider the following system of $N(N + 1)/2$ linear inequalities with N unknowns $x_1, \ldots, x_N$:*

$$\sum_{j=k}^{l} x_j \leq b_{k,l}, \qquad 1 \leq k \leq l \leq N,$$

subject to the constraint $x_1 + \cdots + x_N = 0$. The necessary and sufficient condition for this system to have a solution is that $\sum_n b_{k_n,l_n} \geq 0$ for every simple covering $\mathcal{P} = \{[k_n, l_n]\}_n$ of $[1, N]$.

Proof. The necessity part is easy. In fact, if $\sum_n b_{k_n,l_n} < 0$ for some simple covering $\mathcal{P} = \{[k_n, l_n]\}_n$, then adding up the inequalities

$$\sum_{j=k}^{l} x_j \leq b_{k,l}$$

corresponding to $[k_n, l_n] \in \mathcal{P}$ yields $x_1 + \cdots + x_N < 0$, which contradicts the constraint $x_1 + \cdots + x_N = 0$, and so the system has no solution.

To prove the sufficiency part, we proceed somewhat analogously to the proof of Lemma 4.30, and let $\mathcal{C}$ be the closed convex set in $\mathbb{R}^N$ consisting of vectors $x = (x_1, \ldots, x_N)$ such that $x_k + \cdots + x_l \leq b_{k,l}$ for all $1 \leq k \leq l \leq N$ (without the constraint $x_1 + \cdots + x_N = 0$). It is clear that $\mathcal{C}$ contains all x with sufficiently large (in absolute value) negative coordinates. Therefore, the linear functional

$$L(x) = \langle x, L \rangle = x_1 + \cdots + x_N$$

maps $\mathcal{C}$ onto some interval $(-\infty, \lambda_0]$. In the above, the symbol L stands for the vector $(1, \ldots, 1) \in \mathbb{R}^N$ as well as the associated functional, and $\langle \cdot, \rangle$ is the inner product of $\mathbb{R}^N$.

Number the $N(N + 1)/2$ different intervals of integers $[k, l]$ by the index j, and let e^j be the vector $(0, \ldots, 0, 1, \ldots, 1, 0, \ldots, 0)$ with 1's precisely on the interval $[k, l]$, and write b_j for the bound $b_{k,l}$; in both instances, we assume $[k, l]$ and j are in correspondence. We recall from the proof of Lemma 4.30 the assertion of the *duality theorem of Linear Programming* [127, p. 28]:

$$\max \left\{ \langle x, L \rangle : x \in \mathcal{C} \right\} = \min \left\{ \sum_j \theta_j b_j : \theta_j \in \mathbb{R}_+ \text{ for all } j, \sum_j \theta_j e^j = L \right\}.$$

Also, the vectors

$$\theta = (\theta_1, \dots, \theta_{N(N+1)/2}) \in \mathbb{R}_+^{N(N+1)/2}$$

over which the minimum is taken on the right hand side form a closed convex lower-dimensional polyhedron $\mathcal{S}$ contained in the cube $[0, 1]^{N(N+1)/2}$. Using a covering type argument, we showed in the proof of Lemma 4.30 that the polyhedron $\mathcal{S}$ is the (closed) convex hull of all its "edge points" ϵ, which are characterized as those points $\epsilon = (\epsilon_1, \dots, \epsilon_{N(N+1)/2}) \in \mathcal{S}$ that have $\epsilon_j \in \{0, 1\}$ for all j. In particular, the above minimum is achieved by one of the finitely many "edge points". The assumption $\sum_n b_{k_n, l_n} \geq 0$ for every simple covering $\mathcal{P} = \{[k_n, l_n]\}_n$ of $[1, N]$ means that for any "edge point" θ, we have $\sum_j \theta_j b_j \geq 0$. By the duality theorem, the maximum λ_0 on the left hand side is then ≥ 0, too. Since the whole interval $(-\infty, \lambda_0]$ is attained by the functional L on $\mathcal{C}$, it follows that there exists a point $x \in \mathcal{C}$ with $L(x) = 0$. The proof is complete.   ∎

LEMMA 7.8 *For a premeasure $\mu \in \kappa B^+$, the following conditions are equivalent:*

(1) μ is κ-absolutely continuous.

(2) There is a positive constant C with the property that for every $\varepsilon > 0$ there exists some positive real number M such that the system

$$\begin{cases} x_{k.l} \leq M \kappa(I_{k.l}), \\ \mu(I_{k.l}) + x_{k.l} \leq \min\left\{C \kappa(I_{k.l}), \varepsilon\right\}, \\ x_{k.l} = \sum_{s=k}^{l-1} x_{s.s+1}, \\ x_{0.N} = 0, \end{cases}$$

with unknowns $x_{k.l}$, $0 \leq k < l \leq N$, is consistent for all positive integers N. Here,

$$I_{k.l} = \{e^{i\theta} : 2\pi k/N \leq \theta < 2\pi l/N\}$$

for $0 \leq k < l \leq N$.

Proof. If μ is κ-absolutely continuous, then $x_{k.l} = \mu(I_{k.l})$, $1 \leq k < l \leq N$, satisfy the system in (2).

If the system in (2) is consistent for every $N = 1, 2, 3, \dots$, then for every N we pick a solution $x^N = \{x_{k.l}^N\}_{k.l}$ of the system and form a measure μ_x having constant density $x_{s.s+1}/|I_{s.s+1}|_s$ on each $I_{s.s+1}$; the density is with respect to normalized arc length measure ds. Using the Helly-type selection principle for premeasures and effecting the transition to the limit over a subsequence $N_1 < N_2 < N_3 < \cdots$, we obtain a premeasure x satisfying

$$x(I) \leq M \kappa(I), \qquad \mu(I) + x(I) \leq \min\{C \kappa(I), \varepsilon\},$$

for all open arcs $I \subset \mathbb{T}$ that do not contain $\zeta = 1$; this condition on I can be removed if $(1 + \log 2) C$ and 2ε are substituted for C and ε, respectively. This shows that μ is κ-absolutely continuous. ∎

We continue the proof of Theorem 7.5, and now supply the sufficiency part.

Thus, we assume that μ is not κ-absolutely continuous. Using Lemma 7.8 we see that for every positive constant C there exists an $\varepsilon > 0$ such that no matter how large M is, the system in (2) of Lemma 7.8 has no solution for some N. Combining Lemmas 7.7 and 7.8, with unknowns $x_{k,l} + \mu(I_{k,l})$, we conclude that for such a combination of C, ε, and M, there is a *simple covering* of $\mathbb{T}$ by a finite number of disjoint half-closed arcs $\{I_v\}_v$ such that

$$\sum_v \min \{\mu(I_v) + M \kappa(I_v), C \kappa(I_v), \varepsilon\} < 0.$$

Let $\{I_v'\}_v$ be those arcs from $\{I_v\}_v$ for which

$$\min \{\mu(I_v) + M \kappa(I_v), C \kappa(I_v), \varepsilon\} = \mu(I_v) + M \kappa(I_v),$$

and let $\{I_v''\}_v = \{I_v\}_v \setminus \{I_v'\}_v$. Clearly, $\mu(I_v') < 0$ if $M > C$. Setting $F_M = \cup_v I_v'$, we find that

$$\mu(F_M) < -M \kappa(F_M) - C \sum_v \{\kappa(I_v'') : |I_v''|_s < \delta\} - \varepsilon \sum_v \{1 : |I_v''|_s \geq \delta\},$$

where δ is defined by the equation

$$C \delta \log \frac{e}{\delta} = \varepsilon.$$

Let $C = 3\|\mu\|^+$ in the rest of the proof. For $M > 2C$, the collection $\{I_v'' : |I_v''|_s > \delta\}$ is nonempty; otherwise, the displayed inequalities in the previous two paragraphs would imply that

$$\begin{aligned}
0 &= \mu(\mathbb{T}) = \mu(F_M) + \mu(\mathbb{T} \setminus F_M) \\
&\leq -2C \sum_v \kappa(I_v') - C \sum_v \kappa(I_v'') + \|\mu\|^+ \sum_v \kappa(I_v'') \\
&= -6\|\mu\|^+ \sum_v \kappa(I_v') - 2\|\mu\|^+ \sum_v \kappa(I_v'') < 0,
\end{aligned}$$

which is a contradiction.

We can also show that as $M \to +\infty$,

$$\sum_v \kappa(I_v'') = O(1)$$

and

$$\kappa(F_M) = \sum_v \kappa(I_v') \to 1.$$

In fact, for any partition of $\mathbb{T}$, and in particular for $\{I_v\}_v$, we have

$$\begin{aligned}
\sum_v |\mu(I_v)| &= \sum_v \mu(I_v)^+ + \sum_v \mu(I_v)^- \\
&= 2 \sum_v \mu(I_v)^+ \leq 2\|\mu\|^+ \sum_v \kappa(I_v),
\end{aligned}$$

so that

$$\mu(F_M) \geq -2\|\mu\|^+ \sum_\nu \kappa(I_\nu).$$

The desired estimates then follow from this and our earlier estimate about $\mu(F_M)$. Note that above, we adhere to the standard convention to write $t^+ = \max\{t, 0\}$ and $t^- = \max\{-t, 0\}$.

Applying Lemma 7.6, we can extract a sequence $M_n \to +\infty$ as $n \to +\infty$ such that F_{M_n} converges to a Beurling-Carleson set F_∞. To simplify the notation, we write F_n in place of F_{M_n}.

To summarize the situation, the assumption that μ is not κ-absolutely continuous implies the existence of a sequence $\{F_n\}_n$ of sets with each F_n composed of a finite number of closed arcs such that

(i) $\kappa(F_n) \to 0$ and *a fortiori* $|F_n|_s \to 0$ as $n \to +\infty$,

(ii) $\kappa(\mathbb{T} \setminus F_n) \leq A$ for some constant A and all n,

(iii) $\mu(F_n) \leq -3 \|\mu\|^+ \left[\kappa(F_n) + \sum_k \left\{ \kappa(I_{k,n}) : |I_{k,n}|_s < \delta \right\} \right] - \varepsilon$,

where $\{I_{k,n}\}_k$ are the complementary arcs of F_n, and δ and ε are some positive constants. Moreover, there is a Beurling-Carleson set F_∞ such that for every $\rho > 0$ the ρ- neighborhood of F_∞, F_∞^ρ, contains all but a finite number of F_n, and F_∞ is contained in all but a finite number of F_n^ρ.

We are going to show that $\sigma_\mu(F_\infty) < 0$, where σ_μ is the κ-singular part of μ. We actually show that the contrary assumption $\sigma_\mu(F_\infty) = 0$ would lead to a contradiction.

Thus, we assume $\sigma_\mu(F_\infty) = 0$. Since

$$\sigma_\mu(F_\infty) = \lim_{\rho \to 0^+} \mu(F_\infty^\rho),$$

we can replace F_n by $F_n \setminus F_\infty^{\rho_n}$ in (i)–(iii) and choose ρ_n so small that (i)–(iii) still hold, although perhaps with a smaller ε. Therefore, we can choose a sequence $\{\rho_n\}_n$ of positive numbers, decreasing to 0, as well as a sequence $\{F_n\}_n$ of sets (composed of a finite number of closed arcs) such that

$$F_n \subset F_\infty^{\rho_n} \setminus \overline{F_\infty^{\rho_{n+1}}}$$

and

$$\mu(F_n) \leq -3 \|\mu\|^+ [\kappa(F_n) + \kappa(G_n)] - \varepsilon,$$

where

$$G_n = \left(F_\infty^{\rho_n} \setminus \overline{F_\infty^{\rho_{n+1}}} \right) \setminus F_n.$$

Let $\mathcal{I}_n$, $\mathcal{J}_n$, and $\mathcal{K}_n$ denote the systems of arcs I of which F_n, G_n, and $F_\infty^{\rho_n}$ are composed, respectively. Let $\mathcal{I}_0$ be the system of arcs that form $\mathbb{T} \setminus F_\infty^{\rho_1}$, and let

$$S_n = \left(\bigcup_{k=1}^n \mathcal{I}_k \right) \cup \left(\bigcup_{k=1}^n \mathcal{J}_k \right) \cup \mathcal{K}_{n+1}.$$

Adding up the estimates about $\mu(F_n)$ in the previous paragraph, and keeping in mind that $C = 3\|\mu\|^+$, we get

$$
\sum_{I \in \mathcal{I}_0} |\mu(I)| \;+\; \sum_{I \in \mathcal{S}_n} |\mu(I)| \geq \sum_{\nu=1}^{n} |\mu(F_\nu)|
$$

$$
\geq\; C \left[\sum_{\nu=1}^{n} \kappa(F_n) + \sum_{\nu=1}^{n} \kappa(G_n) \right] + n\varepsilon
$$

$$
=\; C \sum_{I \in \mathcal{S}_n} \kappa(I) - C \sum_{I \in \mathcal{K}_{n+1}} \kappa(I) + n\varepsilon
$$

$$
=\; C \left[\sum_{I \in \mathcal{S}_n \cup \mathcal{I}_0} \kappa(I) - \sum_{I \in \mathcal{K}_{n+1}} \kappa(I) - \sum_{I \in \mathcal{I}_0} \kappa(I) \right] + n\varepsilon.
$$

Since

$$
\lim_{n \to +\infty} \sum_{I \in \mathcal{K}_{n+1}} \kappa(I) = 0,
$$

for large enough n, we must have

$$
\sum_{I \in \mathcal{S}_n \cup \mathcal{I}_0} |\mu(I)| \geq 3 \, \|\mu\|^+ \sum_{I \in \mathcal{S}_n \cup \mathcal{I}_0} \kappa(I),
$$

which contradicts an earlier estimate, because $\mathcal{S}_n \cup \mathcal{I}_0$ is a system of non-overlapping arcs covering $\mathbb{T}$.

The proof of Theorem 7.5 is now complete.   ∎

7.4   Cyclicity in $\mathcal{A}^{-\infty}$

Recall that $\mathcal{A}^{-\infty}$ is the space of analytic functions f in $\mathbb{D}$ such that

$$
|f(z)| = O\left(\frac{1}{(1 - |z|)^N} \right) \qquad \text{as } |z| \to 1^-,
$$

for some positive constant $N = N(f)$. If

$$
f(z) = \sum_{n=0}^{+\infty} \widehat{f}(n)\, z^n, \qquad z \in \mathbb{D},
$$

then it is easy to see that $f \in \mathcal{A}^{-\infty}$ if and only if

$$
|\widehat{f}(n)| = O(n^\lambda), \qquad \text{as } n \to +\infty,
$$

for some positive constant $\lambda = \lambda(f)$. In the rest of this section, We shall make no distinction between $\mathcal{A}^{-\infty}$ as a space of analytic functions and $\mathcal{A}^{-\infty}$ as a space of sequences.

We have

$$\mathcal{A}^{-\infty} = \bigcup_{0 < p < +\infty} A^p,$$

and we define the topology of $\mathcal{A}^{-\infty}$ as that of the inductive limit of A^p as $p \to 0^+$.

Let $\mathcal{A}^\infty$ be the space of functions that are infinitely differentiable on $\bar{\mathbb{D}}$ and analytic in $\mathbb{D}$. If

$$f(z) = \sum_{n=0}^{+\infty} \hat{f}(n) \, z^n, \qquad z \in \mathbb{D},$$

then $f \in \mathcal{A}^\infty$ if and only if

$$|\hat{f}(n)| = O(n^{-\lambda}), \qquad \text{as } n \to +\infty,$$

for all positive constants λ. It is well known that the dual space of $\mathcal{A}^{-\infty}$ is $\mathcal{A}^\infty$ under the following duality pairing, for $f \in \mathcal{A}^{-\infty}$ and $g \in \mathcal{A}^\infty$:

$$\langle f, g \rangle = \sum_{n=0}^{+\infty} \hat{f}(n) \overline{\hat{g}(n)} = \lim_{r \to 1^-} \int_{\mathbb{T}} f(r\zeta) \overline{g(r\zeta)} \, ds(\zeta),$$

where $ds(z) = |dz|/(2\pi)$ is normalized arc length measure.

It will be convenient for us to view $\mathcal{A}^{-\infty}$ as a closed subspace of a larger space $C^{-\infty}$ consisting of formal Fourier (or Laurent) series

$$f(\zeta) = \sum_{n=-\infty}^{+\infty} \hat{f}(n) \, \zeta^n, \qquad \zeta \in \mathbb{T},$$

with the coefficients satisfying

$$|\hat{f}(n)| = O(|n|^\lambda), \qquad \text{as } |n| \to +\infty,$$

for some positive constant $\lambda = \lambda(f)$. The space $C^{-\infty}$ consists of all Schwartzian distributions on $\mathbb{T}$. We shall think of it both as a space of distributions and as a space of sequences. As with $\mathcal{A}^{-\infty}$, a topology can be put on $C^{-\infty}$ via an inductive limit.

Let C^∞ be the space of infinitely differentiable functions on $\mathbb{T}$,

$$f(\zeta) = \sum_{n=-\infty}^{+\infty} \hat{f}(n) \, \zeta^n, \qquad \zeta \in \mathbb{T};$$

it is well known that $f \in C^\infty$ if and only if

$$|\hat{f}(n)| = O(|n|^{-\lambda}), \qquad \text{as } |n| \to +\infty,$$

for all positive constants λ. It is classical that C^∞ and $C^{-\infty}$ are each other's dual spaces via the the duality pairing

$$\langle f, g \rangle = \sum_{n=-\infty}^{+\infty} \hat{f}(n) \, \overline{\hat{g}(n)},$$

where $f \in C^{-\infty}$ and $g \in C^{\infty}$.

It is clear that $\mathcal{A}^{\infty}$ can be viewed as a subspace of C^{∞}. Note also that $\mathcal{A}^{\infty}$ is a subspace of $\mathcal{A}^{-\infty}$.

Let $\times$ be the operation of multiplication on functions defined in terms of the usual convolution $*$ of the corresponding coefficient sequences. We are going to use the operation $\times$ in the following cases:

(i) $C^{-\infty} \times C^{\infty} \subset C^{-\infty}$.

(ii) $C^{\infty} \times C^{\infty} \subset C^{\infty}$.

(iii) $\mathcal{A}^{\infty} \times \mathcal{A}^{\infty} \subset \mathcal{A}^{\infty}$.

(iv) $\mathcal{A}^{-\infty} \times \mathcal{A}^{-\infty} \subset \mathcal{A}^{-\infty}$.

It is important to keep in mind that the operation $\times$ may not be the same as pointwise multiplication, even if the latter is well-defined. However, in the case $C^{\infty} \times C^{\infty}$ the operation $\times$ coincides with pointwise multiplication on $\mathbb{T}$, and in the case $\mathcal{A}^{-\infty} \times \mathcal{A}^{-\infty}$ it agrees with pointwise multiplication in $\mathbb{D}$. The case $\mathcal{A}^{-\infty} \times C^{\infty}$ may be treacherous.

Of the above cases only (i) needs a proof.

PROPOSITION 7.9 *Fix $g \in C^{\infty}$. Then $f \mapsto g \times f$ is a well-defined continuous linear operator on $C^{-\infty}$.*

Proof. Since $f \in C^{-\infty}$, there exist positive constants λ and C such that

$$|c_n| \le C \, (|n| + 1)^{\lambda}$$

for all integers n, positive or negative. Since g is in C^{∞}, we can choose another positive constant C' such that

$$|\widehat{g}(n)| \le \frac{C'}{(|n| + 1)^{\lambda+2}}$$

again for all integers n. Using the elementary inequality

$$1 + |n - m| \le (1 + |m|)(1 + |n|)$$

and the definition

$$g \times f(\zeta) = \sum_{n=-\infty}^{+\infty} \left(\sum_{m=-\infty}^{+\infty} \widehat{g}(n - m) \right) \widehat{f}(m) \, \zeta^n,$$

in the sense of formal Fourier series, we then obtain

$$\left| \sum_m \hat{g}(n-m)\, \hat{f}(m) \right| \leq CC' \sum_m \frac{(1+|m|)^\lambda}{(1+|n-m|)^{\lambda+2}}$$

$$= CC' \sum_m \frac{(1+|n-m|)^\lambda}{(1+|m|)^{\lambda+2}}$$

$$\leq CC' \left[\sum_m \frac{1}{(1+|m|)^2} \right] (|n|+1)^\lambda$$

for all integers n, which shows that $g \times f$ is in $C^{-\infty}$ and that the operation is continuous in f, as the sum of reciprocals of squared positive integers converges. ∎

We now use the operation $\times$ and the space C^∞ to construct a class of invariant subspaces for $\mathcal{A}^{-\infty}$.

PROPOSITION 7.10 *For $g \in C^\infty$, define*

$$I_g = \{f \in \mathcal{A}^{-\infty} : g \times f \in \mathcal{A}^{-\infty}\}.$$

Then I_g is a closed invariant subspace of $\mathcal{A}^{-\infty}$. Furthermore, if $g \notin \mathcal{A}^\infty$, then $I_g \neq \mathcal{A}^{-\infty}$ (but it may happen that $I_g = \{0\}$).

Proof. This follows immediately from Proposition 7.9 and the fact that

$$g \times (zf) = z(g \times f) \in \mathcal{A}^{-\infty}$$

if $g \times f \in \mathcal{A}^{-\infty}$. If g is not in $\mathcal{A}^\infty$, then I_g does not even contain the constant function 1. ∎

We need another lemma before we can characterize the cyclic vectors of $\mathcal{A}^{-\infty}$.

LEMMA 7.11 *If $F \subset \mathbb{T}$ is a Beurling-Carleson set, then there exists an outer function $\Phi \in \mathcal{A}^\infty$ with $\Phi(0) = 1$ and $\Phi(z) \neq 0$ for $z \in \overline{\mathbb{D}} \setminus F$, which is flat on F: $\Phi^{(n)}(z) = 0$ for all $n = 0, 1, 2, 3, \ldots$ and $z \in F$.*

Proof. Let $\mathbb{T} \setminus F = \cup_n I_n$, where I_n are the complementary arcs of F. For each n, let $I_n = \cup_j J_{n.j}$ be a partition of I_n into nonoverlapping closed arcs $J_{n.j}$ (that two arcs are nonoverlapping means that their intersection is empty or a single point) satisfying

$$|J_{n.j}|_s = d_{\mathbb{T}}(J_{n.j}, F)/(2\pi)$$

for all j. Relabel the doubly indexed sequence $\{J_{n.j}\}_{n.j}$ as $\{J_\nu\}_\nu$. It is clear that this sequence of arcs has F as "cluster set". For each ν, let $e^{i\theta_\nu}$ be the middle point of J_ν and $a_\nu = r_\nu e^{i\theta_\nu}$ (with $1 < r_\nu < +\infty$) be the point in $\mathbb{C}$ from which J_ν is seen at a right angle. Clearly, $r_\nu - 1$ is approximately $|J_\nu|/2 = \pi |J_\nu|_s$ and

$$\sum_\nu \kappa(J_\nu) = \sum_\nu |J_\nu|_s \log \frac{e}{|J_\nu|_s} \leq C \hat{\kappa}(F) < +\infty$$

for some positive constant C. Pick some real numbers $\lambda_\nu > 0$ such that $\lambda_\nu \to +\infty$ as $\nu \to +\infty$ and

$$\sum_\nu \lambda_\nu \kappa(J_\nu) = \sum_\nu \lambda_\nu |J_\nu|_s \log \frac{e}{|J_\nu|_s} < +\infty,$$

and consider the function

$$\Phi_1(z) = \exp\left[-\sum_\nu \frac{\lambda_\nu \kappa(J_\nu) e^{i\theta_\nu}}{a_\nu - z}\right].$$

It can be seen that Φ_1 is an outer function with $|\Phi_1(z)| < 1$ in $\mathbb{D}$. Furthermore, we have

$$|\Phi_1(z)| \le \exp\left(c \lambda_\nu \frac{e}{\log |J_\nu|_s}\right), \qquad z \in J_\nu,$$

for some positive constant c, and

$$|J_\nu|_s \le \frac{d_{\mathbb{T}}(z, F)}{2\pi} \le 2|J_\nu|_s, \qquad z \in J_\nu.$$

Hence

$$|\Phi_1(z)| = o\big(d_{\mathbb{T}}(z, F)^N\big) \qquad \text{as} \quad d_{\mathbb{T}}(z, F) \to 0,$$

for all positive constants N. On the other hand, for $n = 1, 2, 3, \ldots$, we easily calculate the n-th derivative:

$$\Phi_1^{(n)}(z) = \Phi_1(z)\, \phi_n(z),$$

where

$$|\phi_n(z)| \le \frac{C_n}{d_{\mathbb{T}}(z, F)^{2n}}$$

holds for some positive constant C_n. It follows that the function $\Phi(z) = \Phi_1(z)/\Phi_1(0)$ has all the desired properties. ∎

THEOREM 7.12 *A function $f \in \mathcal{A}^{-\infty}$ is cyclic in $\mathcal{A}^{-\infty}$ if and only if f is nonvanishing in $\mathbb{D}$ and the premeasure $\mu = \mu_f$ in the representation*

$$f(z) = f(0) \exp\left[\int_{\mathbb{T}} \frac{\zeta + z}{\zeta - z} d\mu(\zeta)\right]$$

is κ-absolutely continuous.

Proof. The sufficiency follows directly from Theorem 7.5 (the approximation theorem for premeasures).

To prove the necessity, we assume that the κ-singular part of μ_f, denoted $\sigma = \sigma_f$, is nonzero. Then there exists a Beurling-Carleson set $F \subset \mathbb{T}$ with $-\infty < \sigma(F) < 0$. Let Φ be the function as in Lemma 7.11. Clearly, the powers Φ^k ($k = 2, 3, \ldots$) also have the same properties.

Now, for $k = 1, 2, 3, \ldots$, define

$$\Psi_k(z) = [\Phi(z)]^k \exp\left[-\int_F \frac{\zeta + z}{\zeta - z} d\sigma(\zeta)\right], \qquad z \in \mathbb{D}.$$

Each Ψ_k is analytic in $\mathbb{D}$, and in fact, it extends to a C^∞ function on $\overline{\mathbb{D}} \setminus F$. It also belongs to the Nevanlinna class, but it is not in $\mathcal{A}^{-\infty}$, because its boundary (pre)measure has a positive κ-singular part (namely, the restriction of $-\sigma$ to F).

If we define $\Psi_k(\zeta) = 0$ for $\zeta \in F$ and $k = 1, 2, 3, \ldots$, then Ψ_k becomes a C^∞ function on $\mathbb{T}$. This is because every derivative of the first factor Φ^k in Ψ_k is $o\left[d_{\mathbb{C}}(\zeta, F)^n\right]$ for all $n \geq 0$, whereas the derivatives of the second factor in Ψ_k increases not faster than some negative power of $d_{\mathbb{C}}(\zeta, F)$ as $d_{\mathbb{C}}(\zeta, F) \to 0$.

Let $g_k(z) = f(z)\Psi_k(z)$ for $k = 1, 2, 3, \ldots$ and $z \in \mathbb{D}$. We can write

$$g_k(z) = f(0) \exp\left[\int_{\mathbb{T}} \frac{\zeta + z}{\zeta - z} (d\mu(\zeta) - d\sigma_F(\zeta) + k \log |\Phi(\zeta)| ds(\zeta))\right],$$

where σ_F is the restriction of σ to F, and $ds(\zeta) = |d\zeta|/(2\pi)$ is normalized arc length measure, as usual. The boundary premeasure of g_k is

$$d\mu_{g_k} = d\mu - d\sigma_F + k \log |\Phi(\zeta)| ds(\zeta).$$

Let J be an arbitrary open arc in $\mathbb{T}$. If $J \cap F = \emptyset$, then

$$\mu_{g_k}(J) = \mu(J) + k \int_J \log |\Phi(\zeta)| ds(\zeta) \leq a\kappa(J) + b |J|_s$$

for some positive constants a, b, both independent of J, because μ is κ-bounded above and $\log |\Phi(\zeta)|$ is bounded above. If $J \cap F \neq \emptyset$, we let $\{J_\nu\}_\nu$ be the components of the open set $J \setminus F$. For every c, $0 < c < +\infty$, there exists a constant $b(c)$ such that

$$\log |\Phi(\zeta)| \leq c \log d_{\mathbb{T}}(\zeta, F) + b(c)$$

for all $\zeta \in \mathbb{T}$. Since at least one of the endpoints of each J_ν belongs to F, we obtain by integration

$$\int_{J_\nu} \log |\Phi(\zeta)| ds(\zeta) \leq b(c) |J_\nu|_s - c\kappa(J_\nu)$$

for all indices ν. If we choose $c > a/k$, then

$$\begin{aligned}
\mu_{g_k}(J) &= \sum_\nu \mu(J_\nu) + k \sum_\nu \int_{J_\nu} \log |\Phi(\zeta)| ds(\zeta) \\
&\leq (a - kc) \sum_\nu \kappa(J_\nu) + b(c) |J|_s \leq b(c) |J|_s.
\end{aligned}$$

This shows that μ_{g_k} is κ-bounded above, and so each g_k belongs to $\mathcal{A}^{-\infty}$.

Take $k = 3$, and let

$$S(z) = \exp\left[\int_F \frac{\zeta + z}{\zeta - z} d\sigma(\zeta)\right], \qquad z \in \mathbb{D}.$$

Then

$$\begin{aligned}
f \times \Psi_3 &= f \times (\Phi^3 S^{-1}) = f \times \left[\Phi^2 \times (\Phi S^{-1})\right] \\
&= (f\Phi^2) \times (\Phi S^{-1}) = \left[(f\Phi S^{-1}) \times (\Phi S)\right] \times (\Phi S^{-1}) \\
&= (f\Phi S^{-1}) \times \Phi^2 = f\Phi^3 S^{-1} = f\Psi_3.
\end{aligned}$$

Here we treat ΦS^{-1}, $\Phi^2 S^{-1}$, and $\Phi^3 S^{-1}$ as elements of C^∞; Φ, Φ^2, Φ^3, and ΦS as elements of $\mathcal{A}^\infty$; and $f\Phi S^{-1}$ and $f\Psi_3$ as elements of $\mathcal{A}^{-\infty}$. We also used the associative law for the operation $\times$.

As $f\Psi_3 \in \mathcal{A}^{-\infty}$, and $\Psi_3 \in C^\infty \setminus \mathcal{A}^\infty$ as a boundary function, Proposition 7.10 tells us that $f \in I_{\Psi_3}$ and that I_{Ψ_3} is a proper invariant subspace of $\mathcal{A}^{-\infty}$. This shows that f is not cyclic in $\mathcal{A}^{-\infty}$.   ∎

7.5 Notes

A difficult (and still open) problem in the theory of Bergman spaces is to characterize the cyclic vectors. Theorem 7.2 was conjectured by Korenblum in [88] and later proved by Aleman, Richter, and Sundberg in [7]. The description of cyclic vectors for $\mathcal{A}^{-\infty}$ was given by Korenblum in [84], where it is a corollary of the theorem describing all invariant subspaces of $\mathcal{A}^{-\infty}$ in terms of zeros and boundary premeasures. Partial results on cyclic vectors were obtained in [117] and [102].

Theorem 7.3 is from Brown and Korenblum [31]; however, the proof is simplified by the use of the results on A^p-inner functions in Chapter 3. The combination of Theorems 7.3 and 7.12 seems to be the most powerful tool available to deal with cyclic vectors in the Bergman spaces A^p. However, as will be clear from the next chapter, there exist functions in A^p that are cyclic for $\mathcal{A}^{-\infty}$ but not for A^p.

7.6 Exercises and Further Results

1. Show that every classical outer function in A_α^p is necessarily a cyclic vector in A_α^p.

2. If f and g are functions in A_α^p with $f = \varphi g$, where φ is a classical outer function, then f and g generate the same invariant subspace. Does this remain true if φ is just a cyclic vector in A_α^p?

3. Fix $0 < p < +\infty$ and $-1 < \alpha < +\infty$. Show that there exists a classical inner function φ such that $1/\varphi$ belongs to A_α^p.

4. Let κV be the space of premeasures $\mu = \mu_1 - \mu_2$, where μ_1 and μ_2 belong to κB^+. Prove that $\mu \in \kappa V$ if and only if there exists a constant $C > 0$ such that for any finite partition $\mathbb{T} = \cup_k I_k$, where each $I_k \in \mathcal{B}(\mathbb{T})$ and $I_k \cap I_l = \emptyset$

for $k \neq l$, we have

$$\sum_k |\mu(I_k)| \leq C \sum_k \kappa(I_k).$$

See [35].

5. Recall that for $f \in A^{-\infty}$ we use σ_f to denote the κ-singular measure of f. Show that $\sigma_{fg} = \sigma_f + \sigma_g$ for all f and g in $A^{-\infty}$.

6. Let Z be a zero set for A_α^p and σ be a κ-singular measure on $\mathbb{T}$. Show that

$$I(Z, \sigma) = \{f \in A_\alpha^p : Z \subset Z_f, \sigma \leq \sigma_f\}$$

is an invariant subspace of A_α^p. Here Z_f is the zero set of f, and the inclusion $Z \subset Z_f$ takes multiplicities into account.

7. Suppose $\mu \in \kappa B^+$ is a premeasure and let F be a Beurling-Carleson set. Show that – in the context of the material following after the proof of Theorem 7.4 – the "restriction" of μ to F is a negative Borel measure. See [83].

8. Let F be a Beurling-Carleson set, and F^δ its δ-neighborhood, for positive δ. Show that $\kappa(F^\delta) \to 0$ as $\delta \to 0^+$. See [84].

8

Invertible Noncyclic Functions

A function f in a space X of analytic functions is said to be invertible if $1/f$ also belongs to X. In the classical theory of Hardy spaces, every invertible function in H^p is necessarily cyclic in H^p. This is also true in the $\mathcal{A}^{-\infty}$ theory; an invertible function in $\mathcal{A}^{-\infty}$ is always cyclic in $\mathcal{A}^{-\infty}$.

In this chapter, we construct invertible functions in A^p that are not cyclic there. What makes the construction possible is a delicate combination of growth and decay. The construction is quite challenging technically, which may explain why it took some 30 years for the solution to appear after Shapiro first posed the problem.

The functions we construct will in some sense be extremal in the given space. In particular, the set of points E in the unit circle where the function is "maximal" should be rather massive. Our functions exhibit bad boundary behavior everywhere on the unit circle. However, it is possible to modify our constructions so that the resulting functions extend analytically across any given proper arc of the unit circle.

In Section 8.1, we supply an estimate of Poisson integrals of Borel measures in terms of the size of of the smallest supporting arc, and introduce harmonic measure. Sections 8.1–8.5 are devoted to the technical details of constructing certain real-valued harmonic functions h on $\mathbb{D}$ with specific properties, and the non-cyclic element f of A^p is then obtained from h via the formula

$$f(z) = \exp\left(h(z) + i\widetilde{h}(z)\right), \qquad z \in \mathbb{D},$$

where the tilde indicates the harmonic conjugation operation.

8.1   An Estimate for Harmonic Functions

For $1 \leq p < +\infty$, let $h^p(\mathbb{D})$ denote the Banach space of complex-valued harmonic functions f in the disk $\mathbb{D}$ with finite norm

$$\|f\|_{h^p} = \sup_{0 < r < 1} \left(\frac{1}{2\pi} \int_{-\pi}^{\pi} |f(re^{i\theta})|^p d\theta \right)^{1/p} < +\infty.$$

Also, let $h^\infty(\mathbb{D})$ be the Banach space of complex-valued bounded harmonic functions on $\mathbb{D}$, with norm

$$\|f\|_{h^\infty} = \sup \{|f(z)| : z \in \mathbb{D}\} < +\infty.$$

For a finite complex-valued Borel measure μ on $\mathbb{T}$, its Poisson integral is the function

$$P\mu(z) = \int_{\mathbb{T}} P(z, \zeta) d\mu(\zeta), \qquad z \in \mathbb{D},$$

where

$$P(z, \zeta) = \frac{1 - |z|^2}{|\zeta - z|^2}, \qquad (z, \zeta) \in \mathbb{D} \times \mathbb{T},$$

is the Poisson kernel. It is well known that a harmonic function belongs to $h^1(\mathbb{D})$ if and only if it is the Poisson integral of a finite Borel measure. Moreover, for $1 < p \leq +\infty$, the space $h^p(\mathbb{D})$ coincides with the space of Poisson integrals of $L^p(\mathbb{T})$ functions. We shall need to estimate the size of the Poisson integral of a measure with small support. For a Borel measure μ on $\mathbb{T}$, $\|\mu\|$ stands for its total variation.

THEOREM 8.1 *For $0 \leq t \leq 2\pi$, let $J(t)$ be the closed arc connecting the point 1 with e^{it}, running counterclockwise. If μ is a complex-valued Borel measure supported on $J(\beta)$ for some β, $0 < \beta < 2\pi$, which has $\|\mu\| = 1$ and $\mu(\mathbb{T}) = 0$, then*

$$|P\mu(z)| \leq \beta \frac{1 - |z|^2}{d_{\mathbb{C}}(z, J(\beta))^3}, \qquad z \in \mathbb{D},$$

where $d_{\mathbb{C}}$ is the Euclidean metric. In particular,

$$|P\mu(z)| \leq \frac{2\beta}{(1 - |z|)^2}, \qquad z \in \mathbb{D}.$$

Proof. Let $M(e^{i\theta})$ be the function $\mu(J(\theta))$, which is well defined at 1 because $\mu(\mathbb{T}) = 0$. The function M is supported on $J(\beta)$ with $\|M\|_{L^\infty} \leq \frac{1}{2}$. Integration by parts gives

$$\int_{-\pi}^{\pi} P(z, e^{i\theta}) d\mu(e^{i\theta}) = (1 - |z|^2) \int_{-\pi}^{\pi} M(e^{i\theta}) \frac{d}{d\theta} |e^{i\theta} - z|^{-2} d\theta.$$

Since

$$\left| \frac{d}{d\theta} \, |e^{i\theta} - z|^{-2} \right| \le \frac{2}{|e^{i\theta} - z|^3}, \qquad z \in \mathbb{D},$$

we get

$$\left| \int_{-\pi}^{\pi} P(z, e^{i\theta}) \, d\mu(e^{i\theta}) \right| \le 2 \, (1 - |z|^2) \int_0^{\beta} \frac{\|M\|_{L^{\infty}}}{|e^{i\theta} - z|^3} d\theta$$

$$\le 2\beta \, \frac{1 - |z|^2}{d_{\mathbb{C}}(z, J(\beta))^3}, \qquad z \in \mathbb{D}.$$

The proof is complete.   ∎

We will need to use harmonic measure late on, so we will briefly review the notion here. Let Ω be a bounded planar domain such that the Dirichlet problem can be solved on Ω. In other words, for any continuous function f on the boundary $\partial\Omega$, there is a harmonic function on Ω which extends continuously to $\overline{\Omega}$ and has boundary values f. By the maximum principle, the harmonic function is uniquely determined by its boundary values. We write $H(f)$ for the harmonic extension of f. The mapping H is clearly linear. In view of the maximum principle,

$$|H(f)(z)| \le \sup \big\{ |f(\zeta)| : \zeta \in \partial\Omega \big\}, \qquad z \in \Omega,$$

so that for each $z \in \Omega$, the mapping $f \mapsto H(f)(z)$ is a bounded linear functional of norm 1 on the space $C(\partial\Omega)$ of continuous functions on $\partial\Omega$. By the Riesz representation theorem, there is a Borel measure $d\omega_z$ on $\partial\Omega$ of total variation norm 1 such that

$$H(f)(z) = \int_{\partial\Omega} f(\zeta) \, d\omega_z(\zeta)$$

for all $f \in C(\partial\Omega)$. If we apply this to the constant function 1, we realize that $d\omega_z$ is a probability measure. It is called *harmonic measure*. To indicate properly the dependence on the point z and the domain Ω, we shall write

$$d\omega(z, \cdot, \Omega)$$

in place of $d\omega_z$, and for Borel measurable subsets E of $\partial\Omega$, $\omega(z, E, \Omega)$ is the corresponding mass of E with respect to harmonic measure. Actually, the notion of harmonic measure can be extended to much more irregular domains than those for which the Dirichlet problem can be solved.

Harmonic measure has a well-known interpretation in terms of Brownian motion. The quantity $\omega(z, E, \Omega)$ represents the probability that a Brownian motion starting at the point $z \in \Omega$ will exit the domain Ω on the subset E of $\partial\Omega$. It is assumed that the Brownian motion comes to a halt once the boundary $\partial\Omega$ is reached.

8.2   The Building Blocks

Our basic building blocks in later constructions will be the functions $\Phi_{\alpha,\beta}$, where $0 < \alpha < \beta \leq 2\pi$, which are defined as

$$\Phi_{\alpha,\beta}(z) = \frac{1}{\alpha} \omega(z, I(\alpha), \mathbb{D}) - \frac{1}{\beta} \omega(z, I(\beta), \mathbb{D}), \qquad z \in \mathbb{D}.$$

Here, ω is harmonic measure and $I(\tau)$ is the arc defined by

$$I(\tau) = \left\{ e^{i\theta} : \theta \in \left[-\tfrac{1}{2}\tau, \tfrac{1}{2}\tau \right] \right\}.$$

In the case $\beta = 2\pi$ we write Φ_α in place of $\Phi_{\alpha,2\pi}$. Thus

$$\Phi_\alpha(z) = \frac{1}{\alpha} \omega(z, I(\alpha), \mathbb{D}) - \frac{1}{2\pi}, \qquad z \in \mathbb{D}.$$

We extend the function $\Phi_{\alpha,\beta}$ to the boundary $\mathbb{T}$ by declaring

$$\Phi_{\alpha,\beta}(z) = \begin{cases} 0, & z \in \mathbb{T} \setminus I(\beta), \\ -1/\beta, & z \in I(\beta) \setminus I(\alpha), \\ 1/\alpha - 1/\beta, & z \in I(\alpha). \end{cases}$$

One checks that $\Phi_{\alpha,\beta}(0) = 0$. An elementary solution of the Dirichlet problem (see [49, pp. 41–42]) yields the explicit formula

$$\omega(z, I(\alpha), \mathbb{D}) = \frac{1}{\pi} \arg \left(\frac{e^{i\alpha/2} - z}{e^{-i\alpha/2} - z} \right) - \frac{\alpha}{2\pi}, \qquad z \in \mathbb{D},$$

with a suitable choice of the argument function, so that

$$\Phi_\alpha(z) = \frac{1}{\pi\alpha} \arg \left(\frac{1 - z e^{-i\alpha/2}}{1 - z e^{i\alpha/2}} \right), \qquad z \in \mathbb{D},$$

and $\Phi_{\alpha,\beta} = \Phi_\alpha - \Phi_\beta$. For $0 \leq r < 1$, put

$$Q_{\alpha,\beta}(r) = \max \left\{ \Phi_{\alpha,\beta}(z) : |z| = r \right\},$$

and extend the function continuously to $[0, 1]$ by setting $Q_{\alpha,\beta}(1) = 1/\alpha - 1/\beta$. Note that it is increasing in r, and has the property that the function $Q_{\alpha,\beta}(e^t)$ is convex on $(-\infty, 0]$. For geometric reasons, the above maximum is attained at $z = r$, and an explicit computation yields

$$Q_{\alpha,\beta}(r) = \Phi_{\alpha,\beta}(r) = \frac{2}{\pi\alpha} \arctan \frac{r \sin \frac{1}{2}\alpha}{1 - r \cos \frac{1}{2}\alpha} - \frac{2}{\pi\beta} \arctan \frac{r \sin \frac{1}{2}\beta}{1 - r \cos \frac{1}{2}\beta}.$$

For $0 < \alpha \leq \pi$, the value of $Q_{\alpha,\beta}$ at $r = \cos \frac{1}{2}\alpha$ is readily estimated as follows:

$$\frac{1}{\alpha} - \frac{1}{\beta} - \frac{1}{2\pi} \leq Q_{\alpha,\beta}(r) < \frac{1}{\alpha} - \frac{1}{\beta}, \qquad \cos \tfrac{1}{2}\alpha \leq r < 1.$$

We proceed to get an estimate on a longer interval. For $0 < \alpha < \frac{1}{3}\pi$, we have

$$\cos \tfrac{1}{2}\alpha > 1 - \frac{1}{3} \sin \tfrac{1}{2}\alpha,$$

so that

$$1 < \frac{r \sin \frac{1}{2}\alpha}{1 - r \cos \frac{1}{2}\alpha}, \qquad 1 - \frac{1}{2} \sin \tfrac{1}{2}\alpha \le r < 1,$$

and consequently,

$$\frac{1}{2\alpha} - \frac{1}{\beta} < Q_{\alpha,\beta}(r) < \frac{1}{\alpha} - \frac{1}{\beta}, \qquad 1 - \frac{1}{2} \sin \tfrac{1}{2}\alpha \le r < 1.$$

The second derivative of $Q_{\alpha,\beta}(r)$ is

$$Q''_{\alpha,\beta}(r) = \frac{4}{\pi\alpha} \frac{(\cos \frac{1}{2}\alpha - r) \sin \frac{1}{2}\alpha}{\left(1 + r^2 - 2r \cos \frac{1}{2}\alpha\right)^2} - \frac{4}{\pi\beta} \frac{(\cos \frac{1}{2}\beta - r) \sin \frac{1}{2}\beta}{\left(1 + r^2 - 2r \cos \frac{1}{2}\beta\right)^2}.$$

When $0 < \alpha < \beta < \pi$, we have $Q''_{\alpha,\beta}(r) > 0$ on the interval $0 < r < \cos \frac{1}{2}\alpha$, so that $Q_{\alpha,\beta}$ is convex there. In the special case $\beta = 2\pi$, we write Q_α in place of $Q_{\alpha,2\pi}$,

$$Q_\alpha(r) = \frac{2}{\pi\alpha} \arctan \frac{r \sin \frac{1}{2}\alpha}{1 - r \cos \frac{1}{2}\alpha};$$

its second derivative reduces to

$$Q''_\alpha(r) = \frac{4}{\pi\alpha} \frac{(\cos \frac{1}{2}\alpha - r) \sin \frac{1}{2}\alpha}{\left(1 + r^2 - 2r \cos \frac{1}{2}\alpha\right)^2}.$$

It follows that Q_α is convex on $[0, \cos \frac{1}{2}\alpha]$, and concave on $[\cos \frac{1}{2}\alpha, 1]$. The value at the inflexion point is

$$Q_\alpha(\cos \tfrac{1}{2}\alpha) = \frac{2}{\pi\alpha} \arctan \frac{\sin \frac{1}{2}\alpha \cos \frac{1}{2}\alpha}{1 - \cos^2 \frac{1}{2}\alpha} = \frac{1}{\alpha} - \frac{1}{\pi},$$

so that by convexity $(0 < \alpha < \pi)$,

$$Q_\alpha(r) \le \frac{\pi - \alpha}{\pi\alpha \cos \frac{1}{2}\alpha} r, \qquad 0 \le r \le \cos \tfrac{1}{2}\alpha.$$

For $0 < \alpha \le \frac{1}{2}\pi$, we have

$$\pi - \alpha < (\pi - \tfrac{1}{2}\alpha) \cos \tfrac{1}{2}\alpha,$$

since $1 - \frac{1}{2}\alpha^2 < \cos \alpha$. Therefore,

$$Q_\alpha(r) \le \left(\frac{1}{\alpha} - \frac{1}{2\pi}\right) r, \qquad 0 \le r \le \cos \tfrac{1}{2}\alpha.$$

We also need to estimate the first derivative of $Q_\alpha(r)$,

$$Q'_\alpha(r) = \frac{2}{\pi\alpha} \frac{\sin\frac{1}{2}\alpha}{1 + r^2 - 2r\cos\frac{1}{2}\alpha}.$$

It can be checked that $Q'_\alpha(r)$ attains its maximum at the point $r = \cos\frac{1}{2}\alpha$ (which is the inflection point for $Q_\alpha(r)$), so that for $0 < \alpha \le \frac{1}{2}\pi$,

$$Q'_\alpha(r) \le Q'_\alpha\left(\cos\frac{1}{2}\alpha\right) = \frac{2}{\pi\alpha\sin\frac{1}{2}\alpha} \le \frac{\sqrt{2}}{\alpha^2}, \qquad 0 \le r \le 1.$$

Here, we used the fact that $(\sin x)/x \ge 2^{3/2}\pi^{-1}$ for $0 < x \le \pi/4$. The function $Q'_\alpha(r)$ increases on $[0, \cos\frac{1}{2}\alpha]$ and decreases on $[\cos\frac{1}{2}\alpha, 1]$, so by estimating its values at $r = 0$ and $r = 1$, we see that $(0 < \alpha \le \frac{1}{2}\pi)$

$$\frac{1}{4} < Q'_\alpha(r) \le \frac{\sqrt{2}}{\alpha^2}, \qquad 0 \le r \le 1, \tag{8.1}$$

and

$$\frac{1}{\alpha^2} < Q'_\alpha(r) \le \frac{\sqrt{2}}{\alpha^2}, \qquad \cos\frac{1}{2}\alpha \le r \le 1. \tag{8.2}$$

Define

$$A(r, \alpha) = \arctan\frac{r\sin\frac{1}{2}\alpha}{1 - r\cos\frac{1}{2}\alpha},$$

and

$$B(r, \alpha) = \left(1 + r^2 - 2r\cos\frac{1}{2}\alpha\right)^{1/2}.$$

Then

$$\frac{rQ'_\alpha(r)}{Q_\alpha(r)} = \frac{\sin A(r, \alpha)}{A(r, \alpha)B(r, \alpha)}.$$

Using this identity and the elementary estimate $2/\pi < (\sin x)/x < 1$ for $0 < x < \frac{1}{2}\pi$, we get

$$\frac{1}{2} < \frac{rQ'_\alpha(r)}{Q_\alpha(r)}, \qquad 0 < r \le 1,\ 0 < \alpha \le \frac{1}{2}\pi, \tag{8.3}$$

and

$$\frac{2}{3\alpha} < \frac{rQ'_\alpha(r)}{Q_\alpha(r)} < \frac{3}{\alpha}, \qquad \cos\frac{1}{2}\alpha \le r \le 1,\ 0 < \alpha \le \frac{1}{2}\pi. \tag{8.4}$$

At some point, we shall also need to be able to handle the function $Q_{2\pi-\alpha}(r)$ for small positive angles α. One shows that for $0 < \alpha \le \frac{1}{2}\pi$,

$$\frac{1}{4} < \frac{rQ'_{2\pi-\alpha}(r)}{Q_{2\pi-\alpha}(r)} < 1, \qquad 0 < r \le 1. \tag{8.5}$$

We omit the details.

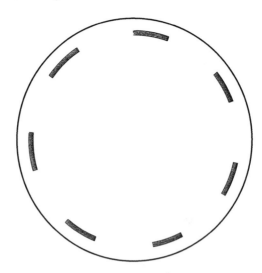

Figure 8.1. The set $E^\sharp(N, \alpha)$

8.3   The Basic Iteration Scheme

We now describe an iteration scheme that produces bounded harmonic functions with certain desired properties. First, we need some notation. For $k = 0, 1, \ldots, N - 1$ and $0 < \alpha < 2\pi$, define arcs $I_k(\alpha, N)$ and $J_k(\alpha, N)$ by

$$I_k(\alpha, N) = \left\{ e^{i\theta} : \theta \in \left(\frac{1}{N} \left(2\pi k - \tfrac{1}{2}\alpha\right), \frac{1}{N} \left(2\pi k + \tfrac{1}{2}\alpha\right) \right) \right\},$$

and

$$J_k(\alpha, N) = \left\{ e^{i\theta} : \theta \in \left(\frac{1}{N} \left(2\pi k + \tfrac{1}{2}\alpha\right), \frac{1}{N} \left(2\pi (k + 1) - \tfrac{1}{2}\alpha\right) \right) \right\},$$

and form the set

$$E^\sharp = E^\sharp(N, \alpha) = \left\{ re^{i\theta} : r_N^\sharp \leq r \leq r_N^*, \quad e^{i\theta} \in \bigcup_j \bar{I}_j(\alpha/2, N) \right\}, \tag{8.6}$$

where $r_N^* = 1 - (N \log N)^{-1}$ and $r_N^\sharp = 1 - 2(N \log N)^{-1}$. The set $E^\sharp$ looks like a union of equidistributed rectangular boxes placed along a concentric circle (see Figure 8.1).

LEMMA 8.2 *Let $u(t) = 1 - \log(1 - t)$ and $v(t) = \beta u(t)$, where β is a positive constant. Suppose $f \in h^\infty(\mathbb{D})$ is real-valued and satisfies*

$$-v(|z|) - A \leq f(z) \leq u(|z|) + B, \qquad z \in \mathbb{D},$$

for some constants A and B. Let $\rho \in (0, 1)$ and $\varepsilon > 0$ be given. Also, fix a parameter α with $0 < \alpha \leq \min\{1, 2\beta\}$. Then for any real parameter ξ, there is a radius $\varrho \in (\rho, 1)$ and a constant $C = C(\alpha)$ such that for sufficiently large positive *integers N we can find a real-valued function g ∈ h∞(𝔻) satisfying the following:*

$$|f(z) - g(z)| \leq \varepsilon, \qquad\qquad |z| \leq \rho,$$
$$-v(|z|) - A - \varepsilon \leq g(z) \leq u(|z|) + B + \varepsilon, \qquad \rho < |z| < \varrho,$$
$$-v(|z|) \leq g(z) \leq u(|z|) - \xi + C, \qquad \varrho \leq |z| < 1,$$

and

$$u(|z|) - \xi - C \leq g(z), \qquad z \in E^\sharp(N, \alpha).$$

Furthermore, the function g can be made to satisfy

$$\int_\mathbb{D} \exp\left(g(z)\right) dA(z) \leq C e^{-\xi} + e^\varepsilon \int_\mathbb{D} \exp\left(f(z)\right) dA(z)$$

as well.

Proof. We may assume that f extends continuously to the closed disk $\overline{\mathbb{D}}$. If it is not, we can replace f by its dilate f_r, where $f_r(z) = f(rz)$ for $z \in \mathbb{D}$, with $0 < r < 1$ sufficiently close to 1.

Let Φ_α and Q_α be as in the previous section. For any positive integer N, let $\phi_N(z) = \Phi_\alpha(z^N)$ and $q_N(r) = Q_\alpha(r^N)$. Then ϕ_N is harmonic in the unit disk and $q_N(r)$ is the maximum of $\phi_N(z)$ on the circle $|z| = r$. The function $\phi_N(z)$ extends continuously to the unit circle except for a finite set of points, and equals $1/\alpha - 1/(2\pi)$ on $\cup_k I_k(\alpha, N)$, and $-1/(2\pi)$ on $\cup_k J_k(\alpha, N)$. As a matter of notation, let us agree to write $\bar{I}_k(\alpha, N)$ and $\bar{J}_k(\alpha, N)$ for the closures of the respective arcs.

Step 1. For any real t, define

$$\Lambda_N(t) = \sup \left\{ \frac{q_N(r)}{u(r) - t} : \sigma(t) \leq r < 1 \right\}, \tag{8.7}$$

where

$$\sigma(t) = \max \left\{ \frac{1}{2}, 1 - e^{-2t} \right\}.$$

Note that $u(r) - t$ is at least 1 for $\sigma(t) \leq r < 1$. By the properties of the functions $u(r)$ and $\sigma(t)$, we also have

$$\frac{1}{2} u(r) \leq u(r) - t, \qquad \sigma(t) \leq r < 1. \tag{8.8}$$

It is clear by inspection that the supremum in (8.7) is attained at some point of $[\sigma(t), 1)$, because the function $u(r)$ tends to $+\infty$ as $r \to 1$. Since

$$\frac{d}{dr} \frac{q_N(r)}{u(r) - t} = \frac{q_N'(r)\left(u(r) - t\right) - q_N(r)u'(r)}{\left(u(r) - t\right)^2},$$

after some simplifications we see that this derivative has the same sign as

$$r^N \frac{Q_\alpha'(r^N)}{Q_\alpha(r^N)} - \frac{r}{N} \frac{u'(r)}{u(r) - t}. \tag{8.9}$$

Suppose the variable t is confined to some given finite interval $[-T, T]$. The first term of (8.9) is greater than $\frac{1}{2}$, by (8.3); for $r = \sigma(t)$, the second term tends to 0 as N grows to infinity, so that for large N, the sign of the quantity in (8.9) is positive, and the supremum in (8.7) is not attained at the left boundary point. So, for N large, any point $r = r_N(t)$ where this maximum is attained is an interior point, and hence, by elementary calculus, we have that

$$r^N \frac{Q'_\alpha(r^N)}{Q_\alpha(r^N)} = \frac{r}{N} \frac{u'(r)}{u(r) - t}, \qquad r = r_N(t).$$

This identity, together with (8.3), (8.4), and (8.8), shows that for large N we can estimate the position of a point where the maximum in (8.7) is attained:

$$1 - \frac{C_1}{N \log N} \le r_N(t) \le 1 - \frac{C_2}{N \log N}, \tag{8.10}$$

where the constants C_1 and C_2 depend only on the parameter α. For example, we may pick C_1 to be 2α and C_2 to be $\alpha/4$.

Step 2. For any large N and real t, put

$$\Lambda_N^*(t) = \frac{\alpha^{-1} - (2\pi)^{-1}}{\log(N \log N) - t}.$$

One then calculates that

$$\left| \frac{1}{\Lambda_N(t)} - \frac{1}{\Lambda_N^*(t)} \right| \le C \tag{8.11}$$

for some positive constant $C = C(\alpha)$ that depends only on α, provided that N is large enough. This is so because $1 - r_N(t)^N$ is comparable to $(\log N)^{-1}$ and $q_N(r_N(t)) = Q_\alpha(r_N(t)^N)$, so that when we plug in the point $r = r_N(t)$ into (8.7), and use the estimate (8.10), the value of the denominator on the right-hand side of (8.7) is close to $\Lambda_N^*(t)^{-1}$, and the numerator is close to 1; see Exercise 1.

By the way the parameter $\Lambda_N(t)$ was defined,

$$\frac{\phi_N(z)}{\Lambda_N(t)} \le u(|z|) - t, \qquad \sigma(t) \le |z| < 1.$$

Combining this with (8.11) we obtain

$$\frac{\phi_N(z)}{\Lambda_N^*(t)} \le u(|z|) - t + C, \qquad \sigma(t) \le |z| < 1, \tag{8.12}$$

with a positive constant C that depends only on α.

The point $r_N^* = 1 - (N \log N)^{-1}$ is similar to the point $r_N(t)$, where the maximum in (8.7) is attained, in that the corresponding distances to the point 1 are comparable. We shall now show that for some positive constant $C = C(\alpha)$, which depends only on α, the following estimate holds for large N:

$$u(|z|) - t - C \le \frac{\phi_N(z)}{\Lambda_N^*(t)}, \qquad z \in E^\sharp(N, \alpha). \tag{8.13}$$

We first observe that by the definition of $r_N(t)$,

$$\frac{\phi_N(z)}{\Lambda_N(t)} = u(|z|) - t, \qquad \frac{z}{r_N(t)} \in E(N),$$

where $E(N)$ is the set of N-th roots of unity consisting of

$$e_k(N) = \exp(2\pi i k/N), \qquad k = 0, \ldots, N - 1.$$

Next, by using (8.1), we see that replacing $\Lambda_N(t)$, $r_N(t)$ with $\Lambda_N^*(t)$, r_N^* carries the cost of introducing a positive constant $C = C(\alpha)$ depending only on α, in the sense that

$$u(|z|) - t - C \le \frac{\phi_N(z)}{\Lambda_N^*(t)}, \qquad \frac{z}{r_N^*} \in E(N).$$

Extending the estimate beyond the set $E(N)$ to the union of rectangular-shaped boxes $E^\sharp(N, \alpha)$ requires some simple estimates of harmonic measure, which are left to the reader; see Exercise 2.

Step 3. Up to this point, the parameter t is confined to a prescribed interval $[-T, T]$, and N is a large integer depending on T and α. We now let T equal the supremum of $|f(z) + \xi|$ on $\mathbb{D}$. Let $e_k(N) = \exp(2\pi i k/N)$ be an N-th root of unity, and set

$$t_k(N, \xi) = f(e_k(N)) + \xi.$$

Then $t_k(N, \xi)$ is confined to the interval $[-T, T]$, and we can consider

$$\mu_k(N, \xi) = \Lambda_N^*(t_k(N, \xi))^{-1} = \frac{\log(N \log N) - f(e_k(N)) - \xi}{\alpha^{-1} - (2\pi)^{-1}}$$

and the associated function

$$\chi_{N,\xi}(z) = \frac{1}{N} \sum_{k=0}^{N-1} \mu_k(N, \xi) \, \Phi_{\alpha/N, 2\pi/N}(\bar{e}_k(N)z), \qquad z \in \mathbb{D}.$$

The points $\bar{e}_k(N) = \exp(-2\pi i k/N)$ are the complex conjugates of the $e_k(N)$. The size of the function $\Phi_{\alpha/N, 2\pi/N}(z)$ is estimated by means of Theorem 8.1; for large N and $0 < \alpha \le \frac{1}{2}\pi$ this results in

$$\left| \chi_{N,\xi}(z) \right| \le \frac{\log N}{N} \frac{8\alpha}{(1 - |z|)^2}, \qquad z \in \mathbb{D}. \tag{8.14}$$

The function $\phi_N(z)$ may be written as

$$\phi_N(z) = \frac{1}{N} \sum_{k=0}^{N-1} \Phi_{\alpha/N, 2\pi/N}(\bar{e}_k(N)z), \qquad z \in \mathbb{D}.$$

We intend to compare $\chi_{N,\xi}(z)$ with the more easily analyzed function $\chi_{N,\xi}^{(k)}(z) = \mu_k(N, \xi) \phi_N(z)$. It, too, enjoys (for large N) the estimate

$$\left| \chi_{N,\xi}^{(k)}(z) \right| \le \frac{\log N}{N} \frac{8\alpha}{(1 - |z|)^2}, \qquad z \in \mathbb{D}.$$

The difference is

$$\chi_{N,\xi}(z) - \chi_{N,\xi}^{(k)}(z) = \frac{1}{N} \sum_{j=0}^{N-1} (\mu_j(N,\xi) - \mu_k(N,\xi)) \Phi_{\alpha/N, 2\pi/N}(\bar{e}_j(N)z)$$

$$= (\alpha^{-1} - (2\pi)^{-1})^{-1}/N$$

$$\times \sum_{j=0}^{N-1} \left(f(e_k(N)) - f(e_j(N)) \right) \Phi_{\alpha/N, 2\pi/N}(\bar{e}_j(N)z).$$

By the uniform continuity of f on $\overline{\mathbb{D}}$, we can fix a positive δ such that $|f(z) - f(w)| \le 1$ whenever $|z - w| \le \delta$. Split up the index set $\{0, 1, \ldots, N-1\}$ into two parts, one, $X(k, N)$, consisting of those j for which $|e_j(N) - e_k(N)| \le \delta$, and the other, $Y(k, N)$, where the opposite occurs. Then, since the various building blocks $\Phi_{\alpha/N, 2\pi/N}(\bar{e}_j(N)z)$ are supported on disjoint arcs of $\mathbb{T}$ for different j, we see that the quantity

$$\frac{1}{N[\alpha^{-1} - (2\pi)^{-1}]} \sum_{j \in X(k,N)} \left| \left(f(e_k(N)) - f(e_j(N)) \right) \Phi_{\alpha/N, 2\pi/N}(\bar{e}_j(N)z) \right|$$

is less than or equal to 1. Summing over the remaining indices in $Y(k, N)$ and noticing that

$$|f(e_k(N)) - f(e_j(N))| \le 2T,$$

we conclude that the quantity

$$\frac{1}{N[\alpha^{-1} - (2\pi)^{-1}]} \sum_{j \in Y(k,N)} \left| \left(f(e_k(N)) - f(e_j(N)) \right) \Phi_{\alpha/N, 2\pi/N}(\bar{e}_j(N)z) \right|$$

is less than or equal to $2T \, \omega(z, \mathbb{T} \setminus L_k, \mathbb{D})$, where ω is harmonic measure, $L_k = L_k(\delta)$ is the arc on $\mathbb{T}$ of points within distance $\frac{1}{2}\delta$ from $e_k(N)$, and N is so large that $2\pi/N$ is considerably smaller than $\frac{1}{2}\delta$. These two estimates lead to

$$\left| \chi_{N,\xi}(z) - \chi_{N,\xi}^{(k)}(z) \right| \le 1 + 2T \, \omega(z, \mathbb{T} \setminus L_k, \mathbb{D}), \qquad z \in \mathbb{D}. \tag{8.15}$$

Let $D_k = D_k(\delta, T)$ be the *lunula* that is the intersection with $\mathbb{D}$ of a disk centered at the point $e_k(N)$ with radius depending only on T and δ, such that $2T \, \omega(z, \mathbb{T} \setminus L_k) \le 1$ for $z \in D_k$; then, by (8.15),

$$\left| \chi_{N,\xi}(z) - \chi_{N,\xi}^{(k)}(z) \right| \le 2, \qquad z \in D_k. \tag{8.16}$$

We may assume that all the D_k are contained in the annulus $\sigma(T) < |z| < 1$, and that the radius of each D_k is at most $\frac{1}{2}\delta$. By (8.12) and (8.13),

$$f(e_k(N)) + \chi_{N,\xi}^{(k)}(z) \le u(|z|) - \xi + C, \qquad \sigma(T) \le |z| < 1,$$

and

$$u(|z|) - \xi - C \le f(e_k(N)) + \chi_{N,\xi}^{(k)}(z), \qquad z \in E^{\sharp}(N, \alpha).$$

Since $\left| f(z) - f\left(e_k(N)\right) \right| \le 1$ for $z \in D_k$, we get from (8.16) that

$$f(z) + \chi_{N,\xi}(z) \le f\left(e_k(N)\right) + \chi_{N,\xi}^{(k)}(z) + 3 \le u(|z|) - \xi + C' \qquad (8.17)$$

for $z \in D_k$, and

$$u(|z|) - \xi - C' \le f\left(e_k(N)\right) + \chi_{N,\xi}^{(k)}(z) - 3 \le f(z) + \chi_{N,\xi}(z) \qquad (8.18)$$

for $z \in D_k \cap E^\sharp(N, \alpha)$; here $C' = C + 3$ with $C = C(\alpha)$ being the constant appearing in (8.12) and (8.13). For large N, the distance between the centers $e_k(N)$ of the *lunulæ* D_k gets much smaller than the radius (which is independent of k and N), so that $\cup_k D_k$ contains an annulus $\vartheta < |z| < 1$, where $\vartheta = \vartheta(\delta, T)$ has $0 < \vartheta < 1$. Moreover, for large N, the set $E^\sharp(N, \alpha)$ will be contained in the annulus $\vartheta < |z| < 1$, by (8.10). So it follows from (8.17) and (8.18) that

$$f(z) + \chi_{N,\xi}(z) \le u(|z|) - \xi + C', \qquad \vartheta < |z| < 1,$$

and

$$u(|z|) - \xi - C' \le f(z) + \chi_{N,\xi}(z), \qquad z \in E^\sharp(N, \alpha).$$

Step 4. Let ρ_0 be the bigger of the two numbers ρ and ϑ. Let $\varrho = \varrho(\rho, \vartheta, \xi, \alpha, f)$, with $\rho_0 < \varrho < 1$, be so close to 1 that

$$|f(z)| \le \frac{1}{2}\beta\, u(|z|) - 2, \qquad \varrho < |z| < 1;$$

after all, f is a bounded function. By (8.14), we have

$$|\chi_{N,\xi}(z)| < \varepsilon, \qquad |z| < \varrho,$$

provided that N is large enough. *We now make the pick*

$$g(z) = f(z) + \chi_{N,\xi}(z), \qquad z \in \mathbb{D};$$

this function meets all the required conditions, save the control from below and the integral estimate. If we can show that

$$-\chi_{N,\xi}(z) \le \frac{1}{2}\beta\, u(|z|) + 2$$

on $\varrho < |z| < 1$, we will be done with the control from below. We turn to estimating the function $\chi_{N,\xi}(z)$ from below on the *lunula* D_k. We have

$$\left| \chi_{N,\xi}(z) - \chi_{N,\xi}^{(k)}(z) \right| \le 2, \qquad z \in D_k,$$

so *we estimate the simpler function instead.* We solve the problem of estimating

$$-\chi_{N,\xi}^{(k)}(z)/v(|z|) = -\mu_k(N, \xi)\phi_N(z)/v(|z|)$$

from above by first noting that along any concentric circle $|z| = r$, the value is the biggest when z^N is real and negative. It is easily checked that

$$-\Phi_\alpha(-w) = \left(2\pi\alpha^{-1} - 1\right)\Phi_{2\pi-\alpha}(w),$$

and since $\phi_N(z) = \Phi_\alpha(z^N)$, we obtain

$$\sup_{z \in \mathbb{D}} \left\{ \frac{-\chi_{N,\xi}^{(k)}(z)}{v(|z|)} \right\} = \frac{2\pi - \alpha}{\alpha \beta \Lambda_N^* \big(t_k(N,\xi)\big)} \sup_{0 < r < 1} \left\{ \frac{Q_{2\pi - \alpha}(r^N)}{u(r)} \right\}. \tag{8.19}$$

The extremal problem on the right-hand side is of the same kind as (8.7), and based on (8.5), one shows with the same methods as were used for problem (8.7) that the point where the above supremum is attained satisfies the analogue of the estimate (8.10), only this time the constants are absolute. When this information is inserted into (8.19), one obtains, using (8.11), that for large N,

$$\sup \left\{ \frac{-\chi_{N,\xi}^{(k)}(z)}{v(|z|)} : z \in \mathbb{D} \right\} \le \frac{\alpha}{4\beta}; \tag{8.20}$$

see Exercise 3. It follows from the restrictions on α that

$$-\chi_{N,\xi}(z) \le \frac{1}{2}\beta \, u(|z|)$$

on $\varrho < |z| < 1$, as desired.

Step 5. We need an additional estimate of the function $g = f + \chi_{N,\xi}$, one that is so accurate that it allows us to say how big the integral

$$\int_{\mathbb{D}} \exp\big(g(z)\big) \, dA(z)$$

is. To this end, we look again at the extremal problem in (8.7) for $t = 0$ and $\frac{1}{2} \le r \le (\cos \frac{1}{2}\alpha)^{1/N}$, and note by the considerations involving the sign of (8.9) that the extremal value is attained at the right endpoint $r = (\cos \frac{1}{2}\alpha)^{1/N}$, at least for large N. This entails that

$$\begin{aligned} q_N(r) &\le \frac{q_N\big((\cos \frac{1}{2}\alpha)^{1/N}\big)}{u\big((\cos \frac{1}{2}\alpha)^{1/N}\big)} u(r) = \frac{Q_\alpha(\cos \frac{1}{2}\alpha)}{u\big((\cos \frac{1}{2}\alpha)^{1/N}\big)} u(r) \\ &= \frac{\alpha^{-1} - \pi^{-1}}{u\big((\cos \frac{1}{2}\alpha)^{1/N}\big)} u(r) \le \frac{\alpha^{-1} - \pi^{-1}}{\log N + C(\alpha)} u(r) \end{aligned}$$

for

$$\frac{1}{2} \le r \le \left(\cos \tfrac{1}{2}\alpha\right)^{1/N},$$

where $C(\alpha)$ is a real-valued constant. In the setting of the estimate (8.17), with the necessary modifications due to ξ, we then arrive at

$$\begin{aligned} f\big(e_k(N)\big) + \mu_k(N,\xi)\phi_N(z) &\le f\big(e_k(N)\big) + \mu_k(N,\xi)q_N(|z|) \\ &\le f\big(e_k(N)\big) + \left(1 - \frac{\alpha}{2\pi}\right) u(|z|) \end{aligned}$$

for large N and for

$$\frac{1}{2} \le |z| \le \left(\cos \tfrac{1}{2}\alpha\right)^{1/N}.$$

If the parameter ϱ selected in Step 4 is sufficiently close to 1, any fixed fraction of $u(|z|)$ will dominate over $f\big(e_k(N)\big) + \xi$ for $\varrho \le |z| < 1$, so that we can get

$$f\big(e_k(N)\big) + \mu_k(N, \xi)\phi_N(z) \le \left(1 - \frac{\alpha}{8}\right) u(|z|) - \xi \qquad (8.21)$$

for

$$\varrho \le |z| \le \left(\cos \tfrac{1}{2}\alpha\right)^{1/N}.$$

By (8.21) and the appropriate analogue of (8.17) involving ξ,

$$f(z) + \chi_{N,\xi}(z) \le \left(1 - \frac{\alpha}{8}\right) u(|z|) - \xi + 3 \qquad (8.22)$$

for $\varrho \le |z| \le \left(\cos \tfrac{1}{2}\alpha\right)^{1/N}$. From Step 4, we know that $|\chi_{N,\xi}(z)| < \varepsilon$ holds on $|z| < \varrho$. To control $f + \chi_{N,\xi}$ in the remaining annulus

$$\left(\cos \frac{1}{2}\alpha\right)^{1/N} < |z| < 1,$$

we need the following elementary estimates of Q_α, which follow from (8.2) (r_0 is a number in the interval $\cos \tfrac{1}{2}\alpha < r_0 < 1$):

$$Q_\alpha(r) = Q_\alpha(r_0) - \int_r^{r_0} Q'_\alpha(t)\, dt \le Q_\alpha(r_0) - \alpha^{-2}(r_0 - r)$$

for $\cos(\alpha/2) \le r \le r_0$, and

$$Q_\alpha(r) = Q_\alpha(r_0) + \int_{r_0}^r Q'_\alpha(t)\, dt \le Q_\alpha(r_0) + \sqrt{2}\,\alpha^{-2}(r - r_0)$$

for $r_0 \le r \le 1$. With $r_0 = (r_N^*)^N \approx 1 - 1/\log N$, these estimates lead to

$$q_N(r) \le q_N(r_N^*) - \alpha^{-2}\big((r_N^*)^N - r^N\big)$$

for $(\cos(\alpha/2))^{1/N} \le r \le r_N^*$, and

$$q_N(r) \le q_N(r_N^*) + \sqrt{2}\,\alpha^{-2}(r^N - (r_N^*)^N)$$

for $r_N^* \le r \le 1$. For large N, we know that

$$f\big(e_k(N)\big) + \mu_k(N, \xi)\, q_N(r_N^*)$$

is within an additive constant (depending only on α) of

$$u(r_N^*) - \xi = 1 + \log(N \log N) - \xi;$$

just look at how we got (8.17) and (8.18). Using the above estimates of $q_N(r)$, we obtain a constant $C(\alpha)$ such that for large N,

$$\begin{aligned}
f\big(e_k(N)\big) + \mu_k(N, \xi)\, \phi_N(z) &\le f\big(e_k(N)\big) + \mu_k(N, \xi)\, q_N(|z|) \\
&\le u(r_N^*) - \xi \\
&\quad - \frac{\log N}{\alpha\big(1 - \alpha/(2\pi)\big)}\,(1 - |z|^N) + C(\alpha)
\end{aligned}$$

for $(\cos(\alpha/2))^{1/N} \le |z| \le r_N^*$, and

$$f\bigl(e_k(N)\bigr) + \mu_k(N,\xi)\,\phi_N(z) \;\le\; f\bigl(e_k(N)\bigr) + \mu_k(N,\xi)\,q_N(|z|)$$
$$\le\; u(r_N^*) - \xi + C(\alpha)$$

for $r_N^* \le |z| < 1$. By the localization trick of (8.17), involving the *lunulæ*, these estimates lead to

$$f(z) + \chi_{N,\xi}(z) \le u(r_N^*) - \frac{(\log N)(1 - |z|^N)}{\alpha\bigl(1 - \alpha/(2\pi)\bigr)} - \xi + C'(\alpha) \qquad (8.23)$$

for $(\cos(\alpha/2)^{1/N} \le |z| \le r_N^*$, and

$$f(z) + \chi_{N,\xi}(z) \le u(r_N^*) - \xi + C'(\alpha), \qquad r_N^* \le |z| < 1, \qquad (8.24)$$

for some other constant $C'(\alpha)$. It follows from (8.24) and

$$u(r_N^*) = 1 + \log(N \log N)$$

that

$$\int_{r_N^* \le |z| < 1} \exp\bigl(f(z) + \chi_{N,\xi}(z)\bigr)\,dA(z) \le C(\alpha)e^{-\xi},$$

where $C(\alpha)$ is a positive constant. An exercise involving Taylor series shows that for positive real τ,

$$\int_{\mathbb{D}} \exp\bigl(-\tau(1 - |z|^N)\bigr)\,dA(z) \le e^{-\tau} + \frac{4}{N}\frac{1 - e^{-\tau}}{\tau},$$

so that by (8.23),

$$\int_{(\cos(\alpha/2))^{1/N} < |z| < r_N^*} \exp\bigl(f(z) + \chi_{N,\xi}(z)\bigr)\,dA(z) \le C(\alpha)e^{-\xi} \qquad (8.25)$$

if $\alpha(1 - \alpha/(2\pi)) < 1$, where $C(\alpha)$ is a positive constant, possibly different from the earlier one. Since $1 - \alpha/8 < 1$, we get from (8.22) that

$$\int_{\varrho < |z| < (\cos(\alpha/2))^{1/N}} \exp\bigl(f(z) + \chi_{N,\xi}(z)\bigr)\,dA(z) \le C(\alpha)e^{-\xi}, \qquad (8.26)$$

where $C(\alpha)$ is yet another positive constant. Moreover, since $|\chi_{N,\xi}(z)| < \varepsilon$ on $|z| < \varrho$, we obtain

$$\int_{|z| < \varrho} \exp\bigl(f(z) + \chi_{N,\xi}(z)\bigr)\,dA(z) \le e^{\varepsilon} \int_{\mathbb{D}} \exp\bigl(f(z)\bigr)\,dA(z). \qquad (8.27)$$

The last part of the lemma now follows from (8.25), (8.26), and (8.27). ∎

8.4 The Mushroom Forest

Let $N(n)$ be a sequence of positive integers approaching $+\infty$ rapidly, and let $E_n^{\sharp}$ be given by (8.6), with $N = N(n)$. Moreover, let ξ_n be a sequence of nonnegative numbers that tend to $+\infty$ rather slowly. For a Borel subset E of $\mathbb{D}$, let $|E|_A$ denote the normalized area of E (the usual area divided by π). If instead E is a rectifiable curve or a (relative Borel) subset of one, we let $|E|_s$ be the length of E divided by 2π. The latter definition is related to ds, the normalized one-dimensional Lebesgue measure in the complex plane. Suppose h is a subharmonic function about which it is known (1) that it is bounded from above by some unspecified constant, (2) that

$$h(z) \le \gamma\, u(|z|), \qquad z \in \mathbb{D}, \qquad (8.28)$$

holds for some positive constant γ, and (3)

$$\sum_{n=1}^{\infty} e^{-\xi_n} \frac{1}{|E_n^{\sharp}|_A} \int_{E_n^{\sharp}} \exp\left(h(z)\right) dA(z) \le M < +\infty. \qquad (8.29)$$

We wish to estimate the average radial growth of $h(z)$. More to the point, we want to know how quickly the integral mean

$$\int_{-\pi}^{\pi} h^+(re^{i\theta})\, d\theta$$

increases as r approaches 1, where $h^+(z) = \max\{h(z), 0\}$. Fubini's theorem together with the Intermediate Value Theorem of Calculus tells us that there is a radius R_n, $r_{N(n)}^{\sharp} < R_n < r_{N(n)}^*$, such that with $E_n = E_n^{\sharp} \cap (R_n \mathbb{T})$,

$$\frac{1}{|E_n|_s} \int_{E_n} \exp\left(h(z)\right) ds(z) \le \frac{1}{|E_n^{\sharp}|_A} \int_{E_n^{\sharp}} \exp\left(h(z)\right) dA(z),$$

so that

$$\sum_{n=1}^{\infty} e^{-\xi_n} \frac{1}{|E_n|_s} \int_{E_n} \exp\left(h(z)\right) ds(z) \le M < +\infty.$$

A crude estimate of each term leads to

$$\frac{1}{|E_n|_s} \int_{E_n} \exp\left(h(z)\right) ds(z) \le M\, e^{\xi_n}. \qquad (8.30)$$

We note that

$$E_n = \left\{ z \in \mathbb{D} : \frac{z}{R_n} \in \bigcup_j \bar{I}_j(\alpha/2, N(n)) \right\},$$

so that $|E_n|_s = \alpha R_n/(4\pi)$ tends to $\alpha/(4\pi)$ as $n \to +\infty$. Introduce the union of rectangular boxes Σ_n,

$$\Sigma_n = \left\{ re^{i\theta} : R_n \le r < 1, \quad e^{i\theta} \in \bigcup_j \bar{I}_j(\alpha/4, N(n)) \right\},$$

and put a hat on each box to form $\Pi_n = E_n \cup \Sigma_n$. The set Π_n looks like a collection of identical mushrooms, with stems affixed to the ground, the unit circle. Let $\Omega_n = \mathbb{D} \setminus \bigcup_{j=n}^{\infty} \Pi_j$, which is an open subset of $\mathbb{D}$, and let $\Omega_n^{\sharp}$ be the connected

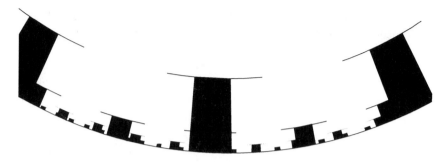

Figure 8.2. The mushroom forest

component that contains the origin (see Figure 8.2 for a graphic illustration of the set $\Omega_n^\natural$), which is easily seen to be simply connected. The boundary $\partial\Omega_n^\natural$ of $\Omega_n^\natural$ consists of a closed subset of the unit circle $\mathbb{T}$, mushroom hats E_j, or parts of them, and stem sides

$$\partial^b \Sigma_j = \left\{ re^{i\theta} : R_j \le r < 1, \quad e^{i\theta} \in \bigcup_j \partial I_j(\alpha/4, N(n)) \right\},$$

or parts of them, as well, for $j = n,\, n+1, n+2, \ldots$; on the right-hand side of the displayed formula, the ∂ is the boundary operation with respect to the topology of $\mathbb{T}$.

Recall that we reserve the symbol ω for harmonic measure; we sometimes write $d\omega(z, \zeta, \Omega)$ and think of it as a measure, where the variable of integration is ζ. Since $h(z)$ is subharmonic and bounded above in $\mathbb{D}$, $h^+(z)$ is subharmonic and bounded as well. If h_n is the harmonic function in $\Omega_n^\natural$ defined by

$$h_n(z) = \int_{\partial\Omega_n^\natural} h^+(\zeta)\, d\omega(z, \zeta, \Omega_n^\natural), \qquad z \in \Omega_n^\natural,$$

then, by the maximum principle, $h^+(z) \le h_n(z)$ on $\Omega_n^\natural$. For any r with $r\overline{\mathbb{D}} \subset \Omega_n^\natural$, the mean-value property for harmonic functions gives

$$\frac{1}{r}\int_{r\mathbb{T}} h^+(z)\, ds(z) \le \frac{1}{r}\int_{r\mathbb{T}} h_n(z)\, ds(z) = h_n(0) = \int_{\partial\Omega_n^\natural} h^+(\zeta)\, d\omega(0, \zeta, \Omega_n^\natural).$$

This calculation suggests that we should estimate $\omega(0, L, \Omega_n^\natural)$ for various Borel subsets L of $\partial\Omega_n^\natural$.

One quickly checks that $\omega(0, L, \Omega_n^\natural)$ is 0 if L is a subset of $\mathbb{T} \cap \partial\Omega_n^\natural$. The *principle of extension of the domain* (see any book on Harmonic Measure) states that the harmonic measure of a piece of the boundary of a region with respect to a fixed interior point gets larger if the region is expanded in such a way that the boundary piece remains on the boundary. If L is a Borel subset of

$$Y_{j,k} = R_k \bar{I}_j\left(\tfrac{1}{2}\alpha, N(k)\right) \subset E_k \cap \partial\Omega_n^\natural$$

for some $k = n, n + 1, n + 2, \ldots$, we replace $\Omega_n^{\natural}$ with $\mathbb{D} \setminus Y_{j,k}$, and see that

$$\omega(0, L, \Omega_n^{\natural}) \leq \omega(0, L, \mathbb{D} \setminus Y_{j,k}) \leq C(\alpha) \, |L|_s, \tag{8.31}$$

for some positive constant $C(\alpha)$. The remaining type of boundary parts is formed by the stem sides. So, let L be a subset of $\partial^b \Sigma_k \cap \partial \Omega_n^{\natural}$ for some $k = n, n+1, n+2, \ldots$, and suppose for simplicity that it is a subset of a single stem side of one mushroom. Then, if we remove all the other mushrooms, the harmonic measure of L increases, but it is still quite small. We can visualize this by thinking of harmonic measure as arising from Brownian motion: to reach L, the particle starting at the origin first has to reach some point of the opening between the hat and the unit circle, and second, it must then also hit the stem, and in particular, the part that lies on L. The hat and the stem define a "boxed" region of dimension $\frac{1}{8}\alpha \, N(k)^{-1}$ by $1 - R_k$, so that an estimate of the second process using harmonic measure for the boxed region shows that

$$\omega(0, L, \Omega_n^{\natural}) \leq C(\alpha) \, N(k)^{-v(\alpha)} \, |L|_s, \tag{8.32}$$

for some positive constants $C(\alpha)$ and $v(\alpha)$; $v(\alpha) = \alpha/6$ should work.

The estimates (8.30) and (8.31) will be used to control

$$\int_{\partial \Omega_n^{\natural}} h^+(\zeta) \, d\omega(0, \zeta, \Omega_n^{\natural}) \tag{8.33}$$

on the hats of the mushrooms, and (8.28) and (8.32) to control it on the stems. Not all mushrooms are so lucky as to form part of the boundary of $\Omega_n^{\natural}$, as many are contained in the stems of earlier generations of them, and some are trapped between two bigger intersecting mushrooms. Approximately the proportion $\alpha/(8\pi)$ of those remaining are lost with each new generation, and by jacking up the growth of the $N(n)$, we may safely claim that the proportion is between $\alpha/30$ and $\alpha/20$ each time.

We first do the stems. In generation $k, k = n, n + 1, n + 2, \ldots$, there are $N(k)$ different mushrooms in Π_k, but at most $(1 - \alpha/30)^{k-n} N(k)$ of them make it to $\partial \Omega_n^{\natural}$. The integral of $u(|z|)$ along the two sides of a single mushroom is at most

$$2 \int_{R_k}^1 u(t) \, dt \leq \frac{6}{N(k)},$$

where the estimate holds for large $N(k)$. It follows from (8.28) and (8.32) that the integral (8.33) taken only over the stems is bounded by the series

$$C(\alpha) \gamma \sum_{k=n}^{\infty} \left(1 - \frac{\alpha}{30}\right)^{k-n} N(k)^{-v(\alpha)},$$

which converges quite rapidly.

We turn to the hats. In generation k ($k = n, n+1, \ldots$), there are $N(k)$ different mushrooms in Π_k, but at most $(1-\alpha/30)^{k-n} N(k)$, and at least $(1-\alpha/20)^{k-n} N(k)$, of them make it to $\partial \Omega_n^{\natural}$. Since $\exp(h^+) \leq \exp(h) + 1$, an application of Jensen's inequality shows that

$$\frac{1}{|E_k \cap \partial\Omega_n^{\natural}|_s} \int_{E_k \cap \partial\Omega_n^{\natural}} h^+(z)\,ds(z)$$
$$\leq \log\left(1 + \frac{1}{|E_k \cap \partial\Omega_n^{\natural}|_s} \int_{E_k \cap \partial\Omega_n^{\natural}} \exp\left(h(z)\right)\,ds(z)\right),$$

and together with (8.30) and the fact that

$$\left(1 - \frac{\alpha}{20}\right)^{k-n} \leq \frac{|(E_k \cap \partial\Omega_n^{\natural}|_s}{|E_k|_s} \leq \left(1 - \frac{\alpha}{30}\right)^{k-n},$$

we see that

$$\int_{E_k \cap \partial\Omega_n^{\natural}} h^+(z)\,ds(z) \leq \left(1 - \frac{\alpha}{30}\right)^{k-n} |E_k|_s \log\left(1 + \left(1 - \frac{\alpha}{20}\right)^{n-k} M e^{\xi_k}\right)$$

Since $|E_k|_s = \alpha R_k/(4\pi) < \alpha/(4\pi)$, it follows that

$$\sum_{k=n}^{\infty} \int_{E_k \cap \partial\Omega_n^{\natural}} h^+(z)\,ds(z)$$
$$\leq \frac{\alpha}{4\pi} \sum_{k=n}^{\infty} \left(1 - \frac{\alpha}{30}\right)^{k-n} \log\left(1 + \left(1 - \frac{\alpha}{20}\right)^{n-k} M e^{\xi_k}\right), \qquad (8.34)$$

where the right-hand side converges, provided that

$$\sum_{k=1}^{\infty} \left(1 - \frac{\alpha}{30}\right)^k \xi_k < +\infty.$$

By estimates (8.31) and (8.34), the integral (8.33) is controlled on the hats as well. For the choice $\xi_n = 2\log n$, we get more specifically

$$\frac{1}{r} \int_{rT} h^+(z)\,ds(z) \leq \int_{\partial\Omega_n^{\natural}} h^+(\zeta)\,d\omega(0, \zeta, \Omega_n^{\natural})$$
$$\leq C + C'\xi_n = C + 2C' \log n,$$

where $C = C(\alpha, \gamma, M)$ and $C' = C'(\alpha)$ are positive constants.

We summarize what we have done in this section as the following.

LEMMA 8.3 *Let h be a subharmonic function on $\mathbb{D}$ that is bounded above, and write $\xi_n = 2\log n$. Suppose h satisfies, for positive constants γ and M,*

$$h(z) \leq \gamma\, u(|z|), \qquad z \in \mathbb{D},$$

and

$$\sum_{n=1}^{\infty} e^{-\xi_n} \frac{1}{|E_n^{\natural}|_A} \int_{E_n^{\natural}} \exp\left(h(z)\right)\,dA(z) \leq M.$$

Then if the sequence $N(n)$ increases sufficiently rapidly, there are two positive constants $C = C(\alpha, \gamma, M)$ and $C' = C'(\alpha)$ such that

$$\frac{1}{r} \int_{r\mathbb{T}} h^+(z)\, ds(z) \le C + C'\xi_n, \qquad 0 < r < R_n,$$

for $n = 1, 2, 3, \ldots$, where R_n belongs to the interval $r^\sharp_{N(n)} < R_n < r^*_{N(n)}$.

8.5   Finishing the Construction

We first use Lemmas 8.2 and 8.3 to construct an extremally growing harmonic function.

THEOREM 8.4 *Let u and v be as in Lemma 8.2 and $\alpha = \min\{1, 2\beta\}$. Moreover, let $N(n)$ be an increasing sequence of positive integers approaching $+\infty$, and $E^\sharp_\cup$ be the union over $n = 1, 2, 3, \ldots$ of the sets $E^\sharp_n = E^\sharp(N(n), \alpha)$ appearing in (8.6). Then, if the positive integers $N(n)$ increase sufficiently rapidly, there are an increasing function $u_0 : [0, 1) \to [0, +\infty)$, with $u_0(t) = o\big(u(t)\big)$ as $t \to 1$ and $\lim_{t \to 1} u_0(t) = +\infty$, and a real-valued harmonic function f in $\mathbb{D}$ such that*

(1) $-v(|z|) - 1 \le f(z) \le u(|z|) - u_0(|z|) + C$ for all $z \in \mathbb{D}$.

(2) $u(|z|) - u_0(|z|) - C \le f(z)$ for all $z \in E^\sharp_\cup$.

(3) $\displaystyle\int_{\mathbb{D}} \exp\big(f(z)\big)\, dA(z) \le C$.

(4) $\displaystyle\limsup_{r \to 1^-} \left(\frac{1}{r} \int_{r\mathbb{T}} f^-(z)\, ds(z) - \lambda\, u_0(r) \right) = +\infty$.

Here λ is any positive constant and $f^-(z) = \max\{-f(z), 0\}$. Furthermore, if h is any subharmonic function on $\mathbb{D}$ that is bounded above and satisfies

$$\int_{\mathbb{D}} \exp\big(f(z) + h(z)\big)\, dA(z) \le 1,$$

then

$$\frac{1}{r} \int_{r\mathbb{T}} h^+(z)\, ds(z) \le C\big(1 + u_0(r)\big), \qquad 0 < r < 1.$$

Above, $C = C(\alpha)$ stands for a positive constant.

Proof. We produce iteratively functions $f_n \in h^\infty(\mathbb{D})$, radii r_n, positive constants C_n, and compact subsets $E^{\sharp,\cup}_n$, as follows. Along the way, we will define increasing functions $u_{0,n} : [0, 1) \to [0, +\infty)$ that tend to the desired function u_0 as $n \to +\infty$. We start with $f_0 = 0$, $r_0 = \frac{1}{2}$, $C_0 = 1$, and $E^{\sharp,\cup}_0 = \emptyset$. We also set $u_{0,0}(t) = 0$ on $[0, 1)$. In general, the radius r_n will be chosen such that $r_{n-1} < r_n < 1$, and such that the set $E_{n-1}^{\sharp,\cup}$ is contained in the disk $|z| < r_n$. Moreover, as $n \to +\infty$, we want $r_n \to 1$. Suppose we have f_{n-1} and $E_{n-1}^{\sharp,\cup}$ satisfying

$$-v(|z|) - 1 + 2^{-n+1} \leq f_{n-1}(z) \leq u(|z|) - u_{0,n-1}(|z|) + C - 2^{-n+1}$$

for $z \in \mathbb{D}$,

$$u(|z|) - u_{0,n-1}(|z|) - C + 2^{-n+1} \leq f(z)$$

for $z \in E_{n-1}^{\sharp,\cup}$, and

$$\int_{\mathbb{D}} \exp\left(f_{n-1}(z)\right) dA(z) \leq C_{n-1}.$$

At this point we choose r_n as prescribed above. If $N(n)$ is large enough, Lemma 8.2 – with $\varepsilon = 2^{-n}$, $\rho = r_n$, and $\xi = \xi_n = 2\log n$ – will then deliver a radius $\varrho = \varrho_n$ with $r_n < \varrho_n < 1$, a compact set $E_n^{\sharp} = E^{\sharp}(N(n), \alpha)$ contained in the ring $\varrho_n < |z| < 1$, and a function $g = f_n \in h^{\infty}(\mathbb{D})$ such that (use a slightly different constant than in the lemma)

$$-v(|z|) - 1 + 2^{-n} \leq f_n(z) \leq u(|z|) - u_{0,n-1}(|z|) + C - 2^{-n}$$

for $|z| < \varrho_n$,

$$-v(|z|) \leq f_n(z) \leq u(|z|) - \xi_n - 1 + C$$

for $\varrho_n \leq |z| < 1$,

$$u(|z|) - u_{0,n-1}(|z|) - C + 2^{-n} \leq f_n(z)$$

for $z \in E_{n-1}^{\sharp,\cup}$,

$$u(|z|) - \xi_n - C + 1 \leq f_n(z)$$

for $z \in E_n^{\sharp}$, and

$$\int_{\mathbb{D}} \exp\left(f_n(z)\right) dA(z) \leq C_n.$$

Here

$$C_n = C e^{-\xi_n} + \exp(2^{-n}) C_{n-1}.$$

Declare $u_{0,n}(t) = u_{0,n-1}(t)$ for $0 \leq t < \varrho_n$, and $u_{0,n}(t) = \xi_n = 2\log n$ for $\varrho_n \leq t < 1$. It is readily checked that this defines an increasing function. Also, put

$$E_n^{\sharp,\cup} = E_{n-1}^{\sharp,\cup} \cup E_n^{\sharp}.$$

Part of the above estimates then simplifies to

$$-v(|z|) - 1 + 2^{-n} \leq f_n(z) \leq u(|z|) - u_{0,n}(|z|) + C - 2^{-n}$$

for $z \in \mathbb{D}$, and

$$u(|z|) - u_{0,n-1}(|z|) - C + 2^{-n} \leq f_n(z)$$

for $z \in E_n^{\sharp,\cup}$. As $n \to +\infty$, the functions $f_n \in h^\infty(\mathbb{D})$ converge, uniformly on compact subsets of $\mathbb{D}$, to a harmonic function f in $\mathbb{D}$, and the functions $u_{0,n}$ converge to a function u_0. Since $\sum_n e^{-\xi_n} = \sum_n n^{-2}$ converges, the constants C_n converge to a limit C_∞ as well. Thus, we obtain

$$-v(|z|) - 1 \le f(z) \le u(|z|) - u_0(|z|) + C$$

for $z \in \mathbb{D}$,

$$u(|z|) - u_0(|z|) - C \le f(z)$$

for $z \in E_\cup^\sharp = \cup_{j=1}^{+\infty} E_j^\sharp$, and

$$\int_\mathbb{D} \exp\big(f(z)\big) \, dA(z) \le C_\infty.$$

For rapidly increasing $N(n)$, the radii ϱ_n, being contained between $r_{N(n-1)}^*$ and $r_{N(n)}^*$, tend to 1 very rapidly in n, so that we can make $u_0(t)$ go to $+\infty$ as slowly as we like as $t \to 1$. In particular, we can get $u_0(t) = o\big(u(t)\big)$ as $t \to 1$.

We now turn to the assertion that

$$\limsup_{r \to 1} \left(\frac{1}{r} \int_{r\mathbb{T}} f^-(z) \, ds(z) - \lambda \, u_0(r) \right) = +\infty.$$

Since f is big and positive on $E_\cup^\sharp$, the integrals

$$\frac{1}{r} \int_{r\mathbb{T}} f^+(z) \, ds(z)$$

are correspondingly big for $r = r_{N(n)}^*$, and the order of magnitude is at least a positive constant times $u(r_{N(n)}^*)$. By the mean value theorem for harmonic functions, the integrals where f^+ is replaced by f^- are of the same magnitude, whence the assertion follows.

Finally, we look at the part of the assertion involving the function h. Since f has the bound from below, the integrability of $\exp(f + h)$ on the unit disk forces the subharmonic function to satisfy (8.28), for some $\gamma = \gamma(\alpha, \beta)$, by the mean-value property of subharmonic functions on disks. In fact, if we subtract a suitable absolute constant from h, we can get $\gamma = \beta + 2$. Moreover, by the way the function u_0 was defined in terms of ξ_n, and the control from below on f on $E_\cup^\sharp$, (8.29) holds for some $M = M(\alpha)$. So, we can apply Lemma 8.3 to get the desired estimate, by replacing the ξ_n with the appropriate expression in terms of u_0; see Exercise 4. The proof is complete. ∎

As a consequence of Theorem 8.4, we obtain an extremally growing analytic function in $\mathbb{D}$, which will enable us to construct noncyclic invertible functions in the Bergman spaces.

COROLLARY 8.5 *For any $\beta > 0$, there are an increasing function $u_0 : [0, 1) \to [0, +\infty)$, with $u_0(t) = o\big(-\log(1-t)\big)$ but $u_0(t) \to +\infty$ as $t \to 1^-$, and a function F holomorphic on $\mathbb{D}$, such that:*

(1) $\displaystyle\int_{\mathbb{D}} |F(z)|\, dA(z) < +\infty.$

(2) $(1 - |z|)^{\beta} \le |F(z)| \le C\,(1 - |z|)^{-1} \exp\left(-u_0(|z|)\right)$ *for all $z \in \mathbb{D}$ and some positive constant C.*

(3) $\displaystyle\limsup_{r\to 1}\left((2\pi)^{-1}\int_{-\pi}^{\pi} \log^- |F(re^{i\theta})|\, d\theta - \lambda\, u_0(r)\right) = +\infty$ *for all $\lambda > 0$.*

(4) *If h is any subharmonic function on $\mathbb{D}$ that is bounded above and satisfies*

$$\int_{\mathbb{D}} |F(z)| \exp(h(z))\, dA(z) = M < +\infty,$$

then

$$\frac{1}{2\pi}\int_{-\pi}^{\pi} h^+(re^{i\theta})\, d\theta \le \log^+ M + C\, u_0(r),$$

for $0 < r < 1$ and some constant C that does not depend on h and M.

We can now prove the main result of the chapter. Recall that $H(\mathbb{D})$ is the space of all analytic functions in $\mathbb{D}$, with the usual topology of uniform convergence on compact sets.

THEOREM 8.6 *For any positive exponents p and q, there exists a function f in A^p such that f is not cyclic in A^p but $1/f$ belongs to A^q.*

Proof. Let F be as in Corollary 8.5, with $0 < \beta < p/q$. Then the function $f = F^{1/p}$ is in A^p, and $1/f$ is in A^q, because $(1 - |z|)^{\beta/p} \le |f(z)|$. We need to show that f is noncyclic in A^p. Let g_k be a sequence of functions in H^∞ such that fg_k converges in norm in A^p. By property (4) of Corollary 8.5 applied to the functions $\log |g_k|$, we have

$$\frac{1}{2\pi}\int_{-\pi}^{\pi} \log^+ |g_k(re^{i\theta})|\, d\theta \le C\left(1 + u_0(r)\right), \qquad 0 < r < 1,$$

for a positive constant C. If $g_k \to g$ in $H(\mathbb{D})$, then

$$\limsup_{r\to 1} \frac{1}{2\pi}\int_{-\pi}^{\pi} \left(\log^- |f(re^{i\theta})| - \log^+ |g(re^{i\theta})|\right) d\theta = +\infty$$

by property (3) of Corollary 8.5. This, however, cannot be true if the limit fg is the constant function 1, for then $\log^- |f| = \log^+ |g|$. The proof is complete. ∎

8.6 Two Applications

As consequences of Theorem 8.6 we discuss two problems related to "inner" and "outer" functions in the context of Bergman spaces.

First, recall from Theorems 7.3 and 7.12 that if a function $f \in A^p$ belongs to a slightly smaller space A^q $(q > p)$, then f is cyclic in A^p if and only if f has no zeros in the disk $\mathbb{D}$ and carries no κ-singular measure on the circle $\mathbb{T}$. It is natural to ask whether the extra assumption that $f \in A^q$ with $q > p$ is really needed. The following corollary answers this question.

COROLLARY 8.7 *For any* $0 < p < +\infty$, *there exists a noncyclic function* $f \in A^p$ *such that* f *has no zeros in* $\mathbb{D}$ *and carries no* κ-*singular measure on* $\mathbb{T}$.

Proof. Let f be the function from Theorem 8.6; it does not matter what the value of q is. Since $1/f$ belongs to $A^q \subset \mathcal{A}^{-\infty}$, we see that f is invertible in $\mathcal{A}^{-\infty}$, so that f is cyclic in $\mathcal{A}^{-\infty}$ (see Exercise 6). By Theorem 7.12, the function f is zero-free in $\mathbb{D}$ and carries no κ-singular measure on $\mathbb{T}$. ∎

Next, recall from Theorem 3.34 that every function $f \in A^p$ admits an "inner-outer" factorization, $f = GF$, where G is A^p-inner and F is cyclic in A^p. A natural follow-up question is whether this factorization is unique. The next corollary answers this question in the negative.

COROLLARY 8.8 *For any* $0 < p < +\infty$, *there exists a function* $g \in A^p$ *such that* g *admits two different "inner-outer" factorizations.*

Proof. Let f be the function from Theorem 8.6. It does not matter what q is here. According to Theorems 3.33 and 3.34, the invariant subspace of A^p generated by f yields a unique extremal function $G \in A^p$, and $f/G \in A^p$ is cyclic. It follows that the function $1/G = (f/G)(1/f)$ belongs to $\mathcal{A}^{-\infty}$. Applying Theorem 7.3, we see that if ε is a sufficiently small positive number, then both $G^{-\varepsilon}$ and $G^{1-\varepsilon}$ are cyclic in A^p. Let $g = G^{1-\varepsilon}$. Then

$$g = 1 \cdot G^{1-\varepsilon} = G \cdot G^{-\varepsilon}$$

are two different "inner-outer" factorizations of g in A^p. ∎

8.7 Notes

The results of this chapter, along with their proofs, are taken from the paper [27] by Borichev and Hedenmalm.

Considering the Hardy space situation, where the "largest" functions are outer, and hence cyclic, it is perhaps surprising that "largeness" can imply non-cyclicity. But in spaces determined by growth, say the separable spaces $\mathcal{A}_0^{-\alpha}$, largeness can imply non-cyclicity, as is seen from the following general idea. If a function $f \in \mathcal{A}_0^{-\alpha}$ grows almost as fast as allowed on a sufficiently massive set, and for polynomials q_n we have that $f q_n$ is norm bounded in $\mathcal{A}_0^{-\alpha}$ uniformly in n, then the polynomials q_n will be very much controlled on the massive set, and this implies that we can estimate q_n uniformly in n by a radial function which increases very slowly in toward the boundary in $\mathbb{D}$. Then $f q_n$ cannot tend to the constant function

1, and consequently, f is non-cyclic. The Bergman spaces A^p share a lot of the characteristics of the growth spaces, and the construction here shows that this idea for growth spaces carries over to the Bergman space setting.

In view of Theorem 8.6, it is natural to ask if there exists at all a decreasing radial function $\Phi(z) = \Phi(|z|)$, with $\Phi(r) \to 0$ as $r \to 1^-$, such that the conditions $f \in A^p$ and $\Phi(|z|) \le |f(z)|$ for $z \in \mathbb{D}$ would imply that f is cyclic. In [25], Borichev shows that the answer is affirmative. More specifically, the function

$$\Phi(|z|) = \delta \exp\left[-\left(\log \frac{1}{1-|z|}\right)^{1/2+\varepsilon}\right], \qquad z \in \mathbb{D},$$

has the desired property when δ and ε are positive.

8.8   Exercises and Further Results

1. Fill in the details of the verification of (8.11).

2. Show how to use estimates of harmonic measure so as to obtain the estimate (8.13) from the corresponding estimate on the set $r_N^* E(N)$. Hint: the argument is simplified if one introduces the complex variable $w = z^N$ and recalls that $\phi_N(z) = \Phi_\alpha(z^N) = \Phi_\alpha(w)$.

3. Check that (8.20) holds for sufficiently large N.

4. Fill in the details of how the mushroom forest Lemma 8.3 is applied at the end of the proof of Theorem 8.4.

5. Fix $0 < p < +\infty$. An A^p-inner function G has the estimate

$$|G(z)| \le \frac{1}{(1-|z|^2)^{1/p}}, \qquad z \in \mathbb{D},$$

which we derived in Chapter 3 from the fact that G is a contractive multiplier $H^p \to A^p$. The latter means that G has the stronger property that $|G|^p dA$ is a Carleson measure. Check that the function f in Theorem 8.6 has an estimate of the same kind as the extremal functions, but with a constant. Then show that for the same f, we can modify the construction so that $|f|^p dA$ is a Carleson measure. Hint: a Carleson measure μ is a finite positive Borel measure on $\mathbb{D}$ with $\mu(Q) \le C\,\ell(Q)$, for some constant C, where Q is a Carleson square and $\ell(Q)$ is the side length of the square. It is therefore enough to get a radial majorant to f which is in $L^p(\mathbb{D}, dA)$. See [27].

6. Show that if both f and $1/f$ are in $\mathcal{A}^{-\infty}$, then f is cyclic in $\mathcal{A}^{-\infty}$.

7. Show that for every $0 < p < +\infty$, there exists an A^p-inner function G such that $\|f\|_p \le \|fG\|_p$ holds for all $f \in H^\infty$ but not all $f \in A^p$.

8. Show that for every $0 < p < +\infty$, there exists an A^p-inner function G such that $(GA^p) \cap A^p$ is not the same as the invariant subspace of A^p generated by G.

9. Suppose f and g are cyclic vectors in A^2 such that fg belongs to A^2. Is fg necessarily cyclic in A^2?

10. Generalize the constructions in this chapter to the setting of the spaces A^p_α, with $0 < p < +\infty$ and $-1 < \alpha < +\infty$.

11. We now mention an open problem. Let the Bergman-Dirichlet space $BD^2(\mathbb{T})$ consist of formal Laurent series

$$f(z) = \sum_{n=-\infty}^{+\infty} a_n \, z^n,$$

where

$$\|f\|^2_{BD^2} = \sum_{n=0}^{+\infty} \frac{|a_n|^2}{n+1} + \sum_{n=1}^{+\infty} (n+1)\,|a_{-n}|^2 < +\infty.$$

The shift operator $\mathbf{S}f(z) = z\,f(z)$ acts boundedly on this space, and so does its inverse $\mathbf{S}^{-1}$. Closed subspaces invariant under both $\mathbf{S}$ and $\mathbf{S}^{-1}$ are called *bilaterally invariant*. Does there exist a bilaterally invariant subspace J in $BD^2(\mathbb{T})$ such that the intersection $J \cap A^2$, which is an invariant subspace in A^2, is nontrivial in A^2? In radially weighted Bergman spaces, with weights that drop down to zero very quickly near $\mathbb{T}$, the corresponding question has an affirmative answer [28]. To throw further light on the issue, we mention that the invariant subspace $I = J \cap A^2$ then has index 1 and no common zeros in $\mathbb{D}$, and hence is generated by a zero-free A^2-inner function φ. The function φ should also be cyclic in $\mathcal{A}^{-\infty}$, and hence similar to the functions constructed in this chapter (if it exists).

9
Logarithmically Subharmonic Weights

In this chapter, we study weighted Bergman spaces for weights that are logarithmically subharmonic and reproduce for the origin; the latter means that if we integrate a bounded harmonic function against the weight over $\mathbb{D}$, we obtain the value of the harmonic function at the origin. Two important examples of such weights are $\omega(z) = |G(z)|^p$, where G is an A^p-inner function, and

$$\omega(z) = (\alpha + 1)(1 - |z|^2)^\alpha, \qquad z \in \mathbb{D},$$

where $-1 < \alpha \leq 0$. Not only are these weights interesting in themselves, they also have nice applications to the study of unweighted Bergman spaces. The main result of the chapter is that the weighted biharmonic Green function Γ_ω is positive, provided that the weight is logarithmically subharmonic and reproduces for the origin. As a consequence, we will prove the domination relation $\|G_A f\|_p \leq \|G_B f\|_p$, where f is any function in A^p, and G_A and G_B are contractive zero divisors in A^p with $A \subset B$.

9.1 Reproducing Kernels

Suppose E is any set and $\mathcal{H}$ is a Hilbert space of complex-valued functions on E such that the point evaluation at each point in E is a bounded linear functional on $\mathcal{H}$. Then by the Riesz representation theorem, for each $y \in E$ there exists a function $K_y = K(\cdot, y)$ in $\mathcal{H}$ such that

$$f(y) = \langle f, K_y \rangle, \qquad f \in \mathcal{H}.$$

The function $K(x, y)$, with $(x, y) \in E \times E$, is called the *reproducing kernel* of $\mathcal{H}$.

It is natural to ask what kind of functions $K : E \times E \to \mathbb{C}$ arise as reproducing kernels of Hilbert spaces of functions on E. We will need the following classical characterization of reproducing kernels. We recall that a square matrix $\{A_{j,k}\}_{j,k=1}^N$ is said to be *positive definite* if

$$\sum_{j,k=1}^N A_{j,k}\, w_j\, \overline{w}_k \geq 0, \qquad \{w_j\}_{j=1}^N \in \mathbb{C}^N.$$

If we require strict inequality for all nonzero vectors in $\mathbb{C}^N$, the matrix is called *strictly positive definite*.

PROPOSITION 9.1 *A function $K : E \times E \to \mathbb{C}$ is the reproducing kernel of a Hilbert space of functions on E if and only if K is positive definite, that is, for any finite subset $\{x_1, \dots, x_N\}$ of E the matrix $\{K(x_i, x_j)\}_{i,j=1}^N$ is positive definite.*

Proof. First assume that K is the reproducing kernel of a Hilbert space $\mathcal{H}$ of functions on E. Given any finite set $\{x_1, \dots, x_N\}$ of E and any finite subset $\{c_1, \dots, c_N\}$ of $\mathbb{C}$ we consider the function

$$f(x) = c_1 K(x, x_1) + \cdots + c_N K(x, x_N), \qquad x \in E.$$

Then $f \in \mathcal{H}$, and the reproducing property of K yields

$$\sum_{j,k=1}^N K(x_j, x_k)\, c_j \overline{c}_k = \langle f, f \rangle \geq 0.$$

This shows that the matrix $\{K(x_j, x_k)\}_{j,k=1}^N$ is positive definite.

Next assume that a function $K : E \times E \to \mathbb{C}$ is positive definite. Let $\mathcal{H}_0$ be the vector space of functions of the form

$$c_1 K(x, x_1) + \cdots + c_N K(x, x_N), \qquad x \in E,$$

where N is any positive integer, $c_1, \dots, c_N$ are arbitrary complex constants, and $x_1, \dots, x_N$ are arbitrary points in E. If

$$f(x) = c_1 K(x, x_1) + \cdots + c_N K(x, x_N)$$

and

$$g(x) = d_1 K(x, y_1) + \cdots + d_M K(x, y_M)$$

are two functions in $\mathcal{H}_0$, we define

$$\langle f, g \rangle = \sum_{j=1}^N \sum_{k=1}^M c_j \overline{d}_k\, K(x_j, y_k).$$

That K is positive definite implies that $\langle \cdot, \cdot \rangle$ defined above is an inner product on $\mathcal{H}_0$, because it is easy to check that any element in $\mathcal{H}_0$ of norm 0 assumes the value 0 at all points. Let $\mathcal{H}$ be the completion of $\mathcal{H}_0$ with respect to this inner product. Then every point evaluation is a bounded linear functional on $\mathcal{H}$, and $\mathcal{H}$ is a space of functions on E, because any element of the completion which vanishes as a function on E must be orthogonal to all the spanning vectors, as the latter are correspond to point evaluations. The kernel K is the reproducing kernel of the Hilbert space of functions $\mathcal{H}$.   ∎

Let K be the reproducing kernel of a Hilbert space of functions on E. For any two points x and y in E, the matrix

$$\begin{pmatrix} K(x, x) & K(x, y) \\ K(y, x) & K(y, y) \end{pmatrix}$$

is positive definite. In particular, the above matrix is self-adjoint, so that

$$K(y, x) = \overline{K(x, y)}.$$

Moreover, a positive definite matrix has a nonnegative determinant, so that

$$|K(x, y)|^2 \leq K(x, x)K(y, y), \qquad x, y \in E.$$

This will be referred to as the Cauchy-Schwarz inequality for reproducing kernels. If $\mathcal{H}$ is a separable Hilbert space, so that it has a countable orthonormal basis $\{e_n\}_{n=1}^{+\infty}$, we can represent the reproducing kernel as a series:

$$K(x, y) = \sum_{n=1}^{+\infty} e_n(x)\overline{e_n(y)}, \qquad (x, y) \in E \times E.$$

It is worth noting that it does not matter which particular orthonormal basis is used.

We will be concerned with the case $E = \mathbb{D}$. Also, we will be interested in only Hilbert spaces of analytic functions in $\mathbb{D}$ on which the point evaluations are bounded linear functionals. In terms of kernel functions, we will study functions $K : \mathbb{D} \times \mathbb{D} \to \mathbb{C}$ such that $K(z, w)$ is analytic in z (and hence conjugate analytic in w).

More specifically, we will be concerned with weighted Bergman spaces. We first specify the kind of weights to be used. Therefore, throughout the chapter, we use ω to denote a function from $\mathbb{D}$ to $[0, +\infty)$ such that

(a) ω is logarithmically subharmonic, that is, $\log \omega$ is subharmonic on $\mathbb{D}$.

(b) ω is reproducing for the origin, that is,

$$p(0) = \int_{\mathbb{D}} p(z)\omega(z) \, dA(z)$$

for every polynomial p.

Note that if we take $p = 1$, the constant polynomial, in condition (b) above, then we get

$$\int_{\mathbb{D}} \omega(z) \, dA(z) = 1.$$

For a holomorphic function f on the unit disk, we consider the norm

$$\|f\|_\omega = \left(\int_{\mathbb{D}} |f(z)|^2 \omega(z) \, dA(z) \right)^{1/2},$$

and let $B^2(\omega)$ consist of all such f for which the above norm is finite. For each $0 < p < +\infty$, the space $B^p(\omega)$ is defined in a similar way.

PROPOSITION 9.2 *The point evaluations on the space $B^2(\omega)$ are uniformly bounded on compact subsets of $\mathbb{D}$.*

Proof. Fix an interior point $z_0 \in \mathbb{D}$, and for $0 < r < 1 - |z_0|$, let $\mathbb{D}(z_0, r)$ be the Euclidean disk

$$\mathbb{D}(z_0, r) = \{z \in \mathbb{C} : |z - z_0| \le r\}.$$

For a holomorphic function f on $\mathbb{D}$, the function $|f|^p \omega$ is subharmonic, and therefore, by the sub-mean value property,

$$|f(z_0)|^2 \omega(z_0) \le \frac{1}{r^2} \int_{\mathbb{D}(z_0, r)} |f(z)|^2 \omega(z) \, dA(z) \le r^{-2} \|f\|_\omega^2,$$

so that

$$|f(z_0)|^2 \le \frac{1}{r^2 \omega(z_0)} \|f\|_\omega^2.$$

Taking logarithms, we obtain

$$\log |f(z_0)| \le \log \frac{\|f\|_\omega}{r} + \frac{1}{2} \log \frac{1}{\omega(z_0)}.$$

The point $z_0 \in \mathbb{D}$ is arbitrary, the left-hand side is subharmonic, and the right-hand side is superharmonic. So we obtain the estimate

$$\log |f(z_0)| \le \log \frac{\|f\|_\omega}{r} + \frac{1}{2r} \int_{\partial \mathbb{D}(z_0, r)} \log \frac{1}{\omega(z)} \, ds(z),$$

where $ds(z) = |dz|/(2\pi)$; the subharmonicity of $\log \omega$ implies that it is integrable on circles in $\mathbb{D}$ such as $\partial \mathbb{D}(z_0, r)$. Introducing the Poisson kernel in these calculations allows us to get a uniform estimate on compact subsets. ■

Note that the above proof does not use the assumption that ω is reproducing for the origin. Also, the proof works for all $B^p(\omega)$ with $0 < p < +\infty$.

Let $A^2(\omega)$ be the closure of the polynomials in $B^2(\omega)$. Although for most weights that come to mind these two spaces coincide, there are weights for which $A^2(\omega)$ is strictly smaller than $B^2(\omega)$; see Exercise 13. Being a closed subspace of $B^2(\omega)$, the space $A^2(\omega)$ is a Hilbert space of analytic functions on $\mathbb{D}$ with locally uniformly bounded point evaluations.

The reproducing kernel for $A^2(\omega)$ will be denoted by K_ω. Since ω reproduces for the origin, we have $K_\omega(z, 0) = K_\omega(0, w) = 1$ for all z and w in $\mathbb{D}$. Before we can obtain deeper properties of K_ω, we need to know how the operator of multiplication by z acts on $A^2(\omega)$.

Let S denote the operator of multiplication by z on $A^2(\omega)$. Thus $Sf(z) = zf(z)$, $z \in \mathbb{D}$, for all $f \in A^2(\omega)$. The operator S is obviously bounded on $A^2(\omega)$ with $\|S\| = 1$.

PROPOSITION 9.3 *For any two functions* $f, g \in A^2(\omega)$, *we have the inequality*

$$\|Sf + g\|_\omega^2 \le 2\left(\|f\|_\omega^2 + \|Sg\|_\omega^2\right).$$

Proof. It is enough to obtain the inequality when f and g are polynomials. We prove the result under the additional assumption that the weight ω is C^∞ up to the boundary. The general case then follows from a simple application of Theorem 9.14 (whose proof will be independent of the present result).

For any $\lambda \in \mathbb{C} \setminus \{0\}$, we have

$$
\begin{aligned}
0 \;\le\; & \Delta_z\!\left(\left|g(z) - \lambda^{-1}z^3 f(z)\right|^2 \omega(z)\right) \\
=\; & \Delta_z\!\left(|g(z)|^2\,\omega(z)\right) - 2\,\mathrm{Re}\left(\bar\lambda^{-1}\Delta_z\!\left(g(z)\bar z^3\,\bar f(z)\,\omega(z)\right)\right) \\
& + |\lambda|^{-2}\,\Delta_z\!\left(|z^3 f(z)|^2\,\omega(z)\right),
\end{aligned}
$$

for $z \in \mathbb{D}$, where the first inequality holds because the product of a logarithmically subharmonic function and the modulus square of a holomorphic function is again logarithmically subharmonic, and in particular, subharmonic. Setting $\lambda = z^2$, we obtain

$$
\begin{aligned}
0 \;\le\; & \Delta_z\!\left(|g(z)|^2\,\omega(z)\right) - 2\,\mathrm{Re}\left(\bar z^{-2}\Delta_z\!\left(g(z)\bar z^3\,\bar f(z)\,\omega(z)\right)\right) \\
& + |z|^{-4}\,\Delta_z\!\left(|z^3 f(z)|^2\,\omega(z)\right).
\end{aligned}
$$

None of the three terms on the right-hand side has any singularity at the origin, even though it may seem so to the inexperienced eye. We are going to integrate the above inequality, term by term, against the positive measure $(1 - |z|^2)^2\, dA(z)$.

By Green's formula,

$$\int_{\mathbb{D}} (1 - |z|^2)^2\, \Delta_z\!\left(|g(z)|^2\,\omega(z)\right) dA(z) = \int_{\mathbb{D}} \left(4|z|^2 - 2\right) |g(z)|^2\,\omega(z)\, dA(z).$$

A slightly more sophisticated exercise involving Green's formula shows that if $\mathbb{D}(0, \varepsilon)$ stands for a small circular disk about the origin of radius ε, then

$$
\begin{aligned}
& \int_{\mathbb{D}\setminus\mathbb{D}(0,\varepsilon)} (1 - |z|^2)^2 \left(\bar z^{-2}\,\Delta_z\!\left(g(z)\bar z^3\,\bar f(z)\,\omega(z)\right)\right) dA(z) \\
& = 2\int_{\mathbb{D}\setminus\mathbb{D}(0,\varepsilon)} \bar z\, g(z)\,\bar f(z)\,\omega(z)\, dA(z) + O(\varepsilon),
\end{aligned}
$$

where the normal derivative is taken inward with respect to the disk $\mathbb{D}(0, \varepsilon)$. We apply Green's formula a third time to obtain

$$\int_{\mathbb{D}\setminus\mathbb{D}(0,\varepsilon)} (1 - |z|^2)^2 \left(|z|^{-4} \Delta_z \left(|z^3 f(z)|^2 \, \omega(z) \right) \right) dA(z)$$

$$= \int_{\mathbb{D}\setminus\mathbb{D}(0,\varepsilon)} (4 - 2 |z|^2) \, |f(z)|^2 \, \omega(z) \, dA(z) + O(\varepsilon).$$

Let $\varepsilon \to 0$. We obtain

$$0 \leq \int_{\mathbb{D}} (4 |z|^2 - 2) \, |g(z)|^2 \omega(z) \, dA(z)$$

$$- 4 \, \mathrm{Re} \int_{\mathbb{D}} \bar{z} \, g(z) \, \bar{f}(z) \, \omega(z) \, dA(z)$$

$$+ \int_{\mathbb{D}} (4 - 2|z|^2) \, |f(z)|^2 \omega(z) \, dA(z),$$

which expresses the inequality we are looking for in expanded form. ∎

In addition to the forward shift S, we will also need the *backward shift* T defined by

$$Tf(z) = \frac{f(z) - f(0)}{z}, \qquad z \in \mathbb{D},$$

which we think of as acting on $A^2(\omega)$. It is clear that TS is the identity operator and ST is given by

$$STf(z) = f(z) - f(0), \qquad z \in \mathbb{D}, f \in A^2(\omega).$$

The shift S is a contraction on $A^2(\omega)$, and so is ST, because the reproducing property of the weight ω leads to the norm identity

$$\|f\|_\omega^2 = \|f - f(0)\|_\omega^2 + |f(0)|^2, \qquad f \in A^2(\omega).$$

What we shall actually use later is the following variant of Proposition 9.3.

COROLLARY 9.4 *For any two functions $f, g \in A^2(\omega)$, we have the inequality*

$$\|Sf + Tg\|_\omega^2 \leq 2 \left(\|f\|_\omega^2 + \|g\|_\omega^2 \right).$$

Proof. Just replace g by Tg in the theorem and use the fact that ST is a contraction. ∎

The following structure result is key to our further investigations.

THEOREM 9.5 *The function L_ω defined by*

$$K_\omega(z, \zeta) = \frac{1 - z\bar{\zeta} L_\omega(z, \zeta)}{(1 - z\bar{\zeta})^2}$$

is the reproducing kernel of a Hilbert space of analytic functions on $\mathbb{D}$.

Proof. Solving for L_ω, we find that

$$
\begin{aligned}
L_\omega(z, \zeta) &= \frac{1}{z\zeta}\left(1 - (1 - z\bar\zeta)^2 K_\omega(z, \zeta)\right) \\
&= \frac{1 - K_\omega(z, \zeta)}{z\bar\zeta} + 2 K_\omega(z, \zeta) - z\bar\zeta\, K_\omega(z, \zeta).
\end{aligned}
$$

Since

$$
K_\omega(z, 0) = K_\omega(0, \zeta) = 1, \qquad (z, \zeta) \in \mathbb{D} \times \mathbb{D},
$$

the function $L_\omega(z, \zeta)$ is analytic in z (and conjugate analytic in ζ). According to Proposition 9.1, we need only show that L_ω is positive definite. In other words, we need to show that for any finite subset $\{z_1, z_2, \dots, z_N\}$ of $\mathbb{D}$, the inequality

$$
\sum_{j,k=1}^{N} L_\omega(z_j, z_k)\, w_j\, \bar w_k \geq 0 \tag{9.1}
$$

holds for all sequences $\{w_j\}_{j=1}^{N} \in \mathbb{C}^N$. From the reproducing property of the kernel K_ω, we easily deduce that

$$
L_\omega(z_j, z_k) = \int_{\mathbb{D}} \left(\int_{\mathbb{D}} L_\omega(z, \zeta)\, K_\omega(z_j, z)\, K_\omega(\zeta, z_k)\, \omega(\zeta)\, dA(\zeta) \right) \omega(z)\, dA(z).
$$

If we define a function f by

$$
f(z) = \sum_{j=1}^{N} \bar w_j\, K_\omega(z, z_j), \qquad z \in \mathbb{D},
$$

then (9.1) is equivalent to

$$
\int_{\mathbb{D}} \left(\int_{\mathbb{D}} L_\omega(z, \zeta)\, \bar f(z)\, f(\zeta)\, \omega(\zeta)\, dA(\zeta) \right) \omega(z)\, dA(z) \geq 0. \tag{9.2}
$$

We proceed to show that (9.2) holds for all $f \in A^2(\omega)$.

Let S^* and T^* denote, respectively, the adjoint of operators S and T on the Hilbert space $A^2(\omega)$. Using the formula for L_ω at the beginning of the proof and the fact that $K_\omega(z, 0) = K_\omega(0, w) = 1$ for all z and w in $\mathbb{D}$, we obtain

$$
\begin{aligned}
\int_{\mathbb{D}} & L_\omega(z, \zeta)\, f(\zeta)\, \omega(\zeta)\, dA(\zeta) \\
&= -\frac{1}{z}\, \langle f, T K_\omega(\cdot, z)\rangle_\omega + 2\,\langle f, K_\omega(\cdot, z)\rangle_\omega - z\,\langle f, S K_\omega(\cdot, z)\rangle_\omega \\
&= -\frac{1}{z}\, \langle T^* f, K_\omega(\cdot, z)\rangle_\omega + 2\,\langle f, K_\omega(\cdot, z)\rangle_\omega - z\,\langle S^* f, K_\omega(\cdot, z)\rangle_\omega \\
&= -\frac{1}{z}\, T^* f(z) + 2\, f(z) - z\, S^* f(z), \qquad z \in \mathbb{D},
\end{aligned}
$$

for every $f \in A^2(\omega)$. Since $T^* f(0) = 0$, which is a consequence of the reproducing property of the weight ω, we can condense the above to

$$\int_{\mathbb{D}} L_\omega(z, \zeta) f(\zeta) \omega(\zeta) \, dA(\zeta) = -TT^* f(z) + 2 f(z) - S S^* f(z), \qquad z \in \mathbb{D}.$$

Integrating now with respect to the z variable, we arrive at

$$\int_{\mathbb{D}} \left(\int_{\mathbb{D}} L_\omega(z, \zeta) \bar{f}(z) f(\zeta) \omega(\zeta) \, dA(\zeta) \right) \omega(z) \, dA(z)$$
$$= -\langle TT^* f, f \rangle_\omega + 2 \langle f, f \rangle_\omega - \langle SS^* f, f \rangle_\omega$$
$$= -\|T^* f\|_\omega^2 + 2 \|f\|_\omega^2 - \|S^* f\|_\omega^2,$$

which shows that what we are trying to prove is the operator inequality

$$2 - TT^* - S S^* \geq 0. \tag{9.3}$$

To prove (9.3), we consider the operator

$$R : A^2(\omega) \oplus A^2(\omega) \to A^2(\omega)$$

defined by

$$R(f, g) = 2^{-\frac{1}{2}} \left(Sf + Tg \right), \qquad f, g \in A^2(\omega).$$

An easy calculation shows that the adjoint of R is given by

$$R^*(h) = 2^{-\frac{1}{2}} (S^* h, T^* h).$$

It follows that

$$R R^* f = \frac{1}{2} \left(SS^* f + TT^* f \right), \qquad f \in A^2(\omega).$$

Thus the operator inequality (9.3) can be rewritten as $R R^* \leq 1$. According to Corollary 9.4, the operator R is a contraction, which is equivalent to $RR^* \leq 1$, completing the proof of the theorem. ∎

COROLLARY 9.6 $|L_\omega(z, \zeta)| < 1$ for all z and ζ in $\mathbb{D}$.

Proof. Since $K_\omega(z, z) \geq 0$ and $L_\omega(z, z) \geq 0$ for all $z \in \mathbb{D}$, the identity

$$K_\omega(z, z) = \frac{1 - |z|^2 L_\omega(z, z)}{(1 - |z|^2)^2}, \qquad z \in \mathbb{D},$$

together with the subharmonicity of $L_\omega(z, z)$ shows that $0 \leq L_\omega(z, z) \leq 1$. In fact, unless $L_\omega(z, z)$ is identically 1, we must have the strict inequality $L_\omega(z, z) < 1$.

If $L_\omega(z, z) \equiv 1$, we must also have $L_\omega(z, \zeta) \equiv 1$, because an analytic kernel function is determined by its values along the diagonal. In this case, K_ω becomes the reproducing kernel of the space H^2, which is impossible, since H^2 is not of the type $A^2(\omega)$.

We conclude that $L_\omega(z, z) < 1$ for all $z \in \mathbb{D}$. An application of the Cauchy-Schwarz inequality for reproducing kernels then yields $|L_\omega(z, \zeta)| < 1$ for all z and ζ in $\mathbb{D}$.   ∎

COROLLARY 9.7 *The reproducing kernel K_ω satisfies*

$$\frac{1 - |z\zeta|}{|1 - z\bar{\zeta}|^2} \leq |K_\omega(z, \zeta)| \leq \frac{1 + |z\zeta|}{|1 - z\bar{\zeta}|^2}$$

for all z and ζ in $\mathbb{D}$. In particular, the kernel function K_ω is nonvanishing on $\mathbb{D} \times \mathbb{D}$.

Proof. This is immediate from Corollary 9.6.   ∎

The kernel L_ω has the following boundary behavior.

THEOREM 9.8 *Suppose ω is continuous on $\overline{\mathbb{D}}$. Then $\omega|_{\mathbb{T}} \geq 1$, and the diagonal function $L_\omega(z, z)$ has a continuous extension to $\overline{\mathbb{D}}$. More specifically,*

$$L_\omega(z, z) = 1 - \frac{1}{\omega(z)}, \qquad z \in \mathbb{T}.$$

Proof. For $\lambda \in \mathbb{D}$, let k_λ be the normalized reproducing kernel of A^2, that is,

$$k_\lambda(z) = \frac{1 - |\lambda|^2}{(1 - \bar{\lambda}z)^2}, \qquad z \in \mathbb{D}.$$

Fix a point $\zeta \in \mathbb{T}$. Since $|k_\lambda(z)|^2$ is the real Jacobian of the involutive Möbius map φ_λ, a change of variable combined with the continuity of ω at ζ gives

$$\int_{\mathbb{D}} |k_\lambda(z)|^2 \, \omega(z) \, dA(z) \to \omega(\zeta)$$

as $\lambda \to \zeta$.

On the other hand, the reproducing property

$$k_\lambda(\lambda) = \langle k_\lambda, K_\omega(\cdot, \lambda) \rangle_\omega$$

together with the Cauchy-Schwarz inequality yields

$$\frac{1}{(1 - |\lambda|^2)^2} = |k_\lambda(\lambda)|^2 \leq K_\omega(\lambda, \lambda) \int_{\mathbb{D}} |k_\lambda(z)|^2 \, \omega(z) \, dA(z),$$

so that

$$\begin{aligned}
\frac{1}{\omega(\zeta)} &\leq \liminf_{\lambda \to \zeta} (1 - |\lambda|^2)^2 \, K_\omega(\lambda, \lambda) \\
&= \liminf_{\lambda \to \zeta} \left(1 - |\lambda|^2 \, L_\omega(\lambda, \lambda)\right) \\
&= 1 - \limsup_{\lambda \to \zeta} L_\omega(\lambda, \lambda),
\end{aligned}$$

which leads to half of the desired assertion,

$$\limsup_{\lambda \to \zeta} L_\omega(\lambda, \lambda) \leq 1 - \frac{1}{\omega(\zeta)}, \qquad \zeta \in \mathbb{T}.$$

For the other half, we use the normalized reproducing kernels of $A^2(\omega)$. Thus for $\lambda \in \mathbb{D}$, we let G_λ be the function

$$G_\lambda(z) = K_\omega(\lambda, \lambda)^{-\frac{1}{2}} K_\omega(z, \lambda), \qquad z \in \mathbb{D},$$

which has norm 1 in $A^2(\omega)$. By Corollary 9.7 and the easy fact that $K_\omega(\lambda, \lambda) \to +\infty$ as $|\lambda| \to 1$, the function G_λ tends to 0 uniformly off a fixed neighborhood of the point ζ as λ approaches $\zeta \in \mathbb{T}$. In particular, the measure $|G_\lambda|^2 \, \omega \, dA$ converges to a point mass at ζ as $\lambda \to \zeta$. In view of Corollary 9.7, we can write

$$G_\lambda(\lambda) = \int_{\mathbb{D}} \frac{G_\lambda(z) \, dA(z)}{(1 - \lambda \bar{z})^2}.$$

By the Cauchy-Schwarz inequality,

$$K_\omega(\lambda, \lambda) = |G_\lambda(\lambda)|^2 \leq (1 - |\lambda|^2)^{-2} \int_{\mathbb{D}} |G_\lambda(z)|^2 \, dA(z),$$

so that

$$\int_{\mathbb{D}} |G_\lambda(z)|^2 \, dA(z) \to \frac{1}{\omega(\zeta)} \qquad \text{as } \lambda \to \zeta.$$

It follows that

$$
\begin{aligned}
1 - \liminf_{\lambda \to \zeta} L_\omega(\lambda, \lambda) &= \limsup_{\lambda \to \zeta} \left(1 - |\lambda|^2 \, L_\omega(\lambda, \lambda) \right) \\
&= \limsup_{\lambda \to \zeta} (1 - |\lambda|^2)^2 \, K_\omega(\lambda, \lambda) \\
&\leq 1 - \frac{1}{\omega(\zeta)},
\end{aligned}
$$

and consequently,

$$1 - \frac{1}{\omega(\zeta)} \leq \liminf_{\lambda \to \zeta} L_\omega(\lambda, \lambda), \qquad \zeta \in \mathbb{T}.$$

This proves that

$$\lim_{\lambda \to \zeta} L_\omega(\lambda, \lambda) = 1 - \frac{1}{\omega(\zeta)}.$$

The inequality $\omega(\zeta) \geq 1$ then follows from the fact that $L_\omega(z, z) \geq 0$ for all $z \in \mathbb{D}$.
∎

The *harmonic polynomials* are functions of the form $p + \bar{q}$, where p and q are (analytic) polynomials. Let $HP^2(\omega)$ denote the closure of the harmonic polynomials in $L^2(\mathbb{D}, \omega \, dA)$. We collect the elementary properties of this weighted Bergman space of harmonic functions in the next proposition.

PROPOSITION 9.9 $HP^2(\omega)$ *is a Hilbert space with locally uniformly bounded point evaluations. Let* $A^2_{\omega,0}$ *denote the subspace of* $A^2(\omega)$ *consisting of those functions that vanish at the origin, and* $\overline{A}^2_{\omega,0}$ *its image under complex conjugation. Then the harmonic Bergman space splits as*

$$HP^2(\omega) = A^2(\omega) \oplus \overline{A}^2_{\omega,0},$$

the two subspaces on the right-hand side being orthogonal. Moreover, the kernel function Q_ω *for* $HP^2(\omega)$ *has the form*

$$Q_\omega(z,\zeta) = 2\operatorname{Re} K_\omega(z,\zeta) - 1, \qquad z, \zeta \in \mathbb{D}.$$

Proof. Let p and q be analytic polynomials. If $q(0) = 0$, then by the reproducing property of ω,

$$\langle p, \bar{q} \rangle_\omega = \int_{\mathbb{D}} p(z)\, q(z)\, \omega(z)\, dA(z) = 0,$$

and so $A^2(\omega)$ and $\overline{A}^2_{\omega,0}$ are perpendicular. The Pythagorean theorem then gives

$$\|p + \bar{q}\|^2_\omega = \|p\|^2_\omega + \|q\|^2_\omega.$$

If we take a Cauchy sequence $\{p_j + \bar{q}_j\}_j$ of harmonic polynomials in $L^2(\mathbb{D}, \omega\, dA)$ with $q_j(0) = 0$ for all j, then the above identity shows that $\{p_j\}_j$ is a Cauchy sequence in $A^2(\omega)$, and $\{q_j\}_j$ a Cauchy sequence in $A^2_{\omega,0}$. By the completeness of the spaces $A^2(\omega)$ and $A^2_{\omega,0}$ there are elements $f \in A^2(\omega)$ and $g \in A^2_{\omega,0}$ such that $p_j \to f$ and $q_j \to g$ as $j \to +\infty$. The limit function $h = f + \bar{g}$ is then harmonic in $\mathbb{D}$, and we have

$$\|h\|^2_\omega = \|f + \bar{g}\|^2_\omega = \|f\|^2_\omega + \|g\|^2_\omega.$$

The local boundedness of point evaluations now follows from Proposition 9.2.

The reproducing kernel for $A^2(\omega)$ is K_ω, and for $\overline{A}^2_{\omega,0}$ it is $\overline{K}_\omega - 1$. It follows from the direct-sum decomposition of $HP^2(\mathbb{D}, \omega)$ that Q_ω is the sum of these two kernels.   ∎

The function L_ω is bounded and sesqui-holomorphic on $\mathbb{D}^2$, meaning that the function $L_\omega(z, \bar{\zeta})$ is a holomorphic function of two variables there, and hence it possesses radial boundary values almost everywhere on the torus $\mathbb{T}^2$. It follows that the kernels K_ω and Q_ω, too, have radial boundary values almost everywhere on $\mathbb{T}^2$. The following result will be used later in the proof of the positivity of the weighted biharmonic Green function.

COROLLARY 9.10 *If* ω *is continuous on* $\overline{\mathbb{D}}$, *then we have*

$$Q_\omega(z, \zeta) \leq -\left(\frac{1}{\omega(z)} + \frac{1}{\omega(\zeta)}\right) \frac{1}{|z - \zeta|^2}, \qquad (z, \zeta) \in \mathbb{T} \times \mathbb{T} \setminus \delta(\mathbb{T}),$$

almost everywhere with respect to surface measure, where $\delta(\mathbb{T})$ *denotes the diagonal in* $\mathbb{T} \times \mathbb{T}$.

Proof. Since L_ω is a reproducing kernel, we have

$$|L_\omega(z, \zeta)| \leq L_\omega(z, z)^{\frac{1}{2}} L_\omega(\zeta, \zeta)^{\frac{1}{2}}, \qquad z, \zeta \in \mathbb{D}.$$

Applying Theorem 9.8 and the geometric-arithmetic mean inequality, we obtain

$$
\begin{aligned}
|L_\omega(z, \zeta)| &\leq \left(1 - \frac{1}{\omega(z)}\right)^{\frac{1}{2}} \left(1 - \frac{1}{\omega(\zeta)}\right)^{\frac{1}{2}} \\
&\leq 1 - \frac{1}{2}\left(\frac{1}{\omega(z)} + \frac{1}{\omega(\zeta)}\right)
\end{aligned}
\tag{9.4}
$$

for almost all $(z, \zeta) \in \mathbb{T}^2$. Since

$$K_\omega(z, \zeta) = \frac{1 - z\bar{\zeta} \, L_\omega(z, \zeta)}{(1 - z\bar{\zeta})^2} = \frac{1}{1 - z\bar{\zeta}} + \frac{z\bar{\zeta}}{(1 - z\bar{\zeta})^2} - \frac{z\bar{\zeta}}{(1 - z\bar{\zeta})^2} L_\omega(z, \zeta),$$

it follows that

$$K_\omega(z, \zeta) = \frac{1}{1 - z\bar{\zeta}} - \frac{1}{|z - \zeta|^2} - \frac{1}{|z - \zeta|^2} L_\omega(z, \zeta), \qquad (z, \zeta) \in \mathbb{T}^2 \setminus \delta(\mathbb{T}),$$

for almost all $(z, \zeta) \in \mathbb{T}^2 \setminus \delta(\mathbb{T})$.

The first term on the right-hand side has real part $\frac{1}{2}$. In view of Proposition 9.9, the above representation formula, and inequality (9.4), we see that

$$
\begin{aligned}
Q_\omega(z, \zeta) &= -\frac{2}{|z - \zeta|^2} - \frac{2}{|z - \zeta|^2} \, \mathrm{Re}\, L_\omega(z, \zeta) \\
&\leq -\frac{2}{|z - \zeta|^2} + \frac{2}{|z - \zeta|^2} |L_\omega(z, \zeta)| \\
&= -\left(\frac{1}{\omega(z)} + \frac{1}{\omega(\zeta)}\right) \frac{1}{|z - \zeta|^2}
\end{aligned}
$$

for almost all $(z, \zeta) \in \mathbb{T}^2 \setminus \delta(\mathbb{T})$, as desired. ∎

9.2 Green Functions with Smooth Weights

Throughout this section we assume that ω, in addition to being logarithmically subharmonic and reproducing for the origin, is strictly positive and real-analytic on $\overline{\mathbb{D}}$ (this means that the weight is real-analytic in a neighborhood of $\overline{\mathbb{D}}$).

Let $D \subset \mathbb{D}$ be a simply connected domain with C^∞ boundary. The Green function $\Gamma_{\omega, D}$ for the weighted biharmonic operator $\Delta \omega^{-1} \Delta$ is defined as follows. For fixed $\zeta \in D$, the function $\Gamma_{\omega, D}(\cdot, \zeta)$ is the unique solution to the boundary value problem

$$\Delta_z \, \omega(z)^{-1} \Delta_z \Gamma_{\omega,D}(z, \zeta) \;=\; \delta_\zeta(z), \qquad z \in D,$$
$$\Gamma_{\omega,D}(z, \zeta) \;=\; 0, \qquad z \in \partial D, \qquad (9.5)$$
$$\partial_{n(z)} \Gamma_{\omega,D}(z, \zeta) \;=\; 0, \qquad z \in \partial D.$$

As we integrate back one Laplacian, we see that

$$\Delta_z \Gamma_{\omega,D}(z, \zeta) = \omega(z)\left(G_D(z, \zeta) + H_{\omega,D}(z, \zeta)\right),$$

where G_D is the Green function for Δ on D, and the function $H_{\omega,D}(z, \zeta)$ is harmonic in the first variable z on $\mathbb{D}$. In view of the boundary condition

$$\Gamma_{\omega,D}(\cdot, \zeta)|_{\partial D} = 0,$$

the integral version of this identity is

$$\Gamma_{\omega,D}(z, \zeta) = \int_D G_D(z, \xi)\left(G_D(\xi, \zeta) + H_{\omega,D}(\xi, \zeta)\right) \omega(\xi)\, dA(\xi). \qquad (9.6)$$

An application of Green's formula along with the zero boundary data of $\Gamma_{\omega,D}$ shows that for a harmonic function h that is smooth up to the boundary of D, the kernel $H_{\omega,D}$ has the balayage property

$$\int_D h(z)\left(G_D(z, \zeta) + H_{\omega,D}(z, \zeta)\right) \omega(z)\, dA(z) = 0. \qquad (9.7)$$

We call $H_{\omega,D}$ the *harmonic compensator* (for the Green function G_D with respect to the weight ω). It follows from (9.7) that $H_{\omega,D}(\cdot, \zeta)$ equals the weighted harmonic projection of $-G_D(\cdot, \zeta)$:

$$H_{\omega,D}(z, \zeta) = - \int_D Q_{\omega,D}(z, \xi)\, G_D(\xi, \zeta)\, \omega(\xi)\, dA(\xi), \qquad (9.8)$$

where $Q_{\omega,D}$ is the reproducing kernel for the weighted harmonic Bergman space $HP^2(D, \omega)$, the completion of the harmonic polynomials in $L^2(D, \omega\, dA)$. Taking the Laplacian with respect to ζ, we obtain

$$\Delta_\zeta H_{\omega,D}(z, \zeta) = -\omega(\zeta)\, Q_{\omega,D}(z, \zeta),$$

a nice relationship between the harmonic compensator and the harmonic reproducing kernel.

The remainder of this section is devoted to proving that the weighted biharmonic Green function

$$\Gamma_\omega = \Gamma_{\omega,\mathbb{D}}$$

is positive on $\mathbb{D} \times \mathbb{D}$, a result that turns out to have far-reaching consequences. Unfortunately, the proof depends on the following result. Recall that $ds(z) = |dz|/(2\pi)$.

THEOREM 9.11 *For each $0 < r \le 1$, there is a (unique) simply connected domain $D(r)$ contained in $\mathbb{D}$ and a conformal map $\phi_r : \mathbb{D} \to D(r)$ with the following properties:*

(1) The boundary of $D(r)$, $\partial D(r)$, is a real-analytic Jordan curve.

(2) Each $D(r)$ contains the origin and each ϕ_r preserves the origin.

(3) The domains $D(r)$ are increasing in r with $D(1) = \mathbb{D}$, and their intersection is the point at the origin.

(4) The reproducing property

$$r^2 h(0) = \int_{D(r)} h(z)\omega(z)\, dA(z) \tag{9.9}$$

holds for all bounded harmonic functions h on $D(r)$.

(5) The mapping $(r, z) \mapsto \phi_r(z)$ extends to a holomorphic function of two complex variables on a neighborhood of $(0, 1] \times \overline{\mathbb{D}}$.

(6) For each $0 < r' \le 1$, there is a small open interval J around it such that all the functions ϕ_r, $r \in J$, extend as conformal maps to one and the same neighborhood of $\overline{\mathbb{D}}$.

(7) For each $0 < r' < 1$, we have

$$\frac{1}{r^2 - (r')^2}\, 1_{D(r)\backslash D(r')}(z)\, dA(z) \rightarrow \frac{d\varpi_{r'}(z)}{\omega(z)}, \tag{9.10}$$

as $r \to r'^{+}$, in the weak-star topology of the Borel measures. Here, ϖ_r is harmonic measure on $\partial D(r)$; we do not use the notation ω for harmonic measure here, to avoid confusion with the weight.

(8) The evolution equation

$$\frac{d\phi_r}{dr}(z) = rz\, \phi_r'(z) \int_{\mathbb{T}} \frac{\zeta + z}{\zeta - z} \frac{ds(\zeta)}{\omega(\phi_r(\zeta))\, |\phi_r'(\zeta)|^2} \tag{9.11}$$

holds for all $0 < r < 1$ and $z \in \mathbb{D}$.

The proof of this theorem is too technical and depends on the theory of weighted Hele-Shaw flows, and is therefore omitted. The interested reader is referred to [69].

We will call the domain $D(r)$ an ω-mean value disk of radius r. The reproducing property (4) above is the most fundamental; in fact, it uniquely determines the domain $D(r)$.

Let $\Gamma_{\omega,r}$ denote the weighted biharmonic Green function $\Gamma_{\omega,D(r)}$. Similarly, let G_r be the classical Green function for $D(r)$, and let $H_{\omega,r}$ be the harmonic compensator corresponding to the weight ω and the domain $D(r)$. We shall derive a variational formula, originally found by Hadamard in 1908, which describes the development of $\Gamma_{\omega,r}$ as r increases quantitatively [50, pp. 515–641]. Since ω is real-analytic on $\overline{\mathbb{D}}$ and the ω-mean value disks $D(r)$ depend on r very smoothly, we conclude that the Green function $\Gamma_{\omega,r}$ extends real-analytically to a neighborhood of the set

$$\overline{D}(r) \times \overline{D}(r) \setminus \delta(\overline{D}(r)),$$

where

$$\delta(\overline{D}(r)) = \big\{(z, z) : z \in \overline{D}(r)\big\}$$

is the diagonal. In particular, for fixed $\zeta \in \overline{D}(r)$, the function $\Gamma_{\omega,r}$ solves the differential equation

$$\Delta \omega^{-1} \Delta \Gamma_{\omega,r}(\cdot, \zeta) = \delta_\zeta$$

on a neighborhood of $\overline{D}(r) \setminus \{\zeta\}$.

We consider two parameter values r and r', with $0 < r < r' < 1$, and introduce the expression

$$F_{r,r'}(\xi, \zeta, z) = \big(G_r(\xi, z) + H_{\omega,r}(\xi, z)\big)\big(G_{r'}(\xi, \zeta) + H_{\omega,r'}(\xi, \zeta)\big).$$

By (9.6) and (9.7),

$$\Gamma_{\omega,r}(z, \zeta) = \int_{D(r)} F_{r,r'}(\xi, \zeta, z)\,\omega(\xi)\,dA(\xi)$$

for $(z, \zeta) \in D(r) \times D(r)$, and

$$\Gamma_{\omega,r'}(z, \zeta) = \int_{D(r')} F_{r,r'}(\xi, \zeta, z)\,\omega(\xi)\,dA(\xi)$$

for $(z, \zeta) \in D(r') \times D(r')$.

Since $D(r) \subset D(r')$, we have

$$\Gamma_{\omega,r'}(z, \zeta) - \Gamma_{\omega,r}(z, \zeta) = \int_{D(r') \setminus D(r)} F_{r,r'}(\xi, \zeta, z)\,\omega(\xi)\,dA(\xi)$$

for $(z, \zeta) \in D(r) \times D(r)$. If r' is sufficiently close to r, we can actually take $(z, \zeta) \in D(r') \times D(r')$. It follows from (9.10) that as $r' \to r$,

$$\frac{d}{dr}\Gamma_{\omega,r}(z, \zeta) = 2r \int_{\partial D(r)} H_{\omega,r}(\xi, z)\,H_{\omega,r}(\xi, \zeta)\,d\varpi_r(\xi). \qquad (9.12)$$

Here, we have used the fact that the Green function G_r vanishes when one of the variables is on the boundary $\partial D(r)$.

We want to turn the differential equation (9.12) into an integral equation. Note that when one of the variables z and ζ is on the boundary $\partial D(r)$ and the other is in the interior $D(r)$, we have $\Gamma_{\omega,r}(z, \zeta) = 0$. If we integrate (9.12) with respect to r, the following formula emerges:

$$\Gamma_{\omega,r}(z, \zeta) = \int_{\max\{R(z),R(\zeta)\}}^{r} \int_{\partial D(\varrho)} H_{\omega,\varrho}(\xi, z)\,H_{\omega,\varrho}(\xi, \zeta)\,d\varpi_\varrho(\xi)\,2\varrho\,d\varrho,$$

$$(9.13)$$

for $(z, \zeta) \in D(r) \times D(r)$. Here, $R(z)$ stands for the parameter value of ϱ for which the boundary of $D(\varrho)$ reaches the point z:

$$R(z) = \inf\big\{\varrho : z \in D(\varrho)\big\}.$$

The integral formula (9.13) will be referred to as *Hadamard's variational formula*; it clearly shows that the positivity of the weighted biharmonic Green function will be established once we are able to show that the harmonic compensators are positive. To this end, we proceed to show that the harmonic compensators can be written as an integral in terms of the Poisson kernel and the weighted harmonic reproducing kernel.

Let P_r be given by

$$P_r(z, \zeta) = -\frac{1}{2} \partial_{n(\zeta)} G_r(z, \zeta), \qquad (z, \zeta) \in D(r) \times \partial D(r),$$

the normal derivative being taken with respect to the boundary $\partial D(r)$ in the interior direction. This function serves as a Poisson kernel on $D(r)$. For instance, we have the identity

$$d\varpi_r(z) = P_r(z_0, z) \, ds(z), \qquad z \in \partial D(r).$$

Using arguments similar to the proof of (9.12), we can show that

$$\frac{d}{dr} G_r(z, \zeta) = -2r \int_{\partial D(r)} P_r(z, \xi) \, P_r(\zeta, \xi) \, \frac{d\varpi_r(\xi)}{\omega(\xi)},$$

for $(z, \zeta) \in D(r) \times D(r)$. In integral form this becomes

$$G_r(z, \zeta) = -\int_{\max\{R(z), R(\zeta)\}}^{r} \int_{\partial D(\varrho)} P_\varrho(z, \xi) \, P_\varrho(\zeta, \xi) \, \frac{d\varpi_\varrho(\xi)}{\omega(\xi)} \, 2\varrho \, d\varrho$$

for $(z, \zeta) \in D(r) \times D(r)$. Combining this with equation (9.8), we get

$$
\begin{aligned}
H_{\omega, r}(\zeta, z) &= \int_{D(r)} \int_{\max\{R(z), R(\eta)\}}^{r} \int_{\partial D(\varrho)} Q_{\omega, r}(\zeta, \eta) \\
&\quad \times P_\varrho(z, \xi) \, P_\varrho(\eta, \xi) \, \frac{d\varpi_\varrho(\xi)}{\omega(\xi)} \, 2\varrho \, d\varrho \, \omega(\eta) \, dA(\eta),
\end{aligned}
$$

which transforms to

$$
\begin{aligned}
H_{\omega, r}(\zeta, z) &= \int_{R(z)}^{r} \int_{\partial D(\varrho)} \int_{D(\varrho)} Q_{\omega, r}(\zeta, \eta) \, P_\varrho(\eta, \xi) \, \omega(\eta) \, dA(\eta) \\
&\quad \times P_\varrho(z, \xi) \, \frac{d\varpi_\varrho(\xi)}{\omega(\xi)} \, 2\varrho \, d\varrho, \tag{9.14}
\end{aligned}
$$

where $(z, \zeta) \in D(r) \times D(r)$.

As a consequence of formula (9.14), we see that if

$$\int_{D(\varrho)} Q_{\omega, r}(\zeta, \eta) \, P_\varrho(\eta, \xi) \, \omega(\eta) \, dA(\eta) \geq 0 \tag{9.15}$$

for $(\xi, \zeta) \in \partial D(\varrho) \times D(r)$, where $0 < \varrho < r < 1$, then the harmonic compensator $H_{\omega, r}$ is positive on $D(r) \times D(r)$. Note that the function $Q_\omega(\zeta, \cdot)$ is harmonic on $\overline{D}(r)$, and in particular, bounded there. Since the Poisson kernel $P(\cdot, \xi)$ is area summable on $D(\varrho)$, we see that the integral in (9.15) makes sense.

We are going to prove the following result, which is seemingly stronger than, but actually equivalent to, inequality (9.15).

LEMMA 9.12 *Fix ϱ and r such that $0 < \varrho < r < 1$. Let h be a positive harmonic function on $D(\varrho)$, and define*

$$h_r(z) = \int_{D(\varrho)} Q_{\omega,r}(z, \xi) \, h(\xi) \, \omega(\xi) \, dA(\xi), \qquad z \in D(r).$$

Then h_r is positive on $D(r)$.

Proof. It suffices to obtain the result under the proviso that h is harmonic and strictly positive on $\overline{D}(\varrho)$. Since $Q_{\omega,r}(0, \cdot) = r^{-2}$, a consequence of the reproducing property of the domain $D(r)$, the value of the function h_r at the center point 0 is

$$h_r(0) = \frac{1}{r^2} \int_{D(\varrho)} h(\xi) \, \omega(\xi) \, dA(\xi) = \frac{\varrho^2}{r^2} h(0),$$

which is positive. We split the proof into three parts.

Part 1: continuity of h_r in r. The function h_r is the orthogonal projection of $h \, 1_{D(\varrho)}$, interpreted to vanish on $D(r) \backslash D(\varrho)$, onto the weighted harmonic Bergman $HP^2(D(r), \omega)$. From the smoothness of the harmonic compensator $H_{\omega,r}$ in the r variable alluded to above, and the corresponding fact for the weighted harmonic Bergman kernel $Q_{\omega,r}$ as deduced from the identity

$$Q_{\omega,r}(z, \zeta) = -\omega(z)^{-1} \Delta_\zeta H_{\omega,r}(z, \zeta),$$

it is immediate that $h_r(z)$ is real-analytic in the coordinates (z, r) in a neighborhood of the set

$$\{(z, r) : z \in \overline{D}(r), \ r \in (\varrho, 1]\} \cup \{(z, r) : z \in D(r), \ r \in [\varrho, 1]\}.$$

We need to investigate the continuity of $h_r(z)$ near the left endpoint $r = \varrho$. By the reproducing property of the domains $D(\varrho)$,

$$\int_{D(\varrho)} Q_{\omega,r}(z, \xi) \, \omega(\xi) \, dA(\xi) = \varrho^2 \, Q_{\omega,r}(z, 0) = \frac{\varrho^2}{r^2}, \qquad z \in D(r),$$

and hence

$$h_r(z) - \frac{r^2}{\varrho^2} h(z) = \int_{D(\varrho)} Q_{\omega,r}(z, \xi) \left(h(\xi) - h(z)\right) \omega(\xi) \, dA(\xi) \qquad (9.16)$$

for $z \in D(r)$, provided that r is so close to ϱ that h is defined as a harmonic function on $D(r)$. Since

$$h(\xi) - h(z) = O(|z - \xi|)$$

for z and ξ in some fixed neighborhood of $\overline{D}(\varrho)$, part of the singularity of the kernel $Q_{\omega,r}$ is neutralized by the appearance of this factor on the right-hand side of (9.16). Let ω_r stand for the pulled-back weight on the unit disk,

$$\omega_r(z) = r^{-2} \omega \circ \phi_r(z) \, |\phi_r'(z)|^2,$$

which is reproducing for the origin as well as logarithmically subharmonic. It follows from the conformal invariance of the reproducing property of the weighted harmonic Bergman kernel that

$$r^2 \, Q_{\omega,r}\big(\phi_r(z), \phi_r(\zeta)\big) = Q_{\omega_r}(z, \zeta) = 2\,\mathrm{Re}\,K_{\omega_r}(z, \zeta) - 1 \qquad (9.17)$$

for $(z, \zeta) \in \mathbb{D} \times \mathbb{D}$. Applying Corollary 9.7 to K_{ω_r} and using the relationship between the harmonic and analytic kernels of Proposition 9.9, we obtain

$$r^2 \, \big|Q_{\omega,r}\big(\phi_r(z), \phi_r(\zeta)\big)\big| = \big|Q_{\omega_r}(z, \zeta)\big| \le 1 + \frac{4}{|1 - z\bar{\zeta}|^2} \qquad (9.18)$$

for $(z, \zeta) \in \mathbb{D} \times \mathbb{D}$. Rewriting (9.16) in terms of the variable ζ, $\phi_r(\zeta) = \xi$, we get

$$h_r \circ \phi_r(z) - \frac{r^2}{\varrho^2} h \circ \phi_r(z) \qquad (9.19)$$

$$= \int_{\phi_r^{-1}(D(\varrho))} Q_{\omega_r}(z, \zeta) \big(h(\phi_r(\zeta)) - h(\phi_r(z))\big) \, \omega_r(\zeta) \, dA(\zeta)$$

for $z \in \mathbb{D}$, where $\phi_r^{-1}(D(\varrho)) \subset \mathbb{D}$. Given the estimates mentioned previously, it is easily deduced from this identity that $h_r \circ \phi_r \to h \circ \phi_\varrho$ uniformly on $\mathbb{D}$ as $r \to \varrho$. In fact, if $r \in (\varrho, 1)$ is close enough to ϱ, the function h is well defined and harmonic on $\overline{D}(r)$, so that the above integral also makes sense when we extend the domain of integration to $\mathbb{D}$. And the integral over $\mathbb{D}$ is zero, by the reproducing property of the harmonic kernel. So the right hand side of (9.19) reduces to an integral over the thin "circular" band $\mathbb{D} \setminus \phi_r^{-1}(D(\varrho))$, which is quite small. In particular, since we assume h to be strictly positive on $\overline{D}(\varrho)$, it follows that $h_r \circ \phi_r$ is uniformly (in r) strictly positive on $\mathbb{D}$ for r in some short interval $(\varrho, \varrho + \delta]$, with $\delta > 0$.

Part 2: the derivative of $h_r \circ \phi_r$. The derivative of the composition $h_r \circ \phi_r$ with respect to the parameter r is, by the chain rule,

$$\frac{d}{dr} h_r \circ \phi_r(z) = \frac{\partial h_r}{\partial r} \circ \phi_r(z) + 2\,\mathrm{Re}\left(\frac{\partial h_r}{\partial z} \circ \phi_r(z) \frac{d\phi_r}{dr}(z)\right), \qquad (9.20)$$

where the partial derivatives with respect to r and z correspond to thinking of the function h_r as a function of two variables: $h_r(z) = h(z, r)$. The derivative of ϕ_r with respect to r is supplied by formula (9.11), which simplifies to

$$\frac{d\phi_r}{dr}(z) = \frac{z}{r} \phi_r'(z) \int_{\mathbb{T}} \frac{\zeta + z}{\zeta - z} \frac{ds(\zeta)}{\omega_r(\zeta)} = \frac{z}{r} \phi_r'(z) \, \mathbf{H}_+\left[\frac{1}{\omega_r}\right](z) \qquad (9.21)$$

for $z \in \mathbb{D}$, where the symbol $\mathbf{H}_+$ stands for the Herglotz transform.

To find a way to express the partial derivative $\partial_r h_r$, let r', $\varrho < r' < r$, be so close to r that $h_{r'}$ extends harmonically and boundedly to $D(r)$. Then, from the reproducing property of the weighted harmonic Bergman kernel,

$$h_{r'}(z) = \int_{D(r)} Q_{\omega,r}(z, \xi) \, h_{r'}(\xi) \, \omega(\xi) \, dA(\xi) \qquad (9.22)$$

for $z \in D(r)$. On the other hand, by the reproducing property again,

$$\int_{D(r')} Q_{\omega,r}(z,\xi) \, Q_{\omega,r'}(\xi,\zeta) \, \omega(\xi) \, dA(\xi) = Q_{\omega,r}(z,\zeta)$$

for $(z,\zeta) \in D(r) \times D(r')$, so that

$$\int_{D(r')} Q_{\omega,r}(z,\xi) \, h_{r'}(\xi) \, \omega(\xi) \, dA(\xi)$$

$$= \int_{D(r')} Q_{\omega,r}(z,\xi) \int_{D(\varrho)} Q_{\omega,r'}(\xi,\zeta) \, h(\zeta) \, \omega(\zeta) \, dA(\zeta) \, \omega(\xi) \, dA(\xi)$$

$$= \int_{D(\varrho)} Q_{\omega,r}(z,\zeta) \, h(\zeta) \, \omega(\zeta) \, dA(\zeta) = h_r(z) \qquad (9.23)$$

for $z \in D(r)$. Forming the difference between (9.22) and (9.23), we obtain

$$h_r(z) - h_{r'}(z) = - \int_{D(r) \setminus D(r')} Q_{\omega,r}(z,\xi) \, h_{r'}(\xi) \, \omega(\xi) \, dA(\xi) \qquad (9.24)$$

for $z \in D(r)$. It follows from (9.10) and (9.24) that

$$\frac{\partial h_r}{\partial r}(z) = -2r \int_{\partial D(r)} Q_{\omega,r}(z,\xi) \, h_r(\xi) \, d\varpi_r(\xi), \qquad z \in D(r).$$

Shifting the coordinates back to the unit disk and keeping in mind (9.17), we obtain

$$\frac{\partial h_r}{\partial r} \circ \phi_r(z) = -\frac{2}{r} \int_{\mathbb{T}} Q_{\omega_r}(z,\zeta) \, h_r \circ \phi_r(\zeta) \, ds(\zeta), \qquad z \in \mathbb{D}. \qquad (9.25)$$

By the Poisson integral formula for harmonic functions in $\mathbb{D}$, we have the representation

$$h_r \circ \phi_r(z) = \int_{\mathbb{T}} \frac{1 - |z|^2}{|1 - z\bar{\zeta}|^2} \, h_r \circ \phi_r(\zeta) \, ds(\zeta), \qquad z \in \mathbb{D},$$

which easily yields

$$\phi_r'(z) \frac{\partial h_r}{\partial z} \circ \phi_r(z) = \int_{\mathbb{T}} \frac{\bar{\zeta}}{(1 - z\bar{\zeta})^2} \, h_r \circ \phi_r(\zeta) \, ds(\zeta), \qquad z \in \mathbb{D}. \qquad (9.26)$$

We insert the above representation formulas (9.21), (9.25), and (9.26) into (9.20), and obtain that the derivative $\frac{d}{dr} h_r \circ \phi_r(z)$ is equal to

$$\frac{2}{r} \int_{\mathbb{T}} \left\{ \operatorname{Re}\left(\mathbf{H}_+\left[\frac{1}{\omega_r}\right](z) \, \frac{z\bar{\zeta}}{(1 - z\bar{\zeta})^2} \right) - Q_{\omega_r}(z,\zeta) \right\} h_r \circ \phi_r(\zeta) \, ds(\zeta),$$

where $z \in \mathbb{D}$. Just as in the proof of Corollary 9.10, we notice the appearance of the Kœbe function. Suppose for the moment that for some value of the parameter r, $\varrho < r < 1$, the real-analytic function $h_r \circ \phi_r|_{\mathbb{T}}$ vanishes along with its (tangential) derivative at some point $z_1 \in \mathbb{T}$. Then $h_r \circ \phi_r(z) = O(|z - z_1|^2)$ as z approaches z_1 along $\mathbb{T}$, which counterbalances the singularities of the Kœbe function and the weighted harmonic Bergman kernel, as estimated by (9.18), at least when $z \in \mathbb{D}$

approaches the boundary point z_1 radially. Taking into account the well-known boundary behavior of the Kœbe function, we obtain in the limit that (the real part of the Herglotz transform is the Poisson integral, with well-known boundary values)

$$\frac{d}{dr} h_r \circ \phi_r(z_1) = -\frac{2}{r} \int_{\mathbb{T}} \left\{ \frac{1}{\omega_r(z_1)} \frac{1}{|\zeta - z_1|^2} + Q_{\omega_r}(z_1, \zeta) \right\} h_r \circ \phi_r(\zeta) \, ds(\zeta).$$

If, in addition, $0 \leq h_r \circ \phi_r$ on $\overline{\mathbb{D}}$, then by invoking Corollary 9.10, which states that

$$Q_{\omega_r}(z_1, \zeta) \leq -\left(\frac{1}{\omega_r(z_1)} + \frac{1}{\omega_r(\zeta)} \right) \frac{1}{|\zeta - z_1|^2}, \qquad \zeta \in \mathbb{T} \setminus \{z_1\},$$

we can assert that

$$0 < \frac{2}{r} \int_{\mathbb{T}} \frac{1}{\omega_r(\zeta)} \frac{1}{|\zeta - z_1|^2} h_r \circ \phi_r(\zeta) \, ds(\zeta) \leq \frac{d}{dr} h_r \circ \phi_r(z_1). \tag{9.27}$$

The leftmost inequality holds because $h_r \circ \phi_r$ cannot vanish identically, since we know that $0 < h_r(0) = h_r \circ \phi_r(0)$.

Part 3: the finishing argument. Consider the function

$$\mathbf{h}(r) = \min \left\{ h_r(z) : z \in \overline{D}(r) \right\} = \min \left\{ h_r \circ \phi_r(z) : z \in \overline{\mathbb{D}} \right\}, \qquad \varrho < r < 1,$$

which, by the results of Part 1, extends continuously to the interval $[\varrho, 1)$, and is positive at the left endpoint. We shall demonstrate that $0 < \mathbf{h}(r)$ holds for all $r \in [\varrho, 1)$, which is actually slightly stronger than what is needed. We argue by contradiction and thus assume $\mathbf{h}(r) \leq 0$ for some $r \in (\varrho, 1)$. Forming the infimum over all such r, we find a parameter value $r_1 \in (\varrho, 1)$ with $\mathbf{h}(r_1) = 0$ such that $0 < \mathbf{h}(r)$ holds for all $r \in [\varrho, r_1)$. By the maximum principle, there exists a point $z_1 \in \mathbb{T}$ such that $h_{r_1} \circ \phi_{r_1}(z_1) = 0$ and $0 \leq h_{r_1} \circ \phi_{r_1}$ elsewhere on $\overline{\mathbb{D}}$. The point z_1 is precisely of the type considered in Part 2, so that by (9.27),

$$\frac{d}{dr} h_r \circ \phi_r(z_1) \Big|_{r=r_1} > 0.$$

We immediately see that $h_r \circ \phi_r(z_1) < 0$ for $r, \varrho < r < r_1$, sufficiently close to r_1; and hence $\mathbf{h}(r) < 0$ for such r. This contradicts the minimality of r_1, and completes the proof of the lemma. ∎

Combining the lemma above with equations (9.13) and (9.14), we conclude that for any $r \in (0, 1)$, both $H_{\omega, r}$ and $\Gamma_{\omega, r}$ are positive on $D(r) \times D(r)$. In fact, they are strictly positive on $D(r) \times D(r)$. Consequently, under the assumptions that ω is logarithmically subharmonic, reproducing for the origin, real-analytic on $\overline{\mathbb{D}}$, and strictly positive on $\overline{\mathbb{D}}$, we have the following result.

THEOREM 9.13 *The weighted biharmonic Green function Γ_ω is strictly positive on $\mathbb{D} \times \mathbb{D}$.*

9.3   Green Functions with General Weights

To prove that Theorem 9.13 remains valid without the assumptions that ω is real-analytic and strictly positive on $\overline{\mathbb{D}}$, we need only establish the following two approximation results.

THEOREM 9.14 *For each positive ε, there is another logarithmically subharmonic weight $\widetilde{\omega}$ that is reproducing for the origin such that:*

(1) $\widetilde{\omega}$ is real-analytic on the closed disk $\overline{\mathbb{D}}$.

(2) $\widetilde{\omega}$ is strictly positive on $\overline{\mathbb{D}}$.

(3) $\displaystyle\int_{\mathbb{D}} \left| \omega(z) - \widetilde{\omega}(z) \right| dA(z) < \varepsilon.$

Proof. Let $\mathrm{Aut}\,(\mathbb{D})$ denote the automorphism group of $\mathbb{D}$. Every $\phi \in \mathrm{Aut}\,(\mathbb{D})$ admits a unique factorization $\phi = R_\beta \circ \varphi_\lambda$ with $\beta \in \mathbb{T}$ and $\lambda \in \mathbb{D}$, where R_β is a rotation and

$$\varphi_\lambda(z) = \frac{\lambda - z}{1 - \bar{\lambda}z}, \qquad z \in \mathbb{D}.$$

Thus we can identify $\mathrm{Aut}\,(\mathbb{D})$ with the set $\mathbb{T} \times \mathbb{D}$. Under the representation above, the (invariant) Haar measure on $\mathrm{Aut}\,(\mathbb{D})$ is given by

$$d\phi = \frac{dA(\lambda)}{(1 - |\lambda|^2)^2}\, ds(\beta).$$

Fix a real-analytic function $\Phi : \mathrm{Aut}\,(\mathbb{D}) \to (0, +\infty)$ in the following form:

$$\Phi(\phi) = \Phi_1(\beta)\Phi_2(\lambda), \qquad \phi = R_\beta \circ \varphi_\lambda,$$

where

$$\Phi_2(\lambda) = (N - 1)\,(1 - |\lambda|^2)^N, \qquad \lambda \in \mathbb{D}, \tag{9.28}$$

for some integer $N = 2, 3, 4, \ldots$, so that

$$\int_{\mathbb{D}} \Phi_2(\lambda)\, \frac{dA(\lambda)}{(1 - |\lambda|^2)^2} = 1,$$

and $\Phi_1 : \mathbb{T} \to (0, +\infty)$ is some real-analytic function with

$$\int_{\mathbb{T}} \Phi_1(\beta)\, ds(\beta) = 1.$$

For instance, we can take

$$\Phi_1(\beta) = \frac{1 - \varrho^2}{|1 + \varrho\beta|^2}, \qquad \beta \in \mathbb{T}, \tag{9.29}$$

for some real parameter ϱ with $0 < \varrho < 1$.

It is easy to see that

$$\int_{\text{Aut}(\mathbb{D})} \Phi(\phi)\, h \circ \phi^{-1}(0)\, d\phi = h(0) \tag{9.30}$$

for all bounded harmonic functions h on $\mathbb{D}$. In fact, for $\phi = R_\beta \circ \varphi_\lambda$, we have $\phi^{-1} = \varphi_\lambda \circ R_{\bar\beta}$, so that $\phi^{-1}(0) = \lambda$, and the left-hand side of (9.30) becomes

$$\int_{\mathbb{T} \times \mathbb{D}} \Phi_1(\beta)\Phi_2(\lambda)h(\lambda)\, ds(\beta)\frac{dA(\lambda)}{(1 - |\lambda|^2)^2}$$

$$= (N-1)\int_{\mathbb{T}} \Phi_1(\beta)\, ds(\beta) \int_{\mathbb{D}} (1 - |\lambda|^2)^{N-2}\, h(\lambda)\, dA(\lambda).$$

An application of the mean value theorem then shows that the integral above equals $h(0)$ for all bounded harmonic functions h on $\mathbb{D}$.

We use the function Φ to regularize ω. More specifically, we consider the function

$$\omega_\Phi(z) = \int_{\text{Aut}(\mathbb{D})} \Phi(\phi)\, \omega \circ \phi(z)\, |\phi'(z)|^2 d\phi. \tag{9.31}$$

It is clear that ω_Φ is strictly positive on $\mathbb{D}$. The function ω_Φ is also logarithmically subharmonic, because each individual function $\omega \circ \phi\, |\phi'|^2$ occurring in the integral is, and because the logarithmically subharmonic functions form a cone. If h is bounded harmonic function in $\mathbb{D}$, we use (9.30) to obtain

$$\int_{\mathbb{D}} h(z)\, \omega_\Phi(z)\, dA(z) = \int_{\text{Aut}(\mathbb{D})} \Phi(\phi) \int_{\mathbb{D}} h(z)\, \omega \circ \phi(z)\, |\phi'(z)|^2 dA(z)\, d\phi$$

$$= \int_{\text{Aut}(\mathbb{D})} \Phi(\phi) \int_{\mathbb{D}} h \circ \phi^{-1}(z)\, \omega(z)\, dA(z)\, d\phi$$

$$= \int_{\text{Aut}(\mathbb{D})} \Phi(\phi)\, h \circ \phi^{-1}(0)\, d\phi = h(0),$$

so that ω_Φ is representing for the origin.

A change of variables gives

$$\omega_\Phi(z) = (1 - |z|^2)^{-2} \int_{\text{Aut}(\mathbb{D})} \Phi(\phi \circ \varphi_z)\, \omega \circ \phi(0)|\phi'(0)|^2\, d\phi$$

for $z \in \mathbb{D}$. This clearly shows that ω_Φ is real-analytic in $\mathbb{D}$.

In order for ω_Φ to approximate ω in $L^1(\mathbb{D})$, we just need to choose Φ so that most of its mass is concentrated near the unit element of Aut $(\mathbb{D})$. This will be achieved if the parameter N is sufficiently large in Φ_2 and if the function Φ_1 has most of its mass near the point -1 on the unit circle.

We have thus shown that ω can be approximated in $L^1(\mathbb{D})$ by a logarithmically subharmonic weight that is reproducing for the origin, real-analytic in $\mathbb{D}$, and strictly positive on $\mathbb{D}$. So, in the rest of this proof we will assume that ω itself has all these additional properties.

For $0 < r < 1$, let $\omega_r(z) = \omega(rz)$ be the associated dilation of ω. Each ω_r is logarithmically subharmonic, strictly positive, and real-analytic on the closed disk $\overline{\mathbb{D}}$, but obviously, ω_r is not representing for the origin. However, each ω_r is *subrepresenting*, that is, for all positive bounded harmonic functions h on $\mathbb{D}$ we have

$$\int_{\mathbb{D}} h(z)\,\omega_r(z)\,dA(z) \leq h(0). \tag{9.32}$$

If fact, if $P(z, \zeta)$ denotes the Poisson kernel

$$P(z, \zeta) = \frac{1 - |z|^2}{|1 - z\bar{\zeta}|^2}, \qquad (z, \zeta) \in \mathbb{D} \times \mathbb{T},$$

then for each $z \in \mathbb{D}$ the function

$$\lambda \mapsto \int_{\mathbb{T}} P(\lambda, \alpha)\,\omega(\alpha z)\,ds(\alpha)$$

is harmonic in $\mathbb{D}$ and equals $\omega(\lambda z)$ for $\lambda \in \mathbb{T}$. Since the function $\omega(\lambda z)$ is subharmonic in the variable λ, we must have

$$\omega(\lambda z) \leq \int_{\mathbb{T}} P(\lambda, \alpha)\,\omega(\alpha z)\,ds(\alpha), \qquad (z, \lambda) \in \mathbb{D} \times \mathbb{D}.$$

In particular,

$$\omega_r(z) \leq \int_{\mathbb{T}} P(r, \alpha)\,\omega(\alpha z)\,ds(\alpha), \qquad z \in \mathbb{D}.$$

By the reproducing property of ω,

$$\begin{aligned}
\int_{\mathbb{D}} h(z)\,\omega_r(z)\,dA(z) &\leq \int_{\mathbb{T}\times\mathbb{D}} h(z)\,P(r, \alpha)\,\omega(\alpha z)\,dA(z)\,ds(\alpha) \\
&= \int_{\mathbb{T}} P(r, \alpha) \int_{\mathbb{D}} h(z)\,\omega(\alpha z)\,dA(z)\,ds(\alpha) \\
&= \int_{\mathbb{T}} P(r, \alpha)\,h(0)\,ds(\alpha) = h(0)
\end{aligned}$$

for all positive bounded harmonic functions h in $\mathbb{D}$.

We now complete the subrepresenting weight ω_r by adding a suitable small term which will make the sum representing for the origin but at the same time preserve the other properties of ω_r.

First, consider the harmonic function

$$P^*[\omega_r](z) = \int_{\mathbb{D}} P(z, \zeta)\,\omega_r(\zeta)\,dA(\zeta), \qquad z \in \mathbb{D},$$

where we have extended the Poisson kernel to the interior:

$$P(z, \zeta) = \frac{1 - |z\zeta|^2}{|1 - z\bar{\zeta}|^2}, \qquad (z, \zeta) \in \mathbb{D} \times \mathbb{D}.$$

This is the *sweep* of f, a function we encountered back in Chapter 3. The function $P^*[\omega_r]$ extends harmonically to a neighborhood of the closed unit disk. By the subrepresenting property (9.32) of ω_r, we have $0 < P^*[\omega_r] \leq 1$ throughout $\mathbb{D}$, and hence $0 \leq P^*[\omega_r] \leq 1$ also on $\mathbb{T}$.

Next, let θ be a real parameter with $0 < \theta < 1$, and consider the function

$$H(z) = 1 - \theta\, P^*[\omega_r](z), \qquad z \in \mathbb{D},$$

which is harmonic, bounded above by 1, and positive, on some $\varrho\mathbb{D}$ with $1 < \varrho < +\infty$. The function

$$F(z) = \int_{\mathbb{T}} \frac{(1 - \varrho^{-2})^2}{|1 - \varrho^{-1}z\bar\zeta|^4}\, H(\varrho\zeta)\, ds(\zeta), \qquad z \in \varrho\mathbb{D}, \tag{9.33}$$

is then real-analytic, strictly positive, and logarithmically subharmonic in $\varrho\mathbb{D}$. Moreover, for each z in $\mathbb{D}$ we have

$$
\begin{aligned}
P^*[F](z) &= \int_{\mathbb{D}} P(z, \zeta)\, F(\zeta)\, dA(\zeta) \\
&= \int_{\mathbb{D}} P(z, \zeta) \int_{\mathbb{T}} \frac{(1 - \varrho^{-2})^2}{|1 - \varrho^{-1}\zeta\bar\xi|^4}\, H(\varrho\xi)\, ds(\xi)\, dA(\zeta) \\
&= \int_{\mathbb{T}} \int_{\mathbb{D}} P(z, \zeta) \frac{(1 - \varrho^{-2})^2}{|1 - \varrho^{-1}\zeta\bar\xi|^4}\, dA(\zeta)\, H(\varrho\xi)\, ds(\xi) \\
&= \int_{\mathbb{T}} P(\varrho^{-1}z, \xi)\, H(\varrho\xi)\, ds(\xi) = H(z).
\end{aligned}
$$

It follows that the weight function

$$\widetilde{\omega}(z) = \theta\, \omega_r(z) + F(z)$$

is logarithmically subharmonic, strictly positive, and real-analytic on some neighborhood of $\overline{\mathbb{D}}$. It also satisfies $P^*[\widetilde{\omega}] = 1$, which is equivalent to

$$\int_{\mathbb{D}} h(z)\, \widetilde{\omega}(z)\, dA(z) = h(0),$$

where h is any bounded harmonic function in $\mathbb{D}$. Therefore, $\widetilde{\omega}$ is representing for the origin as well.

If the parameter r is close to 1, then the dilation ω_r is close to ω in $L^1(\mathbb{D})$. Also, if θ is close to 1, the function $\theta\, \omega_r$ still approximates ω well in $L^1(\mathbb{D})$. But this means that $\theta\, P^*[\omega_r](0)$ is close to 1, and since the $L^1(\mathbb{D})$ norm of F equals the difference $1 - P^*[\omega_r](0)$, the modified weight $\widetilde{\omega}$ approximates ω well in $L^1(\mathbb{D})$. The proof is complete. ∎

Note that if $\omega = |G_A|^2$, where G_A is a finite zero divisor in A^2, then the proof above can be greatly simplified. For example, we can take

$$\widetilde{\omega}(z) = (1 - \delta)\, \omega(z) + \delta, \qquad z \in \mathbb{D},$$

for sufficiently small positive δ.

PROPOSITION 9.15 *Let ω and ω_n, for $n = 1, 2, 3, \ldots$, be logarithmically subharmonic weights that reproduce for the origin. If $\omega_n \to \omega$ in the norm of*

$L^1(\mathbb{D})$ as $n \to +\infty$, then $\Gamma_{\omega_n}(z, \zeta) \to \Gamma_\omega(z, \zeta)$ pointwise in $\mathbb{D} \times \mathbb{D}$ as $n \to +\infty$.

Proof. Let v denote a weight of the same general type as ω and ω_n, and recall that by the reproducing property of v, we have the following identity of reproducing kernel functions (see Proposition 9.9):

$$Q_v(z, \zeta) = 2\,\text{Re}\,K_v(z, \zeta) - 1. \qquad (z, \zeta) \in \mathbb{D} \times \mathbb{D}. \qquad (9.34)$$

Let $H_v(\cdot, \zeta)$ be the harmonic compensator function

$$H_v(z, \zeta) = -\int_\mathbb{D} Q_v(z, \eta)\, G(\eta, \zeta)\, v(\eta)\, dA(\eta). \qquad (z, \zeta) \in \mathbb{D} \times \mathbb{D},$$

so that

$$\Gamma_v(z, \zeta) = \int_\mathbb{D} G(z, \xi)\, \big(G(\xi, \zeta) + H_v(\xi, \zeta)\big)\, v(\xi)\, dA(\xi) \qquad (9.35)$$

for $(z, \zeta) \in \mathbb{D} \times \mathbb{D}$. By Corollary 9.7 and the relationship (9.34), we have the estimate

$$\big|Q_v(z, \zeta)\big| \le 1 + \frac{4}{|1 - z\bar{\zeta}|^2}, \qquad (z, \zeta) \in \mathbb{D} \times \mathbb{D}. \qquad (9.36)$$

We shall use this to estimate the size of the kernel H_v. We observe that by the reproducing property of v,

$$\int_\mathbb{D} \frac{1 - |z\eta|^2}{|1 - z\bar{\eta}|^2}\, v(\eta)\, dA(\eta) = 1, \qquad z \in \mathbb{D}, \qquad (9.37)$$

and that by Fatou's lemma, the integral on the left-hand side is bounded by 1 for $z \in \mathbb{T}$. For ζ confined to a compact subset X of $\mathbb{D}$, the Green function $G(\eta, \zeta)$ is comparable to $-(1 - |\eta|^2)$ near the boundary, which allows us to use estimate (9.36) in conjunction with (9.37) to obtain the uniform estimate

$$\big|H_v(z, \zeta)\big| \le C, \qquad z \in \mathbb{D}, \ \zeta \in X, \qquad (9.38)$$

for some positive constant C depending on X, universal for all the weights v. We now show that $H_{\omega_n}(\cdot, \zeta) \to H_\omega(\cdot, \zeta)$ in an appropriate norm. For fixed $\zeta \in \mathbb{D}$, the function $F_v(\cdot, \zeta)$, defined by

$$F_v(z, \zeta) = G(z, \zeta) + H_v(z, \zeta),$$

is perpendicular to bounded harmonic functions with respect to the inner product of $L^2(\mathbb{D}, v)$. Using this fact and the estimate (9.38), we arrive at the identities

$$\int_\mathbb{D} \big|H_\omega(z, \zeta) - H_{\omega_n}(z, \zeta)\big|^2 \omega(z)\, dA(z)$$

$$= \int_\mathbb{D} \big(|F_{\omega_n}(z, \zeta)|^2 - |F_\omega(z, \zeta)|^2\big)\omega(z)\, dA(z)$$

and

$$\int_{\mathbb{D}} \left| H_\omega(z, \zeta) - H_{\omega_n}(z, \zeta) \right|^2 \omega_n(z)\, dA(z)$$

$$= \int_{\mathbb{D}} \left(|F_\omega(z, \zeta)|^2 - |F_{\omega_n}(z, \zeta)|^2 \right) \omega_n(z)\, dA(z).$$

We add these together to obtain

$$\int_{\mathbb{D}} \left| H_\omega(z, \zeta) - H_{\omega_n}(z, \zeta) \right|^2 \left(\omega(z) + \omega_n(z) \right) dA(z)$$

$$= \int_{\mathbb{D}} \left(|F_\omega(z, \zeta)|^2 - |F_{\omega_n}(z, \zeta)|^2 \right) \left(\omega_n(z) - \omega(z) \right) dA(z).$$

The weight ν always satisfies

$$\nu(z) \le (1 - |z|^2)^{-1}, \qquad z \in \mathbb{D};$$

see Exercise 1. By estimate (9.38), the above growth estimate of weights, and the assumed $L^1(\mathbb{D})$ convergence $\omega_n \to \omega$, we have

$$\int_{\mathbb{D}} \left| H_\omega(z, \zeta) - H_{\omega_n}(z, \zeta) \right|^2 \omega(z)\, dA(z) \to 0 \qquad \text{as} \quad n \to +\infty \qquad (9.39)$$

uniformly for $\zeta \in X$; in other words, $H_{\omega_n}(\cdot, \zeta) \to H_\omega(\cdot, \zeta)$ in the norm of $L^2(\mathbb{D}, \omega)$. Now, by (9.35),

$$\Gamma_\omega(z, \zeta) - \Gamma_{\omega_n}(z, \zeta) = \int_{\mathbb{D}} G(z, \xi) \left(H_\omega(\xi, \zeta) - H_{\omega_n}(\xi, \zeta) \right) \omega(\xi)\, dA(\xi)$$

$$+ \int_{\mathbb{D}} G(z, \xi) \left(G(\xi, \zeta) + H_{\omega_n}(\xi, \zeta) \right) \left(\omega(\xi) - \omega_n(\xi) \right) dA(\xi)$$

for $(z, \zeta) \in \mathbb{D} \times \mathbb{D}$, so that the desired result follows from (9.38), (9.39), the growth estimate of weights, and the $L^1(\mathbb{D})$ convergence $\omega_n \to \omega$. ∎

Combining Theorem 9.13, Theorem 9.14, and Proposition 9.15, we have now proved the following result under the standing assumptions that ω is logarithmically subharmonic on $\mathbb{D}$ and reproducing for the origin, but without the assumptions that ω is real-analytic and strictly positive on $\overline{\mathbb{D}}$.

THEOREM 9.16 *The weighted biharmonic Green function Γ_ω is positive on the set $\mathbb{D} \times \mathbb{D}$.*

9.4   An Application

In this section, we use the positivity of the weighted biharmonic Green function to prove an important result about contractive zero divisors of ordinary (that is, unweighted) Bergman spaces.

Fix $0 < p < +\infty$. We write φ_A for the contractive zero divisor in A^p associated with the zero sequence A; we do not use G_A to avoid conflict with notation for the Green function for the Laplacian.

THEOREM 9.17 *Let A and B be two zero sequences for A^p with $A \subset B$. Then*

$$\|\varphi_A f\| \leq \|\varphi_B f\|$$

for all $f \in A^p$, where $\| \cdot \|$ is the norm in A^p.

Proof. For finite sequences A and B, the functions φ_A and φ_B are holomorphic in a neighborhood of $\overline{\mathbb{D}}$. We consider the function $\Phi_{B,A}$ that solves the boundary value problem

$$\Delta \Phi_{B,A}(z) = |\varphi_B(z)|^p - |\varphi_A(z)|^p, \qquad z \in \mathbb{D},$$
$$\Phi_{B,A}(z) = 0, \qquad z \in \mathbb{T}.$$

From an application of Green's formula, we see that the fact that the function $|\varphi_B|^p - |\varphi_A|^p$ annihilates harmonic functions in $L^2(\mathbb{D})$ translates to the additional boundary condition

$$\partial_{n(z)} \Phi_{B,A}(z) = 0, \qquad z \in \mathbb{T}.$$

Dividing the differential equation by $|\varphi_A(z)|^2$, then applying another Laplacian, we find that

$$\Delta \frac{1}{|\varphi_A(z)|^p} \Delta \Phi_{B,A}(z) = \Delta \left| \frac{\varphi_B(z)}{\varphi_A(z)} \right|^p, \qquad z \in \mathbb{D},$$

which is positive on $\mathbb{D}$. In view of the given boundary data, we may write the function $\Phi_{B,A}$ as an integral in terms of the weighted biharmonic Green function $\Gamma_{|\varphi_A|^p}$:

$$\Phi_{B,A}(z) = \int_{\mathbb{D}} \Gamma_{|\varphi_A|^p}(z, \zeta) \, \Delta_\zeta \left| \frac{\varphi_B(\zeta)}{\varphi_A(\zeta)} \right|^p \, dA(\zeta), \qquad z \in \mathbb{D},$$

which is then positive. The importance of the potential function $\Phi_{B,A}$ comes from the fact that Green's formula yields the identity

$$\|\varphi_B f\|_{A^p}^p - \|\varphi_A f\|_{A^p}^p = \int_{\mathbb{D}} \Phi_{B,A}(z) \, \Delta_z |f(z)|^p \, dA(z), \qquad z \in \mathbb{D},$$

for polynomials f. Because we can approximate functions in A^p by polynomials, and because the functions φ_A and φ_B are bounded on $\mathbb{D}$, we have proved the desired result in the case of finite zero sets. Furthermore, setting $g = \varphi_B f$ leads to

$$\left\| \frac{\varphi_A}{\varphi_B} g \right\|_{A^p} \leq \|g\|_{A^p}, \tag{9.40}$$

for all $g \in A^p$ that vanish on B.

If A and B are arbitrary zero sequences, we form finite subsequences $A' \subset A$ and $B' \subset B$ with $A' \subset B'$. Then inequality (9.40) holds with A and B replaced by A' and B', respectively, and with g vanishing on B. Letting A' grow up to A, and B' up to B, then $\varphi_{A'} \to \varphi_A$ and $\varphi_{B'} \to \varphi_B$ in A^p. An application of Fatou's lemma delivers the inequality (9.40) for arbitrary A and B, which easily implies the assertion of the theorem. ∎

9.5 Notes

Proposition 9.1 is a key result in the theory of reproducing kernels, due to Aronszajn [12], Kreĭn, Mercer, Moore, and Schwartz; see [107]. The other results of this chapter, along with their proofs, are taken from the papers [69, 72] by Hedenmalm, Jakobsson, and Shimorin. The assumption on the weight ω that it is logarithmically subharmonic has a natural differential geometric interpretation: it means that the unit disk $\mathbb{D}$ equipped with the (isothermal) Riemannian metric $\sqrt{\omega(z)}|dz|$ has negative Gaussian curvature everywhere (in other words, it is a hyperbolic surface). The differential operator we have studied, $\Delta\omega^{-1}\Delta$, then corresponds to the squared Laplace-Beltrami operator on the Riemannian manifold.

Engliš has made explicit computations of certain weighted biharmonic Green functions [45].

Hadamard's variational formula is from the classical paper on *plaques élastiques encastrées* [50, pp. 515–641], and the version that applies to the Laplacian Δ is of fundamental importance for conformal mapping; see [97, pp. 42–48, pp. 263–265]. It has been used by Löwner [91] and then later by de Branges [30] in his proof of the Bieberbach conjecture.

There exists a vast mathematical literature on Hele-Shaw flows; here, we mention only Richardson's paper [99]. However, these flows have been studied almost exclusively in the context of Euclidean space, with an irregular but nonempty blob of liquid at time $t = 0$.

9.6 Exercises and Further Results

1. If ω is logarithmically subharmonic and reproduces at the origin, then we have the growth estimate

$$\omega(z) \le (1 - |z|^2)^{-1}, \qquad z \in \mathbb{D}.$$

2. The function

$$J_\omega(z, \zeta) = (1 - z\bar{\zeta}) K_\omega(z, \zeta)$$

is the reproducing kernel of a Hilbert space of analytic functions on $\mathbb{D}$.

3. Suppose K_1 and K_2 are two analytic reproducing kernels on $\mathbb{D}$. If

$$K_1(z, z) = K_2(z, z)$$

for all $z \in \mathbb{D}$, then

$$K_1(z, w) = K_2(z, w)$$

for all z and w in $\mathbb{D}$.

4. Prove Corollary 9.7.

5. Show that $K_\omega(z, z) \to +\infty$ as $|z| \to 1^-$, provided that the weight ω on $\mathbb{D}$ is logarithmically subharmonic and area-summable.

6. Let A and B be two zero sequences for A^2 such that $B \setminus A$ consists of a single point $a \in \mathbb{D}$. Then the quotient $f_a = \varphi_B / \varphi_A$ is a bounded holomorphic function on $\mathbb{D}$, and it vanishes only at the point a in $\mathbb{D}$. Moreover, if b_a is the Blaschke factor corresponding to the point a, then $|f_a / b_a| \geq 1$ holds throughout $\mathbb{D}$. In particular, $f_a(\mathbb{D})$ covers the whole disk $\mathbb{D}$.

7. Let B be a zero sequence for A^p and M an invariant subspace in A^p. If M has index 1 and if $M_B \subset M$, then $M = M_A$ for some subsequence A of B. For details, consult [69].

8. Assume that ω is logarithmically subharmonic and reproducing for the origin, and let $\varphi \in A^2(\omega)$ be an $A^2(\omega)$-inner function; these are defined analogously as for the spaces A^p_α. Then $\|f\|_\omega \leq \|\varphi f\|_\omega$ for all polynomials f. In fact,

$$\|\varphi f\|_\omega^2 = \|f\|_\omega^2 + \int_{\mathbb{D} \times \mathbb{D}} \Gamma_\omega(z, \zeta) \, |\varphi'(z)|^2 |f'(\zeta)|^2 \, dA(z) \, dA(\zeta)$$

for all $f \in A^2(|\varphi|^2 \omega)$. Hint: follow the general outline of the proof of Theorem 9.17.

9. Assume that ω and ω' are two logarithmically subharmonic weights that are reproducing for the origin. Suppose that in addition, both are C^∞ on $\overline{\mathbb{D}}$, and that the quotient ω'/ω is subharmonic. Then $\|f\|_\omega \leq \|f\|_{\omega'}$ holds for all $f \in A^2$.

10. Under the same assumptions of the previous exercise, the difference $K_\omega - K_{\omega'}$ is a reproducing kernel on $\mathbb{D} \times \mathbb{D}$. In other words,

$$\frac{L_{\omega'}(z, \zeta) - L_\omega(z, \zeta)}{(1 - z\bar{\zeta})^2}$$

is a reproducing kernel on $\mathbb{D} \times \mathbb{D}$. What about the kernel

$$\frac{L_{\omega'}(z, \zeta) - L_\omega(z, \zeta)}{1 - z\bar{\zeta}} ?$$

The latter is an open problem; see [69].

11. We now mention an open problem. Let ω be a logarithmically subharmonic weight that is reproducing for the origin. Decide whether for each $\alpha \in \mathbb{D} \setminus \{0\}$, the one-point zero divisor in $A^2(\omega)$,

$$\varphi_\alpha(z) = \left(1 - K_\omega(\alpha, \alpha)^{-1}\right) \left(1 - \frac{K_\omega(z, \alpha)}{K_\omega(\alpha, \alpha)}\right), \qquad z \in \mathbb{D},$$

is univalent, and in particular whether $\varphi_\alpha(\mathbb{D})$ is starshaped about the origin. Is the function always bounded by 3 in modulus? The problem about starshapedness would follow if a certain additional property of the weighted biharmonic Green function Γ_ω were known, namely

$$\frac{\partial}{\partial n(z)} \frac{1}{\omega(z)} \Delta_z \Gamma_\omega(z, \zeta) = -2 P(z, \zeta) + \frac{\partial}{\partial n(z)} H_\omega(z, \zeta) \le 0$$

for all $z \in \mathbb{T}$ and $\zeta \in \mathbb{D}$. Here, P is the Poisson kernel, H_ω is the harmonic compensator, and the normal derivative is in the interior direction. Together with the positivity of the Green function Γ_ω, this conjectured property constitutes the *strong maximum principle for biharmonic operators*, as envisaged by Hadamard in his treatise on *plaques élastiques encastrées* [50, pp. 541–545]. See [69] for details.

12. Suppose $0 < p < +\infty$ and $0 < \sigma_1 < \sigma_2 < +\infty$. Let G_1 and G_2 be the extremal functions of the invariant subspaces of A^p generated by the functions S_{σ_1} and S_{σ_2}, respectively, where S_σ is the classical atomic singular inner function with a point mass σ at $z = 1$. Show that $\|G_1 f\|_p \le \|G_2 f\|_p$ for all bounded analytic functions f in $\mathbb{D}$.

13. If $\omega(z) = |G(z)|^2$ for some A^2-inner function G, then $B^2(\omega) = A^2(\omega)$ if and only if G is a zero divisor.

14. Suppose I is an invariant subspace of $A^2(\omega)$, where the weight ω is logarithmically subharmonic and reproduces for the origin. Prove that $I = [I \ominus zI]$, that is, I is generated by $I \ominus zI$. Hint: try to apply Theorem 6.14.

15. Fix the parameter $-1 < \alpha < +\infty$. Show that each function u on $\mathbb{D}$ with $\Delta \omega_\alpha^{-1} \Delta u = 0$ is of the form $u(z) = g(z) + |z|^{2\alpha+2} h(z)$, where g and h are harmonic. This is an Almansi-type representation of weighted biharmonic functions. Use this information to find an explicit formula for the harmonic compensator for the weight ω_α. See [64].

16. Fix the parameter $-1 < \alpha < +\infty$. Use the positivity of the harmonic compensator for the weight ω_α to prove that the Green function Γ_{ω_α} is positive on $\mathbb{D} \times \mathbb{D}$. Hint: apply Hadamard's variational formula with concentric circles about the origin. See [64]. For $0 \le \alpha < +\infty$, ω_α is logarithmically subharmonic, making this a special case of Theorem 9.16.

In the exercises that follow, ω is a C^∞-smooth strictly positive weight on $\overline{\mathbb{D}}$ that need not be logarithmically subharmonic, nor reproducing for the origin. The symbol M_ω stands for the operator of multiplication by ω. We also let ω_1, ω_2, and ν be weights of the same type.

17. Associate with the reproducing kernel K_ω the integral operator

$$K_\omega f(z) = \int_{\mathbb{D}} K_\omega(z, \zeta) f(\zeta) dA(\zeta), \qquad z \in \mathbb{D},$$

and do the same for the weights ω_1 and ω_2 as well. Show that as operators on $L^2(\mathbb{D}, dA)$, we have

$$K_{\omega_2} = K_{\omega_1} + K_{\omega_1}(M_{\omega_1} - M_{\omega_2})K_{\omega_1}$$
$$+ K_{\omega_1}(M_{\omega_1} - M_{\omega_2})K_{\omega_1}(M_{\omega_1} - M_{\omega_2})K_{\omega_1} + \ldots,$$

provided ω_1 and ω_2 are sufficiently close uniformly on $\mathbb{D}$. We mention that an analogous perturbation formula was key to Fefferman's analysis of the Bergman kernel in $\mathbb{C}^n$ [46].

18. Suppose $\omega_t = \omega + t\nu$ for $0 \leq t < +\infty$. Do the perturbation formula of Exercise 17 infinitesimally, to obtain

$$\frac{d}{dt} K_{\omega_t} = -K_{\omega_t} M_\nu K_{\omega_t}.$$

19. Do Exercises 17 and 18 with Q_ω instead of K_ω. The variational formula from Exercise 18 of course takes the form

$$\frac{d}{dt} Q_{\omega_t} = -Q_{\omega_t} M_\nu Q_{\omega_t}.$$

20. It is known that the reproducing kernel K_ω for $A^2(\omega)$ extends to a C^∞-smooth function on $(\overline{\mathbb{D}} \times \overline{\mathbb{D}}) \setminus \delta(\mathbb{T})$, where $\delta(\mathbb{T})$ is the boundary diagonal. The following seems to be an open problem: Find an asymptotic expansion for singular behavior (that is, the behavior modulo C^∞-smooth functions on $\overline{\mathbb{D}} \times \overline{\mathbb{D}}$) of K_ω near $\delta(\mathbb{T})$.

21. Do the same analysis for the kernel Q_ω. The following may be helpful. If ω reproduces for the origin, then we know from Proposition 9.9 that $Q_\omega = 2 \operatorname{Re} K_\omega - 1$. If ω is not reproducing, is it still true that $Q_\omega - 2 \operatorname{Re} K_\omega$ is a C^∞-smooth kernel on $\overline{\mathbb{D}} \times \overline{\mathbb{D}}$?

22. Let $\mathbf{R}$ denote the operation of restriction to the boundary $\mathbb{T}$. Let $HL^2(\mathbb{D})$ denote the subspace of $L^2(\mathbb{D}, dA)$ consisting of harmonic functions. The restriction of a function in $HL^2(\mathbb{D})$ to $\mathbb{T}$ is in the Sobolev space $W^{-1/2}(\mathbb{T})$ of distributions f on $\mathbb{T}$ with (formal) Laurent (or Fourier) series expansion

$$f(z) = \sum_{n=-\infty}^{+\infty} \widehat{f}(n) z^n, \qquad \sum_{n=-\infty}^{+\infty} \frac{|\widehat{f}(n)|^2}{|n| + 1} < +\infty.$$

In fact, $\mathbf{R}$ maps the harmonic subspace $HL^2(\mathbb{D})$ onto $W^{-1/2}(\mathbb{T})$. We should specify that $\mathbf{R}f$ is defined as the distributional limit as $r \to 1$ of the dilated functions $f_r(z) = f_r(z)$, with $z \in \mathbb{T}$ and $0 < r < 1$.

23. Let us take a closer look at the perturbation formula of Exercise 19. Consider the kernel

$$V_t(z, \zeta) = Q_\nu(z, \zeta)\, \nu(\zeta) - Q_{\omega_t}(z, \zeta)\, \omega_t(\zeta),$$

and check that the perturbation formula implies that

$$\frac{d}{dt}\big(\omega_t(z)\,V_t(z,\zeta)\big) = \int_{\mathbb{D}} V_t(\xi,z)\,V_t(\xi,\zeta)\,\nu(\xi)\,dA(\xi).$$

24. Write the variational formula of Exercise 23 in integral form:

$$\omega_t(z)\,V_t(z,\zeta) = \omega(z)\,V_0(z,\zeta) + \int_0^t \int_{\mathbb{D}} V_\theta(\xi,z)\,V_\theta(\xi,\zeta)\,\nu(\xi)\,dA(\xi)\,d\theta.$$

Let P be the Poisson transform (or Poisson solver):

$$Pf(z) = \int_{\mathbb{T}} P(z,\zeta)\,f(\zeta)\,ds(\zeta), \qquad z \in \mathbb{D},$$

where $P(z,\zeta)$ is the Poisson kernel. Using the harmonicity of $V_t(z,\zeta)$ in the first variable z, we can then write

$$V_t(z,\zeta) = P\left[\frac{\omega}{\omega_t}\,V_0(\cdot,\zeta)\right](z)$$
$$+ \int_0^t \int_{\mathbb{D}} P\left[\frac{V_\theta(\xi,\cdot)}{\omega_t}\right](z)\,V_\theta(\xi,\zeta)\,\nu(\xi)\,dA(\xi)\,d\theta.$$

This equation is amenable to treatment with the classical Picard process from the theory of Ordinary Differential Equations [79]. As a result, if we assume $V_0(z,\zeta) \geq 0$ on $\mathbb{D} \times \mathbb{T}$, we get $V_t(z,\zeta) \geq 0$ on $\mathbb{D} \times \mathbb{T}$ for all $0 < t < +\infty$. See [66]; the technical details involve the operator $\mathbf{R}$ met in Exercise 22, and studying integral operators on Sobolev spaces, using boundary correspondences as exemplified in Exercise 22.

25. Let us say that the weight ω_2 is more *suppressive* than ω_1 if the harmonic compensators satisfy

$$H_{\omega_2}(z,\zeta) \leq H_{\omega_1}(z,\zeta), \qquad (z,\zeta) \in \mathbb{D} \times \mathbb{D}.$$

Show that if ω_2 is more suppressive than ω_1, or vice versa, we have, for all real parameters $0 < t_1, t_2 < +\infty$,

$$t_1\Gamma_{\omega_1}(z,\zeta) + t_2\Gamma_{\omega_1}(z,\zeta) \leq \Gamma_{t_1\omega_1 + t_2\omega_2}(z,\zeta), \qquad (z,\zeta) \in \mathbb{D} \times \mathbb{D}.$$

This is a *concavity-type property* of the biharmonic Green function in the weight space. Show by a local analysis near $\mathbb{T}$ in the first variable z that the above conclusion leads to

$$t_1\omega_1(z)\,H_{\omega_1}(z,\zeta) + t_2\omega_2(z)\,H_{\omega_2}(z,\zeta) \leq \big(t_1\omega_1(z) + t_1\omega_1(z)\big)\,H_{\omega_t}(z,\zeta),$$

for $z \in \mathbb{T}$ and $\zeta \in \mathbb{D}$. Hint: consider $\omega_t = \omega + t\nu$ as before and find a variational formula for Γ_{ω_t}, analogous to that of Q_{ω_t} in Exercise 19. See [66] for details.

26. In Exercise 25, do we really need the assumptions on the two weights to have the concavity-type property? Is there a counterexample?

References

[1] E. Abakumov and A. Borichev, Shift invariant subspaces with arbitrary indices in weighted l_A^p spaces, preprint, 1998.

[2] A. Abkar, *Invariant subspaces in spaces of analytic functions*, PhD dissertation, Lund University, 2000.

[3] S. Agmon, *Lectures on elliptic boundary value problems,* Van Nostrand Mathematical Studies, No. 2D, Van Nostrand Co., Inc., Princeton-Toronto-London, 1965.

[4] P. Ahern, M. Flores, and W. Rudin, An invariant volume-mean-value property, *J. Funct. Anal.* **111** (1993), 380–397.

[5] A. Aleman, The multiplication operators on Hilbert spaces of analytic functions, Habilitationsschrift, Fernuniversität Hagen, 1993.

[6] A. Aleman, H. Hedenmalm, S. Richter, C. Sundberg, Curious properties of canonical divisors in weighted Bergman spaces, *Proceedings of the conference in honor of B. Ya. Levin*, Tel-Aviv University 1997, to appear.

[7] A. Aleman, S. Richter, C. Sundberg, Beurling's theorem for the Bergman space, *Acta Math.* **177** (1996), 275–310.

[8] A. Aleman, S. Richter, C. Sundberg, The majorization function and the index of invariant subspaces in the Bergman spaces, preprint, 2000.

[9] J. M. Anderson, J. Clunie, and Ch. Pommerenke, On Bloch functions and normal functions, *J. Reine Angew. Math.* **270** (1974), 12–37.

[10] C. Apostol, H. Bercovici, C. Foiaş, and C. Pearcy, Invariant subspaces, dilation theory, and the structure of the predual of a dual algebra I, *J. Funct. Anal.* **63** (1985), 369–404.

[11] J. Arazy and S. Fisher, The uniqueness of the Dirichlet space among Möbius invariant Hilbert spaces, *Illinois J. Math.* **29** (1985), 449–462.

[12] N. Aronszajn, Theory of reproducing kernels, *Trans. Amer. Math. Soc.* **68** (1950), 337–404.

[13] S. Axler, The Bergman space, the Bloch space, and commutators of multiplication operators, *Duke Math. J.* **53** (1986), 315–332.

[14] S. Axler, Bergman spaces and their operators, *Surveys of Some Recent Results in Operator Theory,* Volume **1**, (J.B. Conway and B.B. Morrel, editors), Pitman Research Notes in Math. **171**, 1988, 1–50.

[15] D. Békollé, C. A. Berger, L. A. Coburn, and K. Zhu, BMO in the Bergman metric on bounded symmetric domains, *J. Funct. Anal.* **93** (1990), 310–350.

[16] H. Bercovici, C. Foiaş, and C. Pearcy, *Dual algebras with applications to invariant subspaces and dilation theory,* CBMS Regional Conference Series in Mathematics, 56; published for the Conference Board of the Mathematical Sciences, Washington, D.C., by the American Mathematical Society, Providence, R.I., 1985.

[17] F. A. Berezin, Covariant and contravariant symbols of operators, *Math. USSR-Izv.* **6** (1972), 1117–1151.

[18] F. A. Berezin, The relation between co- and contravariant symbols of operators on classical complex symmetric spaces, *Soviet Math. Dokl.* **19** (1978), 786–789.

[19] S. Bergman, *The kernel function and conformal mapping* (second, revised edition), Mathematical Surveys and Monographs, vol. **5**, American Mathematical Society, Providence, RI, 1970.

[20] B. Berndtsson and J. Ortega-Cerdà, On interpolation and sampling in Hilbert spaces of analytic functions, *J. Reine Angew. Math.* **464** (1995), 109–128.

[21] A. Beurling, On two problems concerning linear transformations in Hilbert space, *Acta Math.* **81** (1949), 239–255.

[22] A. Beurling, *The Collected Works of Arne Beurling* by L. Carleson, P. Molliavin, J. Neuberger, and J. Wermer, Volume 2, Harmonic Analysis, 341–365. Boston, Birkhäuser, 1989.

[23] C. Bishop, Bounded functions in the little Bloch space, *Pacific J. Math.* **142** (1990), 209–225.

[24] G. Bomash, A Blaschke-type product and random zero sets for the Bergman spaces, *Ark. Mat.* **30** (1992), 45–60.

[25] A. A. Borichev, Estimates from below and cyclicity in Bergman-type spaces, *Internat. Math. Res. Notices* **12** (1996), 603–611.

[26] A. Borichev, Invariant subspaces of given index in Banach spaces of analytic functions, *J. Reine Angew. Math.* **505** (1998), 23–44.

[27] A. Borichev and H. Hedenmalm, Harmonic functions of maximal growth: invertibility and cyclicity in the Bergman spaces, *J. Amer. Math. Soc.* **10** (1997), 761–796.

[28] A. Borichev, H. Hedenmalm, and A. Volberg, Large Bergman spaces: invertibility, cyclicity, and subspaces of arbitrary index, preprint, 2000.

[29] P. Bourdon, Similarity of parts to the whole for certain multiplication operators, *Proc. Amer. Math. Soc.* **99** (1987), 563–567.

[30] L. de Branges, A proof of the Bieberbach conjecture, *Acta Math.* **154** (1985), 137–152.

[31] L. Brown and B. Korenblum, Cyclic vectors in $A^{-\infty}$, *Proc. Amer. Math. Soc.* **102** (1988), 137–138.

[32] J. Bruna and D. Pascuas, Interpolation in $A^{-\infty}$, *J. London Math. Soc.* (2) **40** (1989), 452–466.

[33] L. Carleson, Sets of uniqueness for functions regular in the unit circle, *Acta Math.* **87** (1952), 325–345.

[34] R. R. Coifman, R. Rochberg, and G. Weiss, Factorization theorems for Hardy spaces in several variables, *Ann. of Math.* (2) **103** (1976), 611–635.

[35] D. S. Cyphert and J. A. Kelingos, The decomposition of bounded κ-variation into differences of κ-decreasing functions, *Studia Math.* **81** (1985), 185–195.

[36] A. E. Djrbashian and F. A. Shamoian, *Topics in the theory of A_α^p spaces,* Teubner-Texte zur Mathematik [Teubner Texts in Mathematics], **105**, BSB B. G. Teubner Verlagsgesellschaft, Leipzig, 1988.

[37] P. L. Duren, *Theory of H^p spaces,* Pure and Applied Mathematics, Vol. **38**, Academic Press, New York–London, 1970.

[38] P. Duren, D. Khavinson, H. S. Shapiro, and C. Sundberg, Contractive zero-divisors in Bergman spaces, *Pacific J. Math.* **157** (1993), 37–56.

[39] P. Duren, D. Khavinson, H. S. Shapiro, and C. Sundberg, Invariant subspaces and the biharmonic equation, *Michigan Math. J.* **41** (1994), 247–259.

[40] P. Duren, D. Khavinson, and H. S. Shapiro, Extremal functions in invariant subspaces of Bergman spaces, *Illinois J. Math.* **40** (1996), 202–210.

[41] P. L. Duren, B. W. Romberg, and A. L. Shields, Linear functionals on H^p spaces with $0 < p < 1$, *J. Reine Angew. Math.* **238** (1969), 32–60.

[42] M. Engliš, *Toeplitz operators on Bergman-type spaces,* Kandidatska disertacni prace (PhD thesis), MU CSAV, Praha, 1991.

[43] M. Engliš, Functions invariant under the Berezin transform, *J. Funct. Anal.* **121** (1994), 233–254.

[44] M. Engliš, A Loewner-type lemma for weighted biharmonic operators, *Pacific J. Math.* **179** (1997), 343–353.

[45] M. Engliš, Weighted biharmonic Green functions for rational weights, *Glasg. Math. J.* **41** (1999), 239–269.

[46] C. Fefferman, The Bergman kernel and biholomorphic mappings of pseudoconvex domains, *Invent. Math.* **26** (1974), 1–65.

[47] F. Forelli and W. Rudin, Projections on spaces of holomorphic functions on balls, *Indiana Univ. Math. J.* **24** (1974), 593–602.

[48] D. Gale, H. W. Kuhn, and A. W. Tucker, Linear programming and the theory of games, *Activity Analysis of Production and Allocation,* Cowles Commission Monograph No. 13. John Wiley & Sons, Inc., New York, N. Y.,; Chapman & Hall, Ltd., London, 1951; pp. 317–329.

[49] J. Garnett, *Bounded analytic functions*, Pure and Applied Mathematics, **96**, Academic Press, Inc., New York–London, 1981.

[50] J. Hadamard, *Œuvres de Jacques Hadamard, Vols. 1–4,* Editions du Centre National de la Recherche Scientifique, Paris, 1968.

[51] P. Halmos, Shifts on Hilbert spaces, *J. Reine Angew. Math.* **208** (1961), 102–112.

[52] J. Hansbo, Reproducing kernels and contractive divisors in Bergman spaces, *Zap. Nauchn. Sem. S.-Peterburg. Otdel. Mat. Inst. Steklov. (POMI)* **232** (1996), Issled. po Linein. Oper. i Teor. Funktsii, 24, 174–198.

[53] G. H. Hardy and J. E. Littlewood, Some properties of fractional integrals, *Math. Z.* **34** (1932), 403–439.

[54] W. W. Hastings, A Carleson measure theorem for Bergman spaces, *Proc. Amer. Math. Soc.* **52** (1975), 237–241.

[55] V. P. Havin, N. K. Nikolski (editors), *Linear and complex analysis, problem book 3, Part II*, Lecture Notes in Mathematics 1574, Springer-Verlag, Berlin, 1994.

[56] W. K. Hayman, *Subharmonic functions, Vol. 2*, London Mathematical Society Monographs, **20**, Academic Press, Inc. [Harcourt Brace Jovanovich, Publishers], London, 1989.

[57] W. K. Hayman, On a conjecture of Korenblum. Analysis (Munich) **19** (1999), 195–205.

[58] W. K. Hayman and P. B Kennedy, *Subharmonic functions, Vol. 1*, London Mathematical Society Monographs, **9**, Academic Press [Harcourt Brace Jovanovich, Publishers], London-New York, 1976.

[59] H. Hedenmalm, A factorization theorem for square area-integrable analytic functions, *J. Reine Angew. Math.* **422** (1991), 45–68.

[60] H. Hedenmalm, A factoring theorem for a weighted Bergman space, *St. Petersburg Math. J.* **4** (1993), 163–174.

[61] H. Hedenmalm, An invariant subspace of the Bergman space having the codimension two property, *J. Reine Angew. Math.* **443** (1993), 1–9.

[62] H. Hedenmalm, Spectral properties of invariant subspaces in the Bergman space, *J. Funct. Anal.* **116** (1993), 441–448.

[63] H. Hedenmalm, A factoring theorem for the Bergman space, *Bull. London Math. Soc.* **26** (1994), 113–126.

[64] H. Hedenmalm, A computation of the Green function for the weighted biharmonic operators $\Delta|z|^{-2\alpha}\Delta$, with $\alpha > -1$, *Duke Math. J.* **75** (1994), 51–78.

[65] H. Hedenmalm, Open problems in the function theory of the Bergman space, *Festschrift in honour of Lennart Carleson and Yngve Domar* (Uppsala, 1993), 153–169, Acta Univ. Upsaliensis Skr. Uppsala Univ. C Organ. Hist., 58, Uppsala Univ., Uppsala, 1995.

[66] H. Hedenmalm, Boundary value problems for weighted biharmonic operators, *St. Petersburg Math. J.* **8** (1997), 661–674.

[67] H. Hedenmalm, Maximal invariant subspaces in the Bergman space. *Ark. Mat.* **36** (1998), 97–101.

[68] H. Hedenmalm, An off-diagonal estimate of Bergman kernels, *J. Math. Pures Appl.*, to appear.

[69] H. Hedenmalm, S. Jakobsson, and S. Shimorin, A biharmonic maximum principle for hyperbolic surfaces, preprint, 2000.

[70] H. Hedenmalm, B. Korenblum, and K. Zhu, Beurling type invariant subspaces of Bergman spaces, *J. London Math. Soc.* (2) **53** (1996), 601–614.

[71] H. Hedenmalm, S. Richter, and K. Seip, Interpolating sequences and invariant subspaces of given index in the Bergman spaces, *J. Reine Angew. Math.* **477** (1996), 13–30.

[72] H. Hedenmalm and S. Shimorin, Hele-Shaw flow on hyperbolic surfaces, preprint, 2000.

[73] H. Hedenmalm and K. Zhu, On the failure of optimal factorization for certain weighted Bergman spaces, *Complex Variables Theory Appl.* **19** (1992), 165–176.

[74] A. Heilper, The zeros of functions in Nevanlinna's area class, *Israel J. Math.* **34** (1979), 1–11.

[75] A. Hinkkanen, A remark on a maximum principle in Bergman space, preprint, 1999.

[76] C. Horowitz, Zeros of functions in the Bergman spaces, *Duke Math. J.* **41** (1974), 693–710.

[77] C. Horowitz, Factorization theorems for functions in the Bergman spaces, *Duke Math. J.* **44** (1977), 201–213.

[78] C. Horowitz, B. Korenblum, and B. Pinchuk, Sampling sequences for $A^{-\infty}$, *Michigan Math. J.* **44** (1997), 389–398.

[79] E. L. Ince, *Ordinary differential equations,* Dover Publications, New York, 1944.

[80] J. Janas, A note on invariant subspaces under multiplication by z in Bergman space, *Proc. Roy. Irish Acad.* **83A** (1983), 157–164.

[81] D. Khavinson and H. S. Shapiro, Invariant subspaces in Bergman spaces and Hedenmalm's boundary value problem, *Ark. Mat.* **32** (1994), 309–321.

[82] P. Koosis, *Introduction to H_p spaces,* Second edition (with two appendices by V. P. Havin), Cambridge Tracts in Mathematics, **115** Cambridge University Press, Cambridge, 1998.

[83] B. Korenblum, An extension of the Nevanlinna theory, *Acta Math.* **135** (1975), 187–219.

[84] B. Korenblum, A Beurling-type theorem, *Acta Math.* **138** (1977), 265–293.

[85] B. Korenblum, Cyclic elements in some spaces of analytic functions, *Bull. Amer. Math. Soc.* **5** (1981), 317–318.

[86] B. Korenblum, Transformation of zero sets by contractive operators in the Bergman space, *Bull. Sci. Math.* **114** (1990), 385–394.

[87] B. Korenblum, A maximum principle for the Bergman space, *Publ. Mat.* **35** (1991), 479–486.

[88] B. Korenblum, Outer functions and cyclic elements in Bergman spaces, *J. Funct. Anal.* **115** (1993), 104–118.

[89] B. Korenblum and K. Zhu, An application of Tauberian theorems to Toeplitz operators, *J. Operator Theory* **33** (1995), 353–361.

[90] E. LeBlanc, A probabilistic zero set condition for Bergman spaces, *Michigan Math. J.* **37** (1990), 427–438.

[91] K. Löwner, Untersuchungen über schlichte konforme Abbildungen des Einheitskreises, *Math. Ann.* **89** (1923), 103–121.

[92] D. H. Luecking, A technique for characterizing Carleson measures on Bergman spaces, *Proc. Amer. Math. Soc.* **87** (1983), 656–660.

[93] D. H. Luecking, Zero sequences for Bergman spaces, *Complex Variables Theory Appl.* **30** (1996), 345–362.

[94] T. H. MacGregor and M. I. Stessin, Weighted reproducing kernels in Bergman spaces, *Michigan Math. J.* **41** (1994), 523–533.

[95] T. H. MacGregor and K. Zhu, Coefficient multipliers between Bergman and Hardy spaces, *Mathematika* **42** (1995), 413–426.

[96] X. Massaneda, Density Conditions for Interpolation in $A^{-\infty}$, preprint, 1999.

[97] Z. Nehari, *Conformal mapping* (Reprinting of the 1952 McGraw-Hill edition), Dover Publications, Inc., New York, 1975.

[98] M. Renardy, R.C. Rogers, *An introduction to partial differential equations*, Texts in Applied Mathematics, **13**, Springer-Verlag, New York, 1993.

[99] S. Richardson, Hele-Shaw flows with a free boundary produced by the injection of fluid into a narrow channel, *J. Fluid Mech.* **56** (1972), 609–618.

[100] S. Richter, Invariant subspaces in Banach spaces of analytic functions, *Trans. Amer. Math. Soc.* **304** (1987), 585–616.

[101] S. Richter, A representation theorem for cyclic analytic two-isometries, *Trans. Amer. Math. Soc.* **328** (1991), 325–349.

[102] J. Roberts, Cyclic inner functions in Bergman spaces and weak outer functions in H^p, $0 < p < 1$, *Illinois J. Math.* **29** (1985), 25–38.

[103] R. Rochberg, Interpolation by functions in Bergman spaces, *Michigan Math. J.* **29** (1982), 229–236.

[104] L. Rubel and R. Timoney, An extremal property of the Bloch space, *Proc. Amer. Math. Soc.* **75** (1979), 45–49.

[105] W. Rudin, *Function theory in the unit ball of* $\mathbb{C}^n$, Grundlehren der Mathematischen Wissenschaften [Fundamental Principles of Mathematical Science], **241** Springer-Verlag, New York–Berlin, 1980.

[106] W. Rudin, *Functional analysis,* Second edition, International Series in Pure and Applied Mathematics, McGraw-Hill, Inc., New York, 1991.

[107] S. Saitoh, *Theory of reproducing kernels and its applications,* Pitman Research Notes in Mathematics Series, 189, Longman Scientific & Technical, Harlow, 1988.

[108] M. Sakai, Regularity of a boundary having a Schwarz function, *Acta Math.* **166** (1991), 263–297.

[109] A. P. Schuster and K. Seip, A Carleson-type condition for interpolation in Bergman spaces, *J. Reine Angew. Math.* **497** (1998), 223–233.

[110] A. P. Schuster and K. Seip, Weak conditions for interpolation in holomorphic spaces, preprint, 1999.

[111] K. Seip, Regular sets of sampling and interpolation for weighted Bergman spaces, *Proc. Amer. Math. Soc.* **117** (1993), 213–220.

[112] K. Seip, Beurling type density theorems in the unit disk, *Invent. Math.* **113** (1993), 21–39.

[113] K. Seip, On Korenblum's density condition for zero sequences of $A^{-\alpha}$, *J. Anal. Math.* **67** (1995), 307–322.

[114] H. S. Shapiro, Some remarks on weighted polynomial approximation of holomorphic functions, *Math. USSR-Sb.* **2** (1967), 285–294.

[115] J. H. Shapiro, Mackey topologies, reproducing kernels, and diagonal maps on Hardy and Bergman spaces, *Duke Math. J.* **43** (1976), 187–202.

[116] J. H. Shapiro, Zeros of random functions in Bergman spaces, *Ann. Inst. Fourier (Grenoble)* **29** (1979), no. 4, vii, 159–171.

[117] J. H. Shapiro, Cyclic inner functions in Bergman spaces, manuscript, 1980.

[118] A. L. Shields, Weighted shift operators and analytic function theory. *Topics in operator theory,* Math. Surveys, No. **13**, Amer. Math. Soc., Providence, R.I., 1974, pp. 49–128.

[119] A. L. Shields, Cyclic vectors in Banach spaces of analytic functions, *Operators and function theory* (S.C. Power, editor; Lancaster, 1984) NATO Adv. Sci. Inst. Ser. C: Math. Phys. Sci., **153**, D. Reidel, Dordrecht-Boston, Mass., 1985, pp. 315–349.

[120] S. M. Shimorin, Factorization of analytic functions in weighted Bergman spaces, *St. Petersburg Math. J.* **5** (1994), 1005–1022.

[121] S. M. Shimorin, On a family of conformally invariant operators, *St. Petersburg Math. J.* **7** (1996), 287–306.

[122] S. M. Shimorin, The Green function for the weighted biharmonic operator $\Delta(1 - |z|^2)^{-\alpha}\Delta$ and the factorization of analytic functions (Russian), *Zap. Nauchn. Sem. S.-Peterburg. Otdel. Mat. Inst. Steklov. (POMI)* **222** (1995), Issled. po Linein. Oper. i Teor. Funktsii, 23, 203–221.

[123] S. M. Shimorin, Single point extremal functions in weighted Bergman spaces, *Nonlinear boundary-value problems and some questions of function theory, J. Math. Sci.* **80** (1996), 2349–2356.

[124] S. M. Shimorin, The Green functions for weighted biharmonic operators of the form $\Delta w^{-1}\Delta$ in the unit disk, *Some questions of mathematical physics and function theory. J. Math. Sci. (New York)* **92** (1998), 4404–4411.

[125] S. M. Shimorin, Approximate spectral synthesis in Bergman spaces, *Duke Math. J.* **101** (2000), 1–40.

[126] S. M. Shimorin, Wold-type decompositions and wandering subspaces for operators close to isometries; submitted for publication, 1999.

[127] J. Stoer and C. Witzgall, *Convexity and optimization in finite dimensions, I,* Grundlehren der mathematischen Wissenschaften, Band **163** Springer-Verlag, New York-Berlin, 1970.

[128] C. Sundberg, Analytic continuability of Bergman inner functions, *Michigan Math. J.* **44** (1997), 399–407.

[129] J. E. Thomson, Approximation in the mean by polynomials, *Ann. of Math.* (2) **133** (1991), 477–507.

[130] J. E. Thomson, Bounded point evaluations and polynomial approximation, *Proc. Amer. Math. Soc.* **123** (1995), 1757–1761.

[131] D. Vukotić, A sharp estimate for A_α^p functions in $\mathbb{C}^n$, *Proc. Amer. Math. Soc.* **117** (1993), 753–756.

[132] K. Zhu, VMO, ESV, and Toeplitz operators on the Bergman space, *Trans. Amer. Math.* **302** (1987), 617–646.

[133] K. Zhu, Positive Toeplitz operators on weighted Bergman spaces of bounded symmetric domains, *J. Operator Theory* **20** (1988), 329–357.

[134] K. Zhu, Multipliers of BMO in the Bergman metric with applications to Toeplitz operators, *J. Funct. Anal.* **87** (1989), 31–50.

[135] K. Zhu, *Operator Theory in Function Spaces,* Marcel Dekker, New York, 1990.

[136] K. Zhu, Bergman and Hardy spaces with small exponents, *Pacific J. Math.* **162** (1994), 189–199.

[137] K. Zhu, Distances and Banach spaces of holomorphic functions on complex domains, *J. London Math. Soc.* **49** (1994), 163–182.

[138] K. Zhu, Interpolating sequences for the Bergman space *Michigan Math. J.* **41** (1994), 73–86.

[139] K. Zhu, Evaluation operators on the Bergman space, *Math. Proc. Cambridge Philos. Soc.* **117** (1995), 513–523.

[140] K. Zhu, Interpolating and recapturing in reproducing Hilbert spaces, *Bull. Hong Kong Math. Soc.* **1** (1997), 21–33.

[141] K. Zhu, Restriction of the Bergman shift to an invariant subspace, *Quart. J. Math. Oxford* **48** (1997), 519–532.

[142] K. Zhu, A class of operators associated with reproducing kernels, *J. Operator Theory* **38** (1997), 19–24.

[143] K. Zhu, Maximal inner spaces and Hankel operators on the Bergman space, *Integral Equations Operator Theory* **31** (1998), 371–387.

[144] K. Zhu, A sharp estimate for extremal functions, *Proc. Amer. Math. Soc.,* in print.

Index